MW01166650

Here are your

1999 Science Year Cross-Reference Tabs

For insertion in your WORLD BOOK

Each year, SCIENCE YEAR, THE WORLD BOOK ANNUAL SCIENCE SUPPLEMENT, adds a valuable dimension to your WORLD BOOK set. The Cross-Reference Tab System is designed especially to help you link SCIENCE YEAR's major articles to the related WORLD BOOK articles that they update.

How to use these Tabs:

First, remove this page from SCIENCE YEAR.

Begin with the first Tab, **Amphibian**. Take the A volume of your WORLD BOOK set and find the **Amphibian** article. Moisten the **Amphibian** Tab and affix it to that page.

Glue all the other Tabs to the corresponding WORLD BOOK articles, if there is one. Your set's G volume, for example, may not have an article on the **Global Positioning System**. Put that Tab in its correct alphabetical location —near the **Glinka, Mikhail Ivanovich** article.

SCIENCE YEAR

1999

The World Book Annual Science Supplement

A review of Science and Technology During the 1998 School Year

World Book, Inc.

www.worldbook.com

a Scott Fetzer company

Chicago • London • Sydney • Toronto

THE YEAR'S MAJOR SCIENCE STORIES

From the arrival at Mars of the first spacecraft to visit the red planet since 1976 to new findings about the Neanderthals, it was an eventful year in science and technology. On these two pages are the stories the editors chose as the most memorable or important of the year, along with details on where to find information about them in the book.

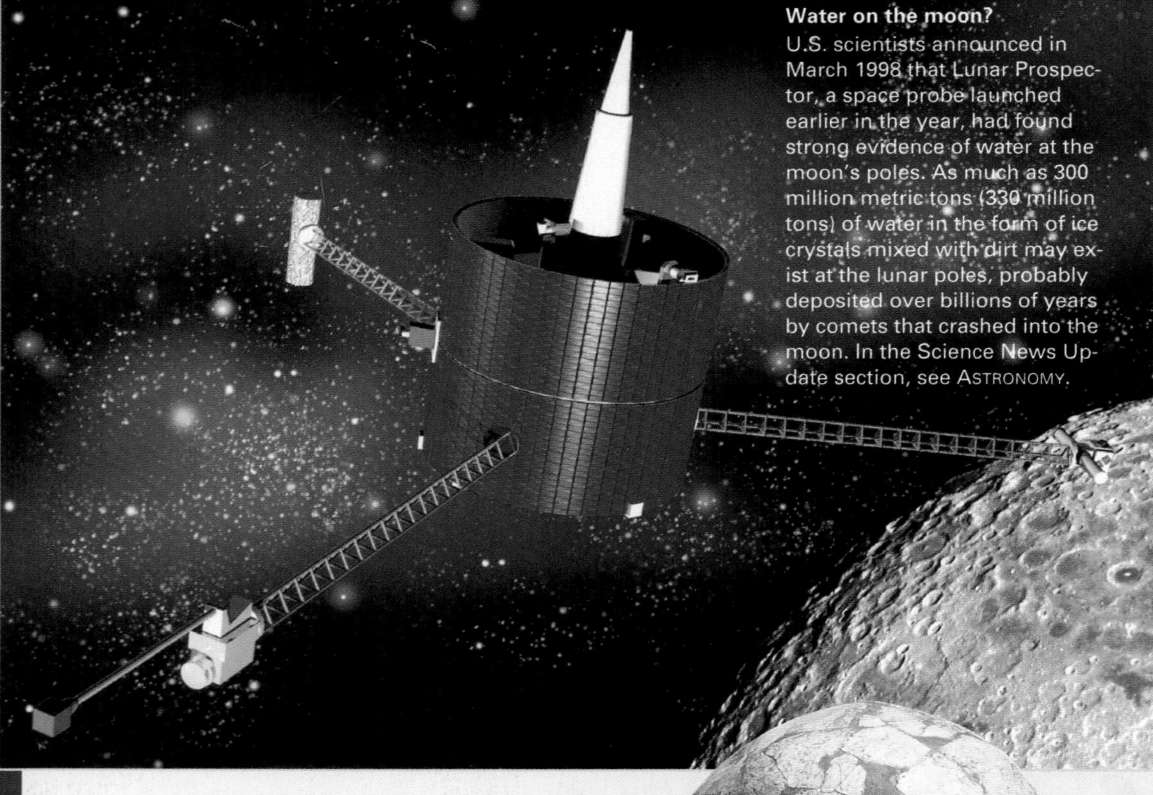

Water on the moon?
U.S. scientists announced in March 1998 that Lunar Prospector, a space probe launched earlier in the year, had found strong evidence of water at the moon's poles. As much as 300 million metric tons (330 million tons) of water in the form of ice crystals mixed with dirt may exist at the lunar poles, probably deposited over billions of years by comets that crashed into the moon. In the Science News Update section, see ASTRONOMY.

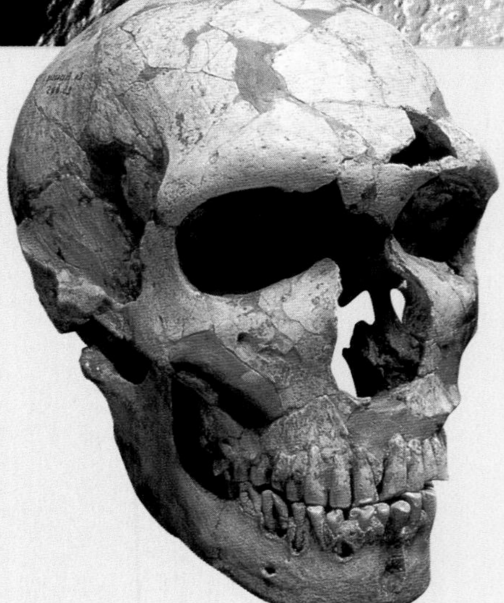

The Neanderthals—probably not our ancestors ▶
A team of German and American anthropologists announced in July 1997 that they had extracted and analyzed genetic material from a fossil of "Neanderthal Man." The analysis indicated that the Neanderthals were most likely not the ancestors of modern human beings. In the Special Reports section, see RETHINKING THE HUMAN FAMILY TREE.

World Book, Inc.
525 W. Monroe
Chicago, IL 60661

ISBN: 0-7166-0599-6
ISSN: 0080-7621
Library of Congress Catalog Number: 65-21776
Printed in the United States of America.

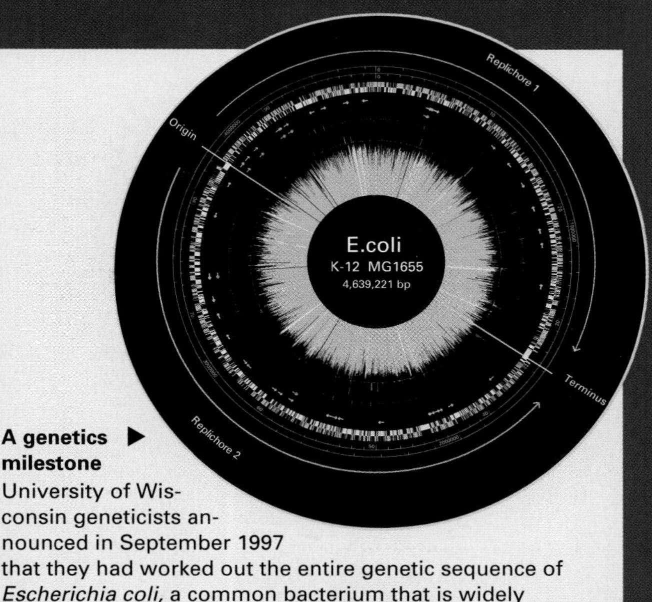

A genetics milestone ▶
University of Wisconsin geneticists announced in September 1997 that they had worked out the entire genetic sequence of *Escherichia coli,* a common bacterium that is widely used in research and sometimes causes food poisoning. In the Science News Update section, see GENETICS.

▲ **Preventing a global flu epidemic**
In January 1998, public-health workers in Hong Kong destroyed an estimated 1.5 million chickens to prevent the spread of the "bird flu," a form of influenza carried by chickens that killed at least eight people. In the Science News Update section, see PUBLIC HEALTH.

An El Nino for the record books
The El Nino of 1997 and 1998 may have been the most severe El Nino of the 1900's. The causes of this periodic phenomenon—a warming of the eastern Pacific Ocean that leads to global changes in weather patterns—are not well understood. So scientists studied the 1997-1998 El Nino intensively to learn as much as possible about it. In the Science News Update section, see OCEANOGRAPHY (Close-Up).

Oppressive air pollution in Southeast Asia
Much of Southeast Asia was shrouded by smog in September and October 1997. The air pollution stemmed from forest fires in Malaysia and Indonesia, most of them deliberately set to clear land for commercial crop production. In the Science News Update section, see ENVIRONMENTAL POLLUTION.

Renewed exploration of Mars ▶
In July 1997, the U.S. space probe Mars Pathfinder landed on Mars, releasing a wheeled robot called Sojourner Rover. Pathfinder beamed many vivid pictures of the Martian surface back to Earth. In September, another U.S. spacecraft, Mars Observer, went into orbit around the red planet. In the Special Reports section, see RETURN TO MARS.

CONTENTS

SPECIAL REPORTS 10

In-depth coverage of significant and timely topics in science and technology.

Page 42

SCIENCE STUDIES 156

Science Versus Crime explains the tools and methods of the forensic scientist and shows how science and technology have become indispensable to crime fighters.

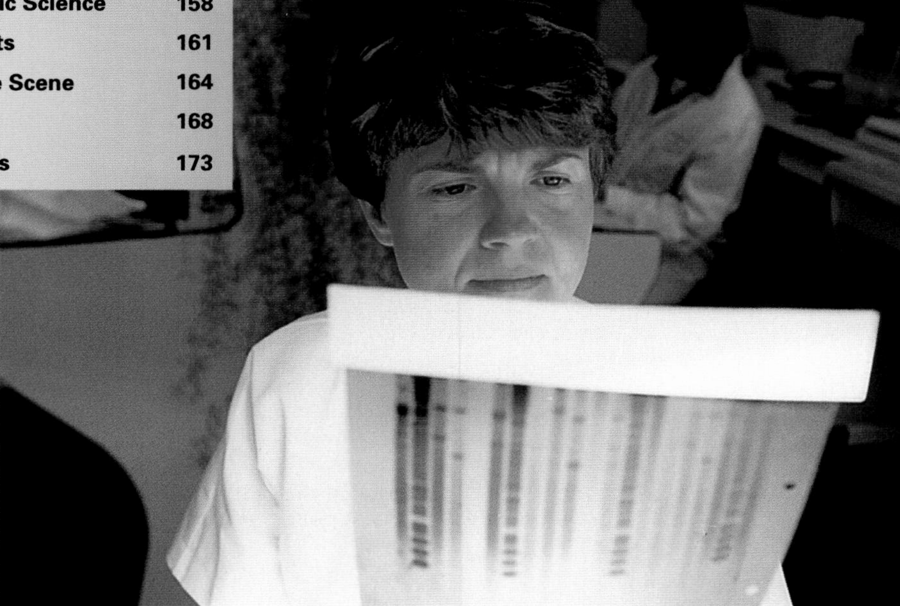

Page 171

SCIENCE NEWS UPDATE

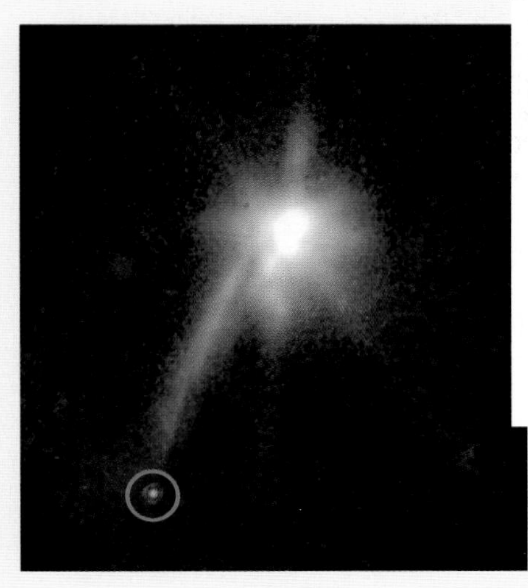

Page 190

SCIENCE YOU CAN USE

WORLD BOOK SUPPLEMENT

INDEX

CROSS-REFERENCE TABS

A tear-out page of cross-reference tabs for insertion in *The World Book Encyclopedia* appears before page 1.

Editorial

Executive Editor
Darlene R. Stille

Managing Editor
David Dreier

Senior Editor
Kristina Vaicikonis

Staff Editors
Tim Frystak
Jay Lenn
Al Smuskiewicz
Peter Uremovic

Contributing Editor
Scott Thomas

Editorial Assistant
Ethel Matthews

Cartographic Services
H. George Stoll, Head
Wayne K. Pichler, Senior Cartographer
Susan E. Ryan, Staff Cartographer

Index Services
David Pofelski, Head
Pam Hori

Art

Executive Director
Roberta Dimmer

Senior Designer, Science Year
Cari L. Biamonte

Senior Designers
Melanie J. Lawson
Brenda B. Tropinski

Photograph Editors
Sandra M. Dyrlund
Senior Photographs Editor
Marc Sirinsky

Production Assistant
Jon Whitney

Library Services
Jon Fjortoft, Head

Production
Daniel N. Bach,
Vice President

Manufacturing/Pre-Press
Sandra Van den Broucke, Director
Barbara Podczerwinski, Manufacturing Manager
Joann Seastrom, Production Manager

Proofreaders
Anne Dillon
Carol Seymour

Text Processing
Curley Hunter
Gwendolyn Johnson

Permissions Editor
Janet Peterson

Publisher
Michael Ross

Editor in Chief
Robert J. Janus

EDITORIAL ADVISORY BOARD

Ralph J. Cicerone is dean of the School of Physical Sciences at the University of California at Irvine. He received the B.S. degree from the Massachusetts Institute of Technology (MIT) in 1965, the M.S. degree from the University of Illinois in 1967, and the Ph.D. degree from the University of Illinois in 1970. He joined the University of California faculty in 1989 after serving for 11 years as director of the Atmospheric Chemistry Division of the National Center for Atmospheric Research in Boulder, Colorado. He was elected to the National Academy of Sciences in 1990. His principal research interest is the chemical composition of the global atmosphere and how it is changing.

Rochelle Easton Esposito is Professor of Molecular Genetics and Cell Biology at the University of Chicago. She received the B.S. degree from Brooklyn College in New York City in 1962 and the Ph.D. degree from the University of Washington in Seattle in 1967. She is a member of the Genetics Society of America—of which she became president in 1996—the American Society of Microbiology, the American Association for the Advancement of Science, and the American Society for Cell Biology. One of her primary research interests is the genetic control of chromosome behavior during cell division.

Alan H. Guth is Victor Weisskopf Professor of Physics at the Massachusetts Institute of Technology in Cambridge, Mass. He received the B.S. degree and the M.S. degree from MIT in 1969 and the Ph.D. degree from MIT in 1972. He is a member of the American Astronomical Society and the National Academy of Sciences, and he is a fellow of the American Physical Society, the American Association for the Advancement of Science, and the American Academy of Arts and Sciences. His main research interests are particle physics and cosmology, and he is the author of the inflationary theory of the early universe.

Neena B. Schwartz is director of the Center for Reproductive Science and William Deering Professor of Biological Sciences at Northwestern University in Evanston, Illinois. She received the B.A. degree from Goucher College in Baltimore in 1948, the M.S. degree from Northwestern in 1950, and the Ph.D. degree in 1953, also from Northwestern. She is a fellow of the American Academy of Arts and Sciences and also holds memberships in several other professional societies. Her primary research interest is the factors that regulate the secretion from the pituitary gland of hormones that stimulate the development of the sex organs.

Donald L. Wolberg is Executive Director of Special Projects at The Academy of Natural Sciences in Philadelphia. He received the B.A. degree in 1968 from New York University in New York City and the Ph.D. degree from the University of Minnesota in 1978. From 1978 to 1996, he served in various capacities at several American institutions of higher learning. From 1988 to 1994, he was chief operating officer and secretary of the Paleontological Society, the largest and oldest professional society for the study of fossils. He has done fossil work in many parts of the world. He is also the creator of "Dinofest"™, an international traveling exhibit and symposium on dinosaurs.

CONTRIBUTORS

Asker, James R., B.A.
Washington Bureau Chief,
Aviation Week & Space Technology magazine.
[Special Report, *The International Space Station; Space Technology*]

Bell, Larry, B.S., M.S.
Vice President for Exhibits,
Museum of Science, Boston.
[Special Report, *Museums That Make Science an Adventure*]

Black, Harvey, B.S., M.S., Ph.D.
Free-Lance Writer.
[Science You Can Use, *Wet Cleaners: the Latest Wave in a Green Revolution*]

Bolen, Eric G., B.S., M.S., Ph.D.
Professor,
Department of Biological Sciences,
University of North Carolina at Wilmington.
[*Conservation*]

Brett, Carlton E., M.S., Ph.D.
Professor,
Department of Earth and Environmental Sciences,
University of Rochester.
[*Fossil Studies; Fossil Studies* (Close-Up)]

Brody, Herb, B.S.
Senior Editor,
Technology Review.
[Science You Can Use, *The Global Positioning System Charts New Territory*]

Cain, Steven A., B.S.
Communication Specialist,
Purdue University School of Agriculture.
[*Agriculture*]

Chiras, Dan, B.A., Ph.D.
Adjunct Professor,
Environmental Policy and Management Program,
University of Denver.
[*Environmental Pollution*]

Cruz-Uribe, Kathryn, B.A., M.A., Ph.D.
Associate Professor of Anthropology,
Northern Arizona University.
[*Anthropology*]

Ferrell, Keith
Free-Lance Writer.
[Computers and Electronics]

Gorant, Jim, B.A., M.A.
Outdoors Editor,
Popular Mechanics magazine
[Science You Can Use, *Golf Ball Technology Slices into a Mystery of the Game*]

Graff, Gordon, B.S., M.S., Ph.D.
Free-Lance Science Writer.
[Special Report, *New Tools of the Art Conservator; Chemistry*]

Hay, William W., B.S., M.S., Ph.D.
Professor of Geological Sciences,
University of Colorado at Boulder.
[*Geology*]

Haymer, David S., M.S., Ph.D.
Professor,
Department of Genetics and Molecular Biology,
University of Hawaii.
[*Genetics*]

Hengartner, Michael O., B.S., Ph.D.
Associate Investigator,
Cold Spring Harbor Laboratory.
[*Genetics* (Close-Up)]

Hester, Thomas R., B.A., Ph.D.
Professor of Anthropology and Director,
Texas Archeological Research Laboratory,
University of Texas at Austin.
[*Archaeology*]

Johnson, Christina S., B.A., M.S.
Free-Lance Science Writer
[*Oceanography*]

Kowal, Deborah, M.A.
Adjunct Assistant Professor,
Emory University Rollins School of Public Health.
[*Public Health*]

Lannoo, Michael J., B.S., M.S., Ph.D.
Associate Professor,
Indiana University School of Medicine.
[Special Report, *Amphibian Mystery*]

Lasser, Robert A., B.A., M.D.
Clinical Associate,
National Institutes of Mental Health.
[*Psychology*]

Lunine, Jonathan I., B.S., M.S., Ph.D.
Professor of Planetary Science,
University of Arizona Lunar and Planetary Laboratory.
[*Astronomy*]

Mack, Alison, A.B., M.A., M.S.
Free-Lance Science and Medical Writer.
[Science You Can Use, *Controlling the Causes of Underarm Odor*]

Maran, Stephen P., B.S., M.A., Ph.D.
Press Officer,
American Astronomical Society and Editor,
The Astronomy and Astrophysics Encyclopedia.
[Special Report, *Return to Mars*]

March, Robert H., A.B., M.S., Ph.D.
Professor of Physics and Liberal Studies,
University of Wisconsin at Madison.
[*Physics*]

Marschall, Laurence A., B.S., Ph.D.
Professor of Physics,
Gettysburg College.
[*Books About Science*]

Maugh, Thomas H., II, Ph.D.
Science Writer,
Los Angeles Times.
[*Biology*]

Moser-Veillon, Phylis B., B.S., M.S., Ph.D.
Professor,
Department of Nutrition and Food Science,
University of Maryland.
[*Nutrition*]

Rickart, Eric A., B.S., M.A., Ph.D.
Curator of Vertebrates,
Utah Museum of Natural History,
University of Utah.
[*Biology* (Close-Up)]

Riley, Thomas N., B.S., Ph.D.
Professor,
School of Pharmacy,
Auburn University.
[*Drugs*]

Roper, Clyde F. E., B.S., M.S., Ph.D.
Zoologist,
National Museum of Natural History,
Smithsonian Institution.
[Special Report, *The Hunt for the Giant Squid*]

Sforza, Pasquale M., B.Ae.E., M.S., Ph.D.
Program Director,
Graduate Engineering and Research Center,
University of Florida.
[*Energy*]

Shipman, Pat, B.A., M.A., Ph.D.
Adjunct Associate
Professor of Anthropology,
Pennsylvania State University.
[Special Report, *Rethinking the Human Family Tree*]

Snow, John T., B.S.E.E., M.S.E.E., Ph.D.
Dean,
College of Geosciences,
University of Oklahoma.
[*Atmospheric Sciences; Oceanography* (Close-Up)]

Snow, Theodore P., B.A., M.S., Ph.D.
Professor of Astrophysics and Director,
Center for Astrophysics and Space Astronomy,
University of Colorado at Boulder.
[*Astronomy*]

Stephenson, Joan, B.S., Ph.D.
Associate Editor, Medical News,
Journal of the American Medical Association.
[Science You Can Use, *Screening Tests for Genetic Disorders*]

Tamarin, Robert H., B.S., Ph.D.
Dean of Sciences,
University of Massachusetts.
[Special Report, *Cloning Isn't Science Fiction Anymore; Ecology*]

Teich, Albert H., B.S., Ph.D.
Director,
Science and Policy Programs,
American Association for the Advancement of Science.
[*Science and Society*]

Thornton, John I., B.A., Ph.D.
Professor Emeritus of Forensic Science,
University of California at Berkeley.
[Science Studies, *Science Versus Crime*]

Trubo, Richard, B.A., M.A.
Free-Lance Writer.
[*Medical Research*]

Webster, Larry, B.S.
Technical Editor,
Car and Driver magazine.
[*Engineering* (Close-Up)]

Wright, Andrew G., B.A.
Associate Editor,
Engineering News-Record.
[*Engineering*]

SPECIAL REPORTS

Feature articles take an in-depth look at significant and timely subjects in science and technology.

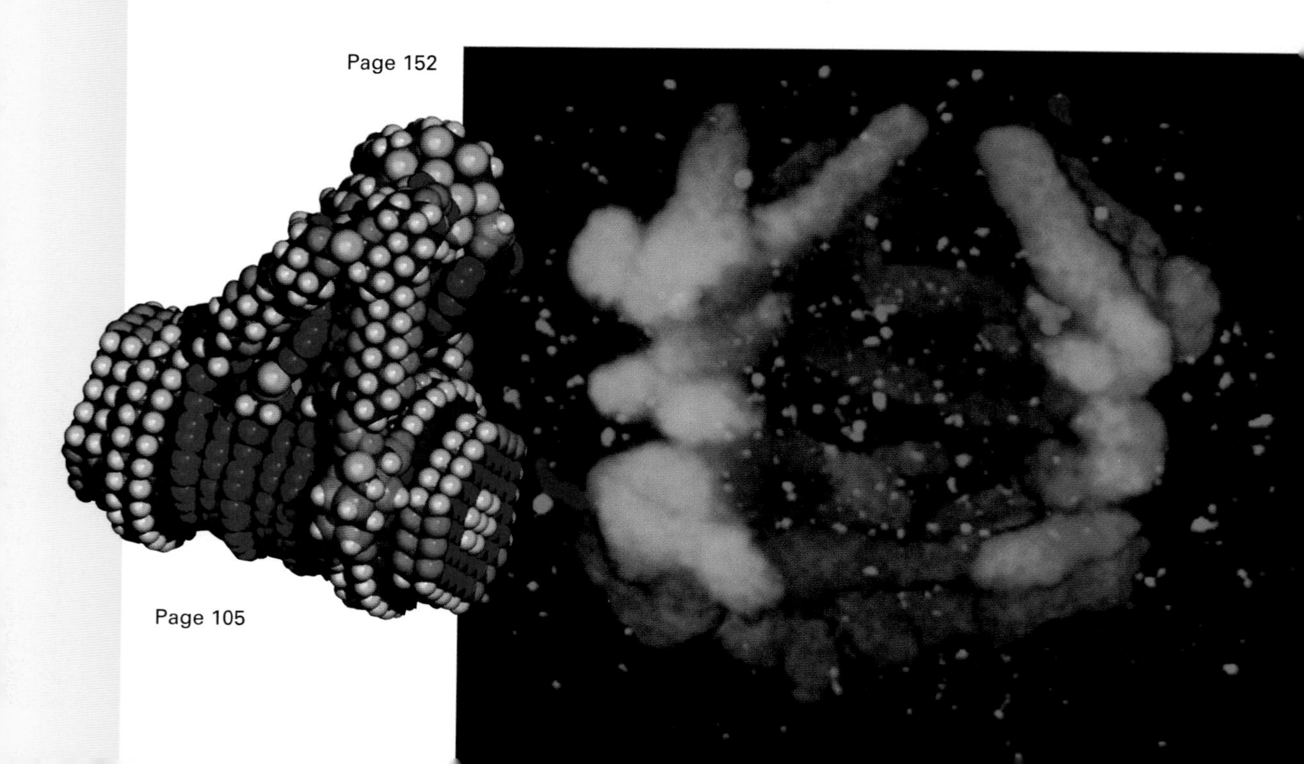

Page 152

Page 105

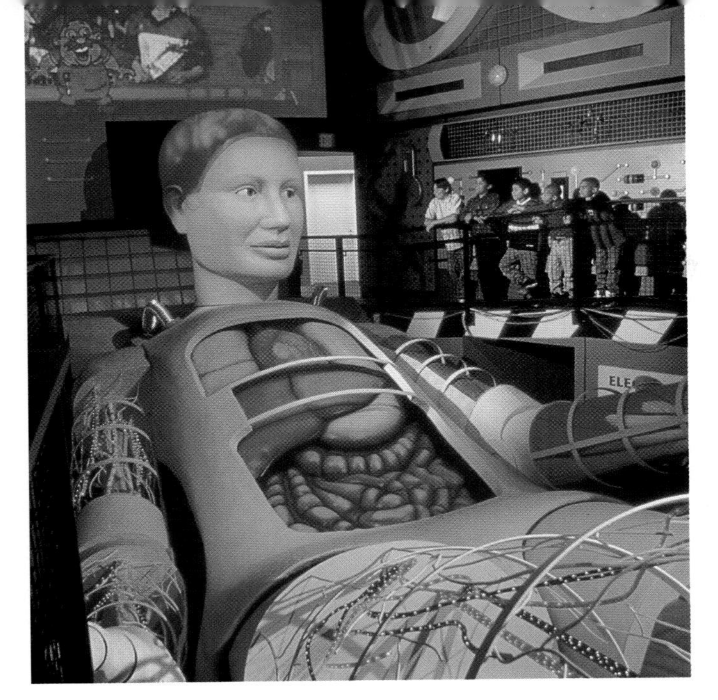

Page 127

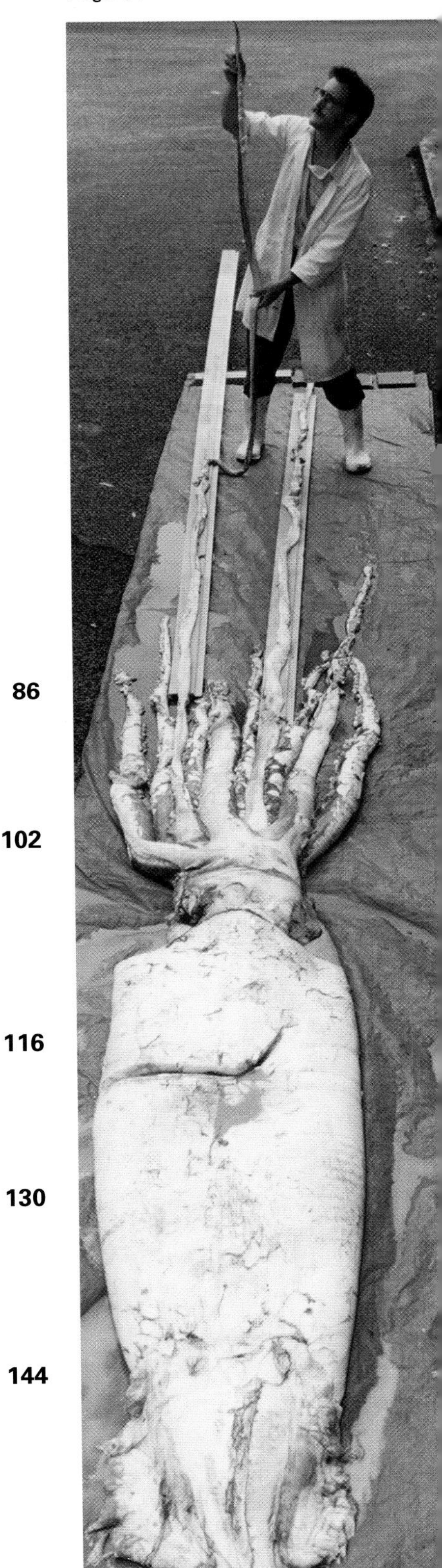

Page 94

Rethinking the Human Family Tree

In 1997, genetic research and new fossil discoveries provided clues to the origins of modern humans.

By Pat Shipman

With the discovery of the "Neanderthal man" in 1856 in the Neander Valley of Germany, scientists had the first fossil evidence of human evolution. In 1997, those renowned fossils were again in the spotlight when an international team of anthropologists and geneticists successfully extracted genetic material from an arm bone of the original Neanderthal specimen. For the first time, scientists could compare Neanderthals to modern humans both *morphologically* (in terms of the shapes and sizes of their bones) and genetically. Based on the new genetic evidence, the researchers concluded that Neanderthals were probably not direct ancestors of modern humans.

That finding had important consequences for ideas about the origins of modern humans, one of the hottest debates in paleoanthropology in the 1990's. One hypothesis, accepted by most anthropologists, is known as the out-of-Africa theory. It asserts that anatomically modern humans, *Homo sapiens,* evolved in Africa about 200,000 years ago, relatively recently in evolutionary terms. According to this scenario, all modern humans are descended from a single African population of *H. sapiens,* which evolved from a species called *H. erectus.* The out-of-Africa theory also holds that Neanderthals were a distinct but dead-end species rather than the direct ancestors of modern humans.

The other leading explanation of human origins is called the multiregional theory. This hypothesis contends that modern human features began to appear 1 million to 2 million years ago in several populations of *H. erectus* as neighboring groups interbred and exchanged genes. In this way, according to the theory, *H. sapiens* emerged at about the same time in populations in Africa, Asia, and Europe. A certain amount of geographic isolation allowed each regional group to evolve the distinctive racial characteristics that are observed in modern humans. In the multiregional scenario, Neanderthals were a regional population and the direct ancestors of modern Europeans.

The genetic research reported in 1997 appeared to favor the out-of-Africa theory, but those findings were not conclusive and left many questions unanswered. Fossil discoveries reported in 1997—including the remains of a possible new human ancestor—added more fuel to the debate. In an effort to arrive at more complete answers, researchers continued to reevaluate the fossil evidence, search for new specimens, and refine their analytical techniques. As they gather additional evidence, the story of human origins will undoubtedly become even more complex but also more firmly grounded.

Opposite page:
A cast of a 20,000-year-old burial site, which was found in Italy in 1939, offers a glimpse of early modern humans. The evolutionary relationship of such anatomically modern humans to earlier humans and prehumans is a central debate in paleoanthropology.

The beginning of human ancestry

Most anthropologists agree that human evolution started some 5 million years ago in Africa with the divergence of a single species of apelike creatures into two separate lineages. One evolutionary line eventually gave rise to our closest living relatives, the African chimpanzees. The other was the hominid lineage, which led to modern humans. *Bipedalism*—the ability to walk on two feet—may have been the feature that distinguished the hominid lineage from the chimpanzee lineage from the beginning,

The author:
Pat Shipman is an adjunct associate professor of anthropology at Pennsylvania State University.

though the fossil evidence supporting that idea is not conclusive. The oldest known hominid species, *Ardipithecus ramidus,* may have climbed in trees, but whether it walked upright remains to be proved.

After *A. ramidus* came at least six species of hominids known as australopithecines (*awe stral oh PITH uh seenz*), all of which were clearly bipedal. The oldest of these was *Australopithecus anamensis,* which lived about 4 million years ago. The first *A. anamensis* fossils were discovered in northern Kenya in 1994 by paleontologist Meave G. Leakey of the National Museums of Kenya in Nairobi, anatomist Alan C. Walker of Pennsylvania State University in State College, and their colleagues.

The various species of *Australopithecus* evolved during a period of about 3 million years. Researchers have not reached a consensus about how those species were related to one another, but they do know what the general anatomical features of the australopithecines were. Besides being bipedal, all the species had brains that were larger than those of apes but smaller than those of modern humans. Some australopithecines were slender, while others, known as robust australopithecines, were bigger-boned. Researchers agree that all of these early hominids lived only in Africa and were largely or completely vegetarian.

The australopithecines died out about 1.3 million years ago. Long before that, however—by about 2 million years ago—one species of *Australopithecus* had evolved into the first *Homo* species, *H. habilis.* Compared with Australopithecus, *H. habilis* was taller and had a larger brain and a less projecting face. Nonetheless, the beginning of the *Homo* lineage has remained somewhat murky. Anthropologist Louis S. B. Leakey of the National Museums of Kenya and colleagues from South Africa and London described the species, which was discovered in 1960 in Tanzania's Olduvai Gorge. They named it *H. habilis* (handy man) because the site included stone tools. At the time, tools were believed to be an exclusive invention of *Homo.* They also chose that name because the fossil skull's brain capacity was about 600 cubic centimeters (37 cubic inches), a size thought to be too large for *Australopithecus.* Later, researchers found stone tools in deposits with larger-brained australopithecines, and the definition of *H. habilis* became blurred.

Terms and concepts:

Australopithecines: Hominids that lived only in Africa from about 4.2 million to 1.3 million years ago.

DNA (deoxyribonucleic acid): The molecule of which genes are made.

Hominids: Human beings and their prehuman ancestors.

Homo antecessor: A possible human ancestor that lived in Europe more than 780,000 years ago.

Homo erectus: A hominid species, which evolved in Africa about 2 million years ago, that is a distant ancestor of modern humans.

Homo sapiens: The species to which modern humans belong.

Mitochondrial DNA (mtDNA): DNA in mitochondria, the energy-producing structures in cells; mtDNA is inherited only from the mother and so can be used to trace genetic lineages.

Multiregional theory: A theory of human evolution holding that modern humans arose in different regions of the world from the ancestors of *H. erectus,* who left Africa 1 million to 2 million years ago.

Mutation: A change in genetic material.

Neanderthals: Hominids that lived mostly in Europe from about 230,000 to 27,000 years ago.

Out-of-Africa theory: A theory of human evolution holding that modern humans evolved about 200,000 years ago from a single African population that descended from *H. erectus.*

Populating the Old World

The primary importance of *H. habilis* is that it apparently occupied an evolutionary position between the australopithecines and a much better understood *Homo* species, *H. erectus,* which first appeared about 1.95 million years ago in east Africa. Some anthropologists believe the oldest of these fossils are distinct enough to be classified as a separate species, *H. ergaster.* These scientists consider *H. ergaster* to be the African ancestor of *H. erectus,* a species originally recognized from Asian specimens.

A nearly complete 1.5-million-year-old skeleton of an approximately 11-year-old *H. erectus* boy—excavated in Nariokotome, Kenya, in 1984 and 1985—clarified anthropologists' understanding of the species. Paleoanthropologist Richard E. Leakey, then of the National Museums of Kenya, Alan C. Walker, and their colleagues estimated that if the boy had lived to

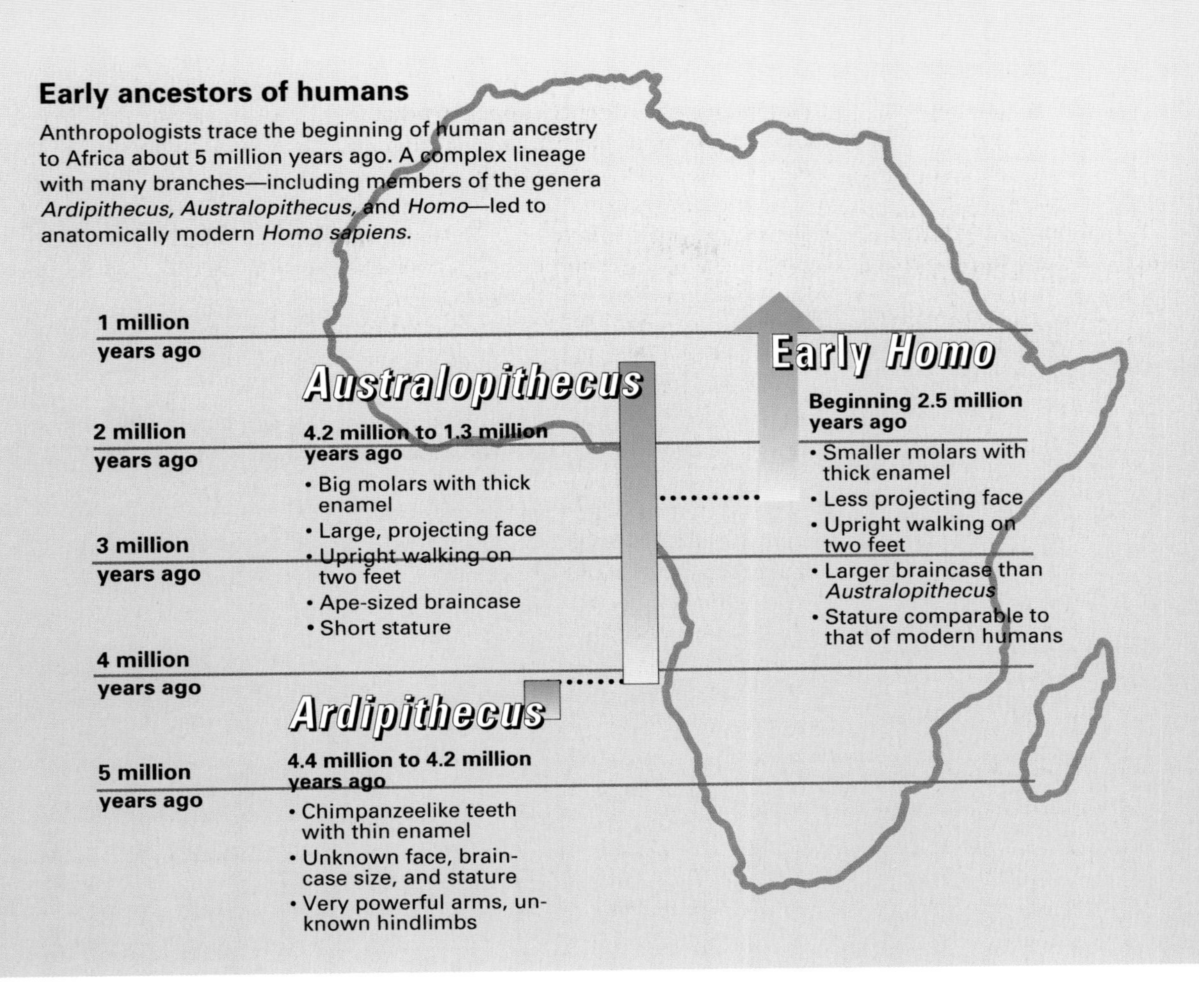

adulthood, he might have topped 1.8 meters (6 feet) in height and weighed about 68 kilograms (150 pounds). Those proportions were unexpectedly modern for an early hominid. Despite his large body, however, the boy had a brain no larger than that of a 1-year-old modern human, about 880 cubic centimeters (54 cubic inches). The brain capacity of modern human adults is about 1,250 cubic centimeters (76 cubic inches).

Most paleoanthropologists agree that *H. erectus* was probably the first hominid to become a full-time hunter. The large body and brain size of these hominids suggest that they consumed more protein, calories, and fat on a regular basis than earlier hominids did. *H. erectus* was also the first hominid species to move out of Africa, about 1 million to 2 million years ago. This expansion into a larger territory may have been necessary because of the switch to a hunting lifestyle. Meat-eaters typically need a large geographic range or they soon kill all available prey.

After the emergence of *H. erectus,* the story of hominid evolution gets more complex, and the out-of-Africa and multiregional theories differ sharply in their interpretation of the fossil record. The out-of-Africa scenario asserts that *Homo* lineages in Asia and Europe that descended from

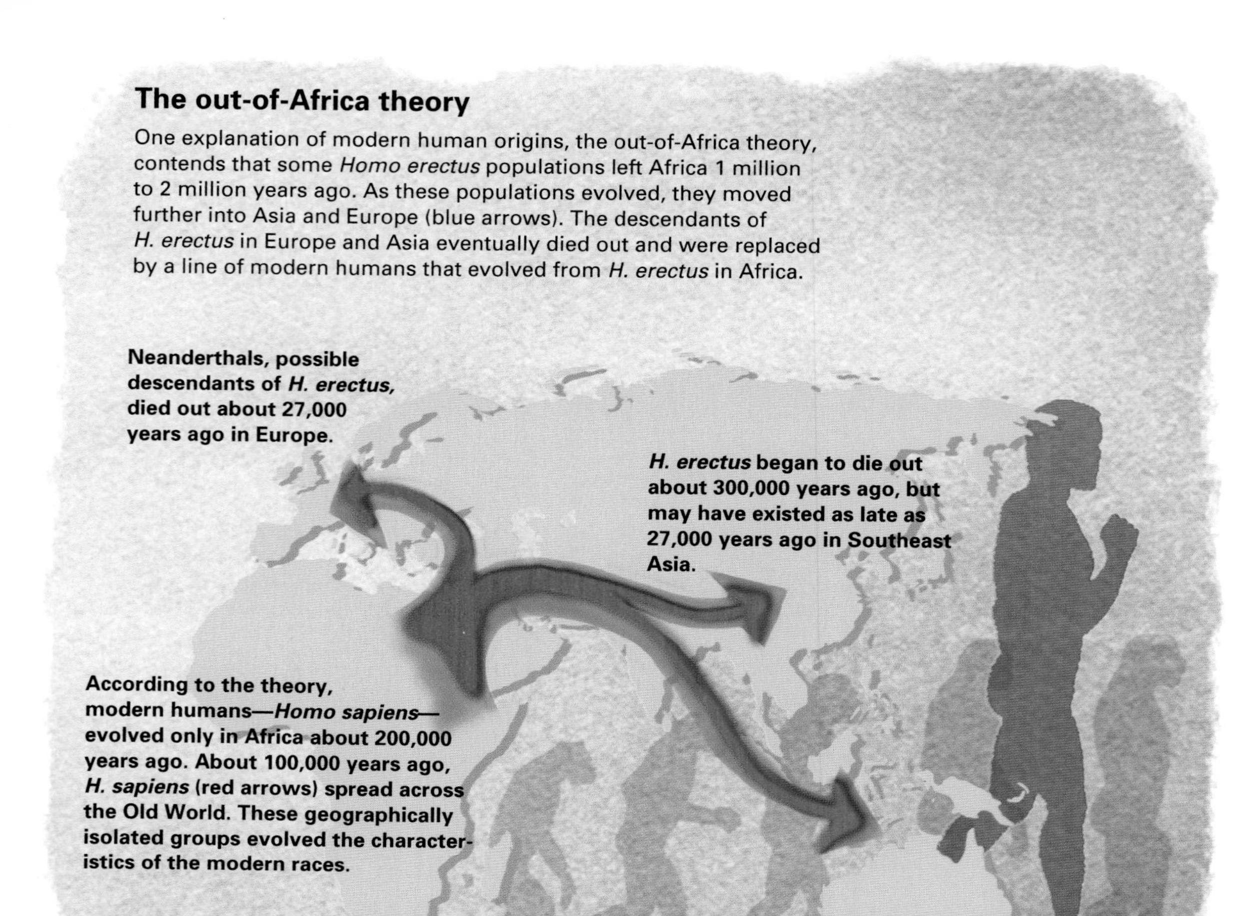

those early *H. erectus* hunters were all evolutionary dead ends that left no modern descendants. Proponents of the theory maintain that an *archaic* (primitive) form of *H. sapiens* evolved about 200,000 years ago in Africa from a single population of *H. erectus*. These early *H. sapiens* evolved into fully modern humans, some of whom left Africa about 100,000 years ago and spread over Europe and Asia. The newcomers replaced the *H. erectus* descendants who had first settled in those regions.

In this scenario, the races of modern humans began to appear sometime after the second exodus from Africa. The theory predicts that evidence of modern humans should appear suddenly and recently in the hominid fossil record. Partial *H. sapiens* skulls found in the 1960's in Ethiopia and in the 1970's in South Africa, and dated at ages between 75,000 and 130,000 years, have been accepted by adherents of the out-of-Africa theory as the oldest-known examples of anatomically modern humans.

Multiregionalists, however, reject those fossils as too fragmentary and too poorly dated. They believe that the migration of *H. erectus* was the only exodus from Africa. As individual populations spread throughout Europe and Asia, the multiregionalists contend, they gradually evolved into mod-

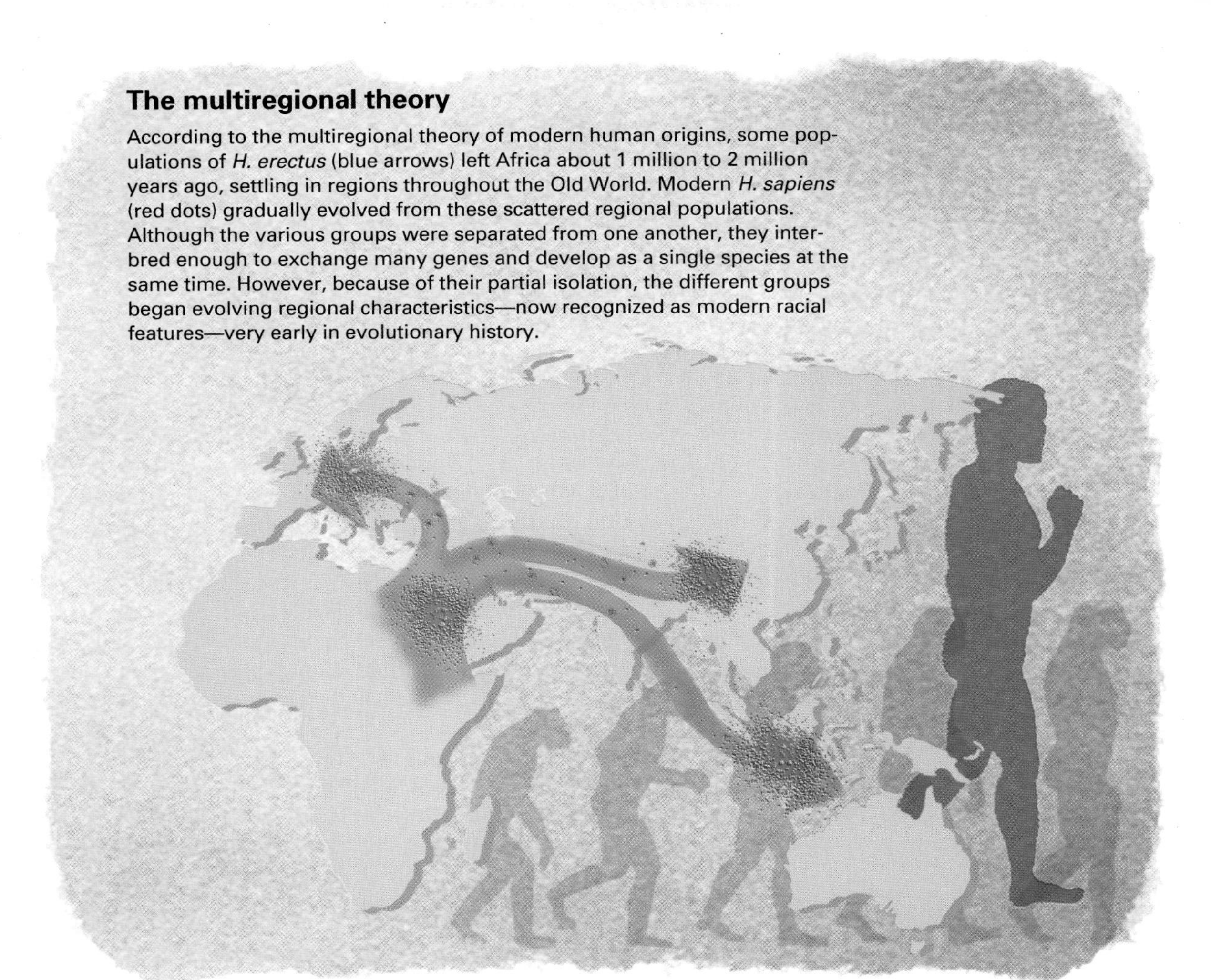

The multiregional theory

According to the multiregional theory of modern human origins, some populations of *H. erectus* (blue arrows) left Africa about 1 million to 2 million years ago, settling in regions throughout the Old World. Modern *H. sapiens* (red dots) gradually evolved from these scattered regional populations. Although the various groups were separated from one another, they interbred enough to exchange many genes and develop as a single species at the same time. However, because of their partial isolation, the different groups began evolving regional characteristics—now recognized as modern racial features—very early in evolutionary history.

ern humans. There was enough contact between the various populations to keep them all part of one widespread species, but each population was sufficiently isolated to evolve regional racial characteristics. According to this scenario, modern Africans, Asians, and Europeans are direct descendants of *H. erectus* populations—or of intermediary hominids between *H. erectus* and *H. sapiens*—that first settled in various regions.

The multiregional theory predicts that the hominid fossil record in many parts of the world will show a seamless, unbroken continuum of local evolution from *H. erectus* to *H. sapiens*. Two prominent multiregionalists, paleoanthropologists Milford H. Wolpoff of the University of Michigan in Ann Arbor and Alan G. Thorne of the Australian National University in Canberra, have taken the theory a step further. They see *H. sapiens* as an extremely old, widespread, and highly variable species, encompassing all hominids after *H. habilis*.

Three questions about the hominid fossil record are central to the debate between the proponents of the out-of-Africa and multiregional theories: Did a particular specimen come from a modern human? If not, does it represent a new species? And how ancient are the various fossils that have been found? Paleoanthropologists use the comparative anatomy of fossils, a variety of dating methods, and genetic analysis to answer these questions and test the two interpretations of modern human origins.

The fossil record: ancient clues about our ancestors

The fossil record is the most concrete evidence of the course of evolution. By comparing the anatomical features of different fossils, such as the size of teeth or the length of an arm, paleoanthropologists assess the degree of variation between specimens. Then they ask how much of a difference makes a fossil a new species. There is no simple answer.

The classic definition of a species is that it is a group of creatures that are "reproductively isolated." That is, the members of a species can breed only among themselves to produce fertile offspring. Applying this criterion to hominid fossils is difficult because researchers have no direct knowledge about the reproductive behavior of beings who lived thousands or millions of years ago. Instead, paleoanthropologists must make inferences about that behavior based on an assessment of the physical features of fossils. If they determine that similarities among specimens are evidence of interbreeding, they conclude that all the fossils came from members of the same species. Unfortunately, scientists must assess resemblances among specimens from very fragmentary remains.

The early European hominids, who lived between about 1 million and 300,000 years ago, provide an example of the problem. These European inhabitants shared many features with those of African and Asian *H. erectus.* The fossil evidence shows that both groups had large bodies, faces that were lightly built by australopithecine standards but heavier than those of modern humans, and fairly large braincases with protruding browridges. But anthropologists do not agree on how to classify these early people. Most multiregionalists refer to the early Europeans as archaic *H. sapiens*—a transitional stage in the evolution from *H. erectus* to modern *H. sapiens.* Most proponents of the out-of-Africa theory, however, maintain that early European fossils represent at least one other *Homo* species, *H. heidelbergensis.* Fossils classified as *H. heidelbergensis* have been dated to between 500,000 and 100,000 years old. The head of the Human Origins Group at the Natural History Museum in London, Christopher Stringer, argues that *H. heidelbergensis* was the direct ancestor of the best-known group of archaic European hominids, the Neanderthals.

Anthropologists are also divided on whether to classify Neanderthals as a distinct species, *Homo neanderthalensis,* or as a subspecies of modern humans, *Homo sapiens neanderthalensis.* Neanderthals lived from about 230,000 to 27,000 years ago. The oldest fossils of anatomically modern humans in Europe are about 40,000 years old. This means that Neanderthals and modern humans coexisted for at least 13,000 years. If these two groups were members of the same species, they could have interbred. If they were two separate species, then Neanderthals did not contribute to the genetic makeup of modern Europeans. Clarifying the relationship of Neanderthals to anatomically modern humans is crucial to resolving the debate over modern human origins.

A common ancestor?

In 1997, a team of researchers classified 780,000-year-old fossils found in Spain as a new *Homo* species, *H. antecessor.* The scientists theorized that the species, which probably descended from *H. erectus,* was a common ancestor to both *H. sapiens* and a lineage that led to *H. heidelbergensis* and *H. neanderthalensis.*

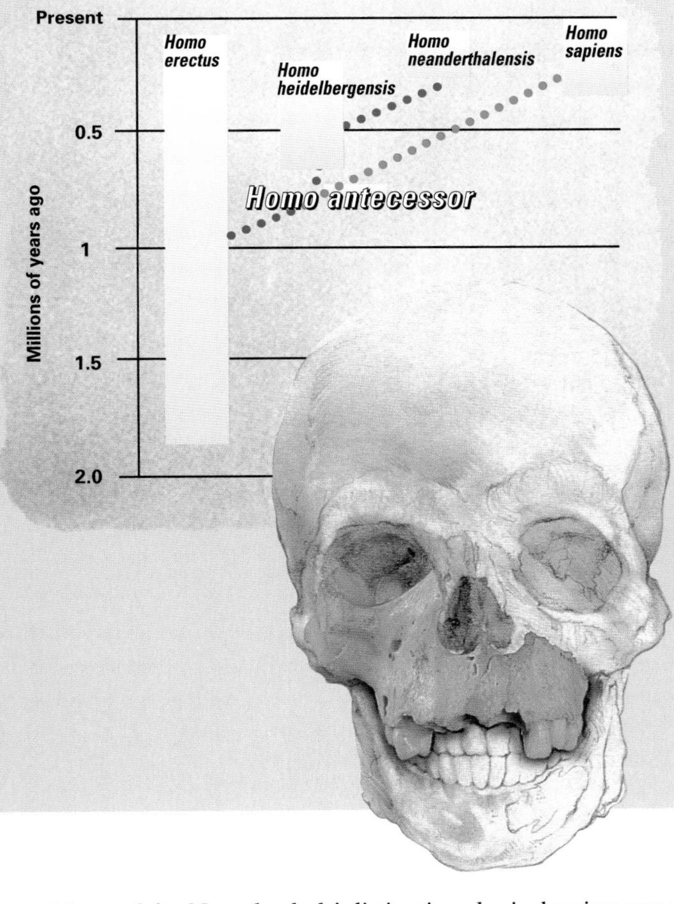

The researchers based their conclusions on their analysis of skull fragments. The lower face, *below left,* closely resembles the structure of a modern human skull. However, the browridge, *below right,* is heavier and more protruding than a human brow. The combination of modern and archaic features suggests that the specimen represents a link between two *Homo* lineages.

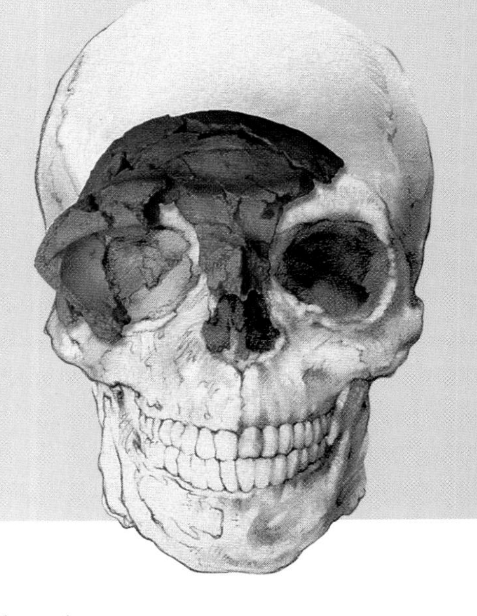

Many of the Neanderthals' distinctive physical traits were adaptations to the harsh climate of Ice Age Europe and, consequently, were much different from those of modern humans. Neanderthals' bodies were short and stout to conserve body heat. Their other features included a low forehead with prominent browridges, a broad nose, a protruding lower face, and a receding chin. However, they had a braincase capacity as large as or slightly larger than that of modern humans. Because modern humans have more slender bodies, their proportion of brain size to body size is about the same. The brain capacity of Neanderthals, the sophistication of their tools, their survival under arctic conditions, and other evidence suggest that they could have had an intelligence comparable to that of the earliest anatomically modern humans.

Multiregionalists believe that Neanderthals and anatomically modern humans in Europe interbred. They base their conclusions on detailed

Not in our gene pool?

In 1997, geneticists in Germany and the United States extracted mitochondrial DNA (mtDNA), the genetic material in a cell's energy-producing structures, from the arm bone of a Neanderthal. Differences between the mtDNA of the Neanderthal and that of modern humans led the researchers to conclude that Neanderthals were not the direct ancestors of living humans. Instead, the scientists said, the two groups belong to separate lineages and probably separate species that diverged from a common ancestor 550,000 to 690,000 years ago.

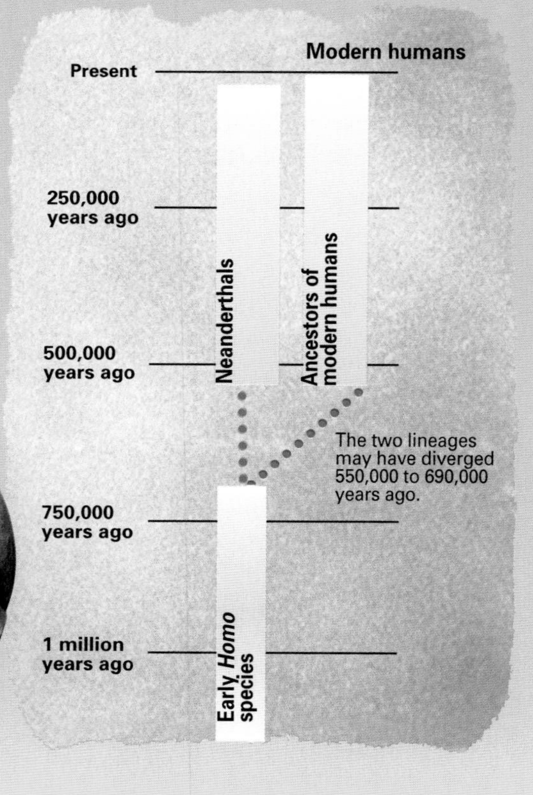

anatomical comparisons of the skulls and skeletons of different human races and of Neanderthals. According to multiregionalists, some unusual anatomical details found in modern Europeans can be traced back directly to Neanderthals, showing an evolutionary link. Most proponents of the out-of-Africa theory disagree. They maintain that the total package of Neanderthal features—from the prominent browridge and low forehead to the massive, large-jointed limb bones—is too distinctive for Neanderthals to have contributed genes to modern humans.

A new *Homo* species?

The debate over this evolutionary relationship broadened in May 1997, when researchers in Spain reported the discovery of fossils that may represent another species of *Homo.* Paleobiologist Jose Maria Bermudez de Castro of the National Museum of Natural Sciences in Madrid and his colleagues assigned the name *H. antecessor* to fossil remains from more than 780,000 years ago. The 80 teeth, skull fragments, and jaws, excavated from 1994 to 1996 in the Gran Dolina caves near Atapuerca, Spain, came from at least six individuals. These fossils are among the oldest known *Homo* specimens in Europe.

Bermudez's team chose to name a new species because skull fragments in the fossil remains suggest a combination of primitive and modern features. The heavy brow resembles those of *H. heidelbergensis* and Neanderthals. The lower face, however, is not protruding and is lightly built, more like the features of modern humans. The researchers concluded that *H. antecessor* may have been a descendant of the African *H. erectus.* Although there is no direct fossil evidence, they speculated that some *H. antecessor* populations moved about 1 million years ago from Africa to Europe, where they evolved into *H. heidelbergensis. H. antecessor* populations remaining in Africa, the researchers contended, led to modern *H. sapiens.*

Essentially, this suggestion introduced a new group of characters in the out-of-Africa scenario. But most researchers in 1998 contended that the evidence was not yet strong enough to name a new species. The most complete *H. antecessor* skull came from an adolescent who had not finished growing. Consequently, most anthropologists agreed that adult skulls were needed to show if these fossils do indeed represent a distinct species and if that species does hold a key place in human ancestry.

Determining the age of fossils

Studying the features is only one way of evaluating the different theories of modern human origins. For a fossil to mean anything, in terms of its place in an evolutionary lineage, it must be accurately dated. Undated fossils are simply remains that may differ in appearance but cannot tell us the order in which various species arose.

Bones and fossils less than about 50,000 years old can be dated directly with the carbon 14 dating method. All living organisms contain carbon 14, a radioactive *isotope* (form) of carbon. Like all radioactive substances, carbon 14 *decays* (breaks down) at a regular rate. As long as the organism is alive, the carbon 14 is renewed, but after the organism dies, it no longer absorbs the atoms. By measuring the amount of carbon 14 remaining in a fossil, scientists can determine how long ago the organism lived. With a specimen older than about 50,000 years, however, too little carbon 14 remains to make a precise determination of its age.

The age of older fossils that were buried above or below layers of volcanic rock can be estimated accurately with a dating technique called the potassium-argon method. During a volcanic eruption, a small amount of radioactive potassium is trapped by the molten lava. Once the lava has cooled and solidified, the radioactive potassium in the rock crystals decays at a regular rate into argon. In the laboratory, a piece of the volcanic rock is reheated, releasing all the trapped gases. Researchers use instruments to analyze the gases and determine the relative proportions of potassium and argon. Once they know the age of the rock sample, the scientists can accurately estimate the age of fossils that originally lay above or below the lava.

Unfortunately, in many parts of the world either volcanic eruptions or potassium—or both—are rare, so researchers are forced to use other, less precise, dating techniques. Because of the uncertainty in dating, re-

Ongoing research

A team of researchers from the National Museum of Natural Sciences in Madrid, Spain, excavate a site at the Gran Dolina caves near Atapuerca, where the *H. antecessor* fossils were discovered. The team hoped to uncover more complete fossil evidence to clarify *H. antecessor*'s place in the lineage of modern humans.

searchers do not always agree on the chronological order of fossils. Consequently, poorly dated specimens leave room for speculation and doubts about how they support one theory or another.

Genetic research: unraveling complex relationships

The newest source of evidence about modern human origins is the genetic study of human populations, a process that requires no fossils and no dating. This technique was pioneered by the late Allan C. Wilson, a biochemist at the University of California at Berkeley. It is based on an analysis of DNA (deoxyribonucleic acid), the molecule that carries the genetic code.

The four molecular subunits of DNA, called bases, are attached in pairs along the two-stranded DNA molecule. The DNA housed in the nuclei of cells carries genes inherited from both parents. But there is another type of DNA found outside of the nucleus in tiny bodies called mitochondria, which are the energy-producing structures in cells. Mitochondrial DNA (mtDNA) is inherited only from an individual's mother, who got it from her mother, and so on. Because a father passes no mtDNA to his children, any changes that occur in mtDNA from one generation to the next must be the result of spontaneous genetic *mutations* (changes). In these occasional chance alterations, one base pair is substituted for another when DNA is being replicated. An individual's mtDNA records most or all of the mutations that have occurred in his or her maternal lineage.

Geneticists can use this record of mutations to determine the relationships among human populations. For example, a researcher might com-

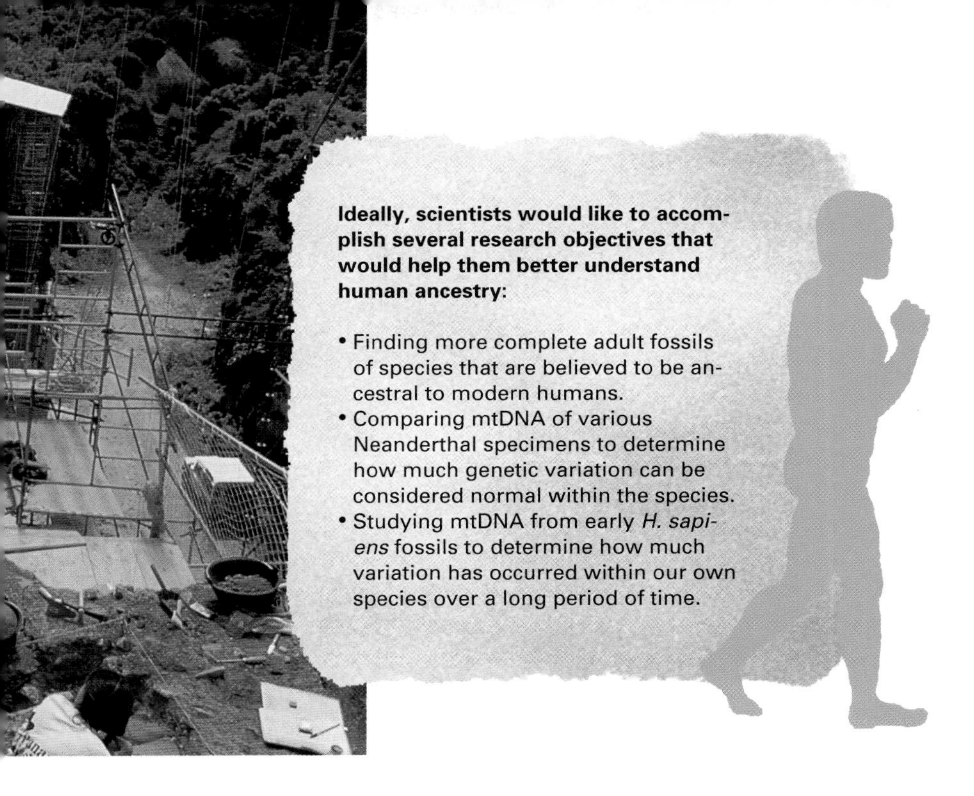

Ideally, scientists would like to accomplish several research objectives that would help them better understand human ancestry:

- Finding more complete adult fossils of species that are believed to be ancestral to modern humans.
- Comparing mtDNA of various Neanderthal specimens to determine how much genetic variation can be considered normal within the species.
- Studying mtDNA from early *H. sapiens* fossils to determine how much variation has occurred within our own species over a long period of time.

pare the mtDNA of an individual from Iceland and an individual from Sudan. The differences between the two mtDNA samples would show how many mutations occurred in each person's maternal lineage since they shared a common maternal ancestor. A picture of this would be like a tree trunk that splits into two branches. If the study included 1,000 people from around the world, the picture would be a sort of evolutionary tree showing how the various maternal lineages connect. Researchers who have conducted such studies have found that the oldest branch of this modern human tree is almost always a maternal lineage in Africa. This conclusion is consistent with the prediction of the out-of-Africa theory that modern humans arose only in Africa.

Geneticists have devised a method for estimating how long ago a particular branching event occurred. The basis for the method is the idea that genetic mutations occur regularly, like the ticking of a clock. First they count the number of mutations that separate the mtDNA of chimpanzees—our closest living relatives—and human mtDNA. Since humans and chimps branched off from their last common ancestor about 5 million years ago, geneticists can use this information to calculate a *mutation rate*—the number of mutations that occur, on average, in a 1,000-year period. Then they count the number of mutations in the mtDNA of two humans from different regional populations. Finally, geneticists use a formula based on the mutation rate to calculate the number of years since those populations shared a maternal ancestor. Most genetic studies have estimated the appearance of modern humans to between 200,000 and 500,000 years ago.

The more recent date supports the out-of-Africa theory, but many anthropologists—even some supporters of the out-of-Africa theory—doubt the accuracy of this dating method. First, the 5 million-year-old starting date is itself an estimation. Second, mutations may not always occur at a steady rate. If mutation rates vary, the estimates of the branching will be wrong. Also, a mutation could be followed by another mutation that changes the sequence back to its original form. Consequently, researchers may be missing some mutations in their count.

Probing our links to the Neanderthals

Although there are problems with using genetic information to date divergences in evolutionary history, most anthropologists still agree that such studies provide valuable insights into the relationships within human ancestry. The use of genetic data was expanded even further in 1997 with the study of Neanderthal mtDNA by geneticist Svante Paabo of the University of Munich in Germany. Paabo and his colleagues were able to extract mtDNA from an arm bone of the 1856 Neanderthal specimen, estimated to be between 30,000 and 100,000 years old. Paabo analyzed half of the mtDNA sample in his laboratory and sent the other half to the laboratory of geneticist Mark Stoneking at Pennsylvania State University.

Each team used stringent procedures to prevent contamination of the mtDNA samples with modern DNA. Although the Neanderthal mtDNA was fragmented, each laboratory found enough overlapping pieces to reconstruct a lengthy sequence of 379 base pairs—a segment of mtDNA called hypervariable region I. When the two laboratories' sequences proved to be identical, the researchers were confident that they had retrieved genuine, untainted Neanderthal mtDNA—a conclusion accepted by many anthropologists.

The researchers compared the Neanderthal mtDNA to the hypervariable region I of each of more than 2,000 modern humans, representing 994 maternal lineages. On average, modern human mtDNA differed from each other by only eight base-pair substitutions. This means that although the modern human species appears to be quite diverse, its genetic variation is relatively small. The average difference between the Neanderthal and modern human sequences, however, was about 27 base-pair substitutions. Paabo's team concluded that this number was too large to account for a normal range of biological variation within a single species. In other words—as proponents of the out-of-Africa theory had long maintained—Neanderthals were too genetically different from us to be the ancestors of modern humans. The researchers also concluded that Neanderthals and modern humans evolved from a common ancestor that lived probably 550,000 to 690,000 years ago.

The findings also contradicted another important argument of multiregionalists. If the multiregional scenario were true, the Neanderthal mtDNA would resemble that of Europeans more closely than the mtDNA of people from other areas of the world. Paabo's team found, however, that the Neanderthal mtDNA differed equally from that of modern Europeans, Native Americans, Australian Aborigines, Pacific Islanders, Africans, and Asians.

Despite the findings on Neanderthals, the debate over modern human origins was not resolved in 1997, because the mtDNA sample represented just one Neanderthal. One individual's mtDNA is inadequate to represent the genetic variability among all Neanderthals, just as a single Neanderthal skeleton cannot possibly reveal a population's variability in body size or skeletal structure. Perhaps, against all likelihood, this particular Neanderthal was an unusual individual rather than a typical member of his species. The mtDNA of another Neanderthal might look more like the mtDNA of modern humans, lending support to the multiregional theory. Anthropologists also wondered how closely the mtDNA of Neanderthal fossils from other geographic regions—Israel, Italy, or Yugoslavia, for example—would resemble the German specimen's mtDNA. And how much genetic variation might there be between Neanderthals separated by tens of thousands of years?

The next steps in research

Obtaining mtDNA from other Neanderthals and from the remains of the first anatomically modern Europeans, who appeared about 40,000 years ago, would allow geneticists to evaluate Neanderthals' place in human evolution from another perspective. A comparative analysis of Neanderthal and early-modern mtDNA might show that some Neanderthals evolved into the ancestors of modern Europeans, while others grew genetically more distinct and died out. Similarly, comparative studies of mtDNA from fossils in Asia and Africa and from the modern humans in those areas would also clarify relationships in the human family tree.

More fossils and more fossil mtDNA are needed for such studies to be carried out. Unfortunately, the genetic evidence is likely to remain scarce, because DNA degrades completely under most conditions in about 100,000 years or less. Retrieving mtDNA from the 1856 Neanderthal specimen was a remarkable accomplishment, and even Paabo's team doubted that mtDNA would be retrieved from many other hominid fossils. Their research was a rare blend of genetics, dating, and fossil studies that provided a tantalizing but limited glimpse of our evolutionary heritage. This imperfect look into the human past, as well as other recent findings, has answered some questions but raised many more. Consequently, the complex picture of the human family tree, whether based on the premises of the out-of-Africa theory or the multiregional theory, will undoubtedly undergo further revisions in the future.

For further reading:

Gore, Rick. "Dawn of Humans: Expanding Worlds." *National Geographic,* May 1997, pp. 84–109.

Gore, Rick. "Dawn of Humans: The First Europeans." *National Geographic,* July 1997, pp. 96–113.

Thorne, Alan G., and Milford H. Wolpoff. "The Multiregional Evolution of Humans." *Scientific American,* April 1992, pp. 76–83.

Kunzig, Robert. "Atapuerca: The Face of an Ancestral Child." *Discover,* December 1997, pp. 88–101.

Tattersall, Ian. *The Last Neanderthal: The Rise, Success, & Mysterious Extinction of Our Closest Human Relatives.* Macmillan Publishing Company, 1995.

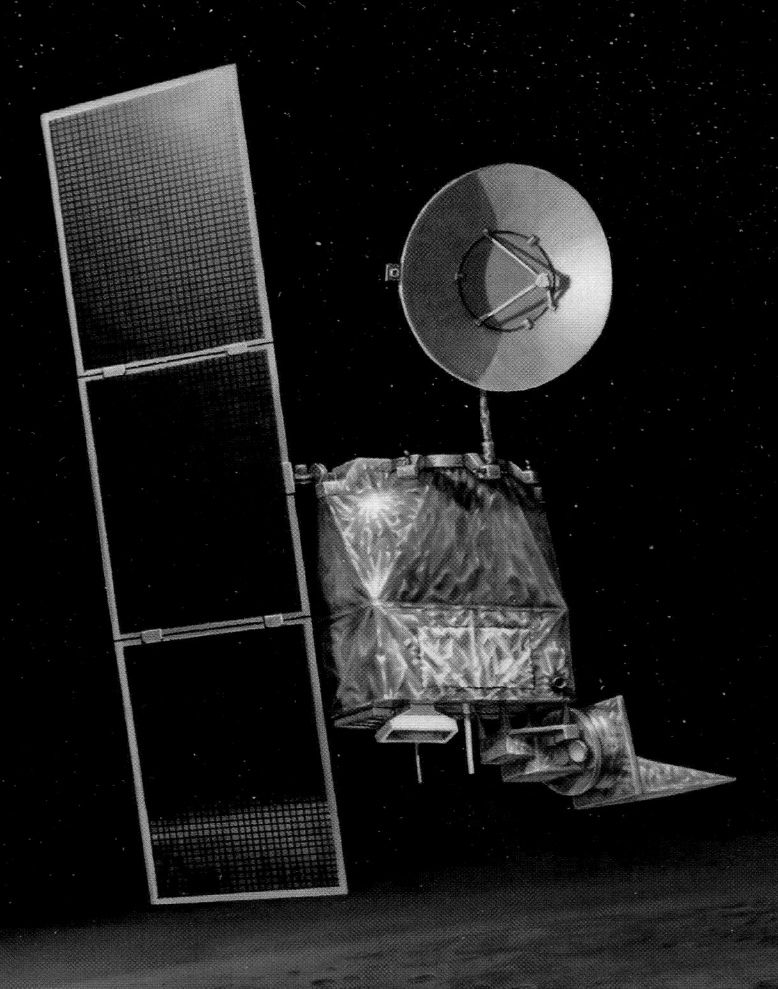

Twenty years after the last probe mission to Mars, a new generation of unmanned spacecraft is bringing Mars exploration into the 21st century.

RETURN TO MARS

By Stephen P. Maran

Parachuting down through the thin, cold air, then bouncing across the desolate landscape shielded by a cluster of giant air bags, Mars Pathfinder made a triumphant—if inelegant—landing on the surface of Mars on July 4, 1997. The arrival of the unmanned probe marked the first visit to Mars by a spacecraft from Earth since 1976. As an audience of millions watched back on Earth, Pathfinder beamed back photographs showing Sojourner Rover, a small six-wheeled exploration robot, rolling down a ramp from the lander onto the Martian soil and slowly forging a path across the rocky plain. Ten weeks later, on Sept. 11, 1997, another spacecraft, Mars Global Surveyor, reached orbit around the red planet to begin an extended mission to scan the surface of Mars and study its upper atmosphere.

Scientists, space enthusiasts, and the general public exulted in the near-perfect missions, which greatly added to our understanding of the fourth planet from the sun. One important new finding was that Mars

Shown in an artist's conception, Mars Surveyor 98, an orbiter and lander-rover combination, was scheduled to reach Mars in late 1999. The mission was designed to document atmospheric and surface changes during an entire Martian year (687 days).

probably once had large bodies of open water. If so, it is possible that some form of life emerged on Mars in the distant past.

As spectacular as the 1997 missions were, however, they were only the beginning. NASA planned to follow up on those successes by launching two more unmanned Mars probes in December 1998 and January 1999, with others to follow in 2001 and 2003. And according to the space agency's long-range plans, these probes could lead to the first manned interplanetary mission—a round-trip voyage to Mars and back to Earth—perhaps as early as 2010. The boldest visionaries believe that human beings will one day colonize Mars. Living in the warmth of an enclosed structure with its own atmosphere and energy sources, they would develop a "New World," much as European pioneers did in the Americas centuries ago. With advanced technology, they might even transform Mars into an Earthlike planet.

Glossary

Carbonates: Compounds formed when water and carbon dioxide chemically erode certain kinds of rock and redeposit the minerals elsewhere to form new rock.

Greenhouse effect: The warming of a planet caused by heat being trapped in a planet's atmosphere by carbon dioxide or other atmospheric gases.

Lander: A probe designed to land on a planet or moon and study the surface conditions.

Orbiter: A probe designed to enter orbit around a planet or moon and make observations over an extended period of time.

Polycyclic aromatic hydrocarbons (PAH's): Carbon-based compounds that both occur naturally in space and can be created by the decay of dead microorganisms.

Rover: A robotic vehicle with wheels or treads that makes observations as it roams across the surface of a planet or moon.

Terraforming: The process of creating an Earthlike environment on another planet.

Early exploration of Mars

Mars has been a source of fascination and speculation throughout history. Even the earliest stargazers took note of the red "star" that, unlike most other heavenly objects, continually changed its position in relation to the constellations. And every two years or so, baffled observers would watch as, week by week, the mysterious object grew steadily brighter in the night sky. The ancient Babylonians regarded the red wanderer as a bad omen and named it Nirgal, after their god of death and pestilence. The ancient Romans associated it with Mars, their god of war. This name has been used for the planet by scientists ever since.

For centuries, astronomers tracked the motions of Mars and other planets across the constellations, carefully refining their calculations in an attempt to understand the solar system. In the early 1600's, the newly invented telescope enabled astronomers to study celestial objects in greater detail. In 1659, the famous Dutch astronomer Christiaan Huygens spotted a dark, roughly V-shaped feature on Mars, which he sketched in a notebook. Huygens observed that the feature (later named Syrtis Major) was moving and realized that Mars must be rotating. He calculated that, like Earth, Mars rotates once every 24 hours. We now know that a day on Mars is 24 hours, 37 minutes long. A century later, in 1779, the German-born British astronomer William Herschel observed that the ice caps at the north and south poles of Mars change color and size from time to time. He correctly concluded from these and other surface changes that Mars has seasons.

About a century after Herschel's discovery, observations by the Italian astronomer Giovanni V. Schiaparelli triggered a fundamental change in the way people thought of Mars. In 1877, Schiaparelli drew a detailed map of the Martian surface based on his studies. Schiaparelli claimed to have seen many fine, straight lines—which he called *canali,* Italian for "grooves" or "channels"—crisscrossing the planet. In the English-speaking world, however, this word was mistranslated as "canals." In addition, he reported seeing a "wave of darkening," in which the areas surrounding the polar caps became darker, then ex-

The author
Stephen P. Maran is Press Officer of the American Astronomical Society in Washington, D.C., and editor of *The Astronomy and Astrophysics Encyclopedia.*

The view from Earth

Telescopes have been aimed at Mars ever since their invention in the early 1600's. Before the development of photography, astronomers sketched what they saw, *left.* By the 1890's, the Italian astronomer Giovanni Schiaparelli had drawn a detailed map of the Martian surface, *above,* that depicted what he described as *canali* ("channels"). The mistranslation of this term as "canals" inspired numerous tales of intelligent beings on Mars. These "Martians" have been depicted in numerous popular science fiction stories ever since, as in this fanciful illustration from an H. G. Wells story, "The Things that Live on Mars," *right.*

In later years, better telescopes and observatory sites provided more detail, *middle,* but interference caused by Earth's atmosphere still obscured many details. Images from telescopes in Earth orbit, such as the Hubble Space Telescope, *below,* are not affected by Earth's atmosphere, but they are limited by the planetwide dust storms that sometimes totally blanket the Martian surface.

The first close looks

Many of Mars's secrets were revealed by unmanned probes in the 1960's and 1970's. Armed with a variety of cameras and other instruments, they gave scientists a close-up view that would otherwise have been impossible.

In 1964, Mariner 4 made the first successful fly-by of Mars. The probe sent back 22 images that provided the best views of Mars available up to that time, *above*. In 1976, the Viking mission sent two orbiters and two landers on an extended mission to study Mars in detail. The orbiters scanned the surface extensively with a variety of instruments, enabling researchers to generate detailed maps of the surface and use computers to enhance the images, *left*. The landers analyzed soil samples and photographed the Martian surface throughout a Martian year (687 Earth days). Some lander images revealed that frosts of water ice sometimes accumulate on Mars during the winter, *below*.

panded toward the equator as the cap shrank. To Schiaparelli, it appeared as though meltwater from the polar ice was flowing in a widening sheet toward the equator.

The idea that there were canals and water on Mars sparked many people's imagination. Some theorized that the supposed canals were so long and straight that they could not be natural. But it was Percival Lowell, an American author, diplomat, and astronomer, who brought worldwide attention to the notion. In 1895, Lowell triggered a storm of controversy when he declared that the canals had been built by Martian engineers to carry water from the polar caps to drought-stricken regions near the equator. Ever since, tales of intelligent Martians have been popular in science fiction literature and films.

The first close-up views of Mars

Many astronomers in the 1900's suspected that the canals on Mars were an optical illusion, but that was difficult to prove because observation of Mars using Earth-based telescopes is difficult. We now know that our view of Mars changes frequently because the planet's surface is often obscured by tremendous dust storms. But even when the dust clears, astronomers simply cannot see much detail. Therefore, a close study of the Martian surface—and proof that the canals do not exist—had to wait until space probes arrived to take pictures from up close.

The first probes to Mars were fly-by missions, each collecting data during one brief pass by the planet. The first successful fly-by of Mars was made in 1964, when the NASA probe Mariner 4 flew within 9,800 kilometers (6,100 miles) of Mars. Unfortunately, the spacecraft's view was limited to the heavily cratered southern hemisphere. Therefore, the 22 photographs taken by Mariner 4 showed a surface that was not much different from that of the Earth's moon. The images included pictures of many craters caused by meteorite impacts but nothing to indicate geological activity. Mars fly-bys in 1969 by Mariners 6 and 7 also showed a long-dead surface dominated by impact craters.

The next generation of probes were orbiters, which circled the planet for extended periods. In 1971, Mariner 9 entered orbit around Mars and revolutionized our understanding of the planet. During its year-long mission, the probe took about 7,000 photographs revealing that Mars is nothing like the moon. Among the surface features discovered by Mariner 9 was the tallest mountain in the solar system, Olympus Mons. This huge volcanic peak, 25 kilometers (15.5 miles) tall, is almost three times as tall as Mount Everest, the highest peak on Earth. Also seen for the first time was Vallis Marineris, an enormous canyon stretching about 4,000 kilometers (2,500 miles), almost as far as the distance from New York City to Los Angeles. It became obvious that Mars was once alive with geological processes. Mariner 9 also confirmed that it is seasonal dust storms, not polar floodwater spreading toward the equator, that change the appearance of the surface.

The next major phase in the exploration of Mars involved landers that touched down on the surface. It began in 1976, with the arrival of

New views from the surface

On July 4, 1997, probes returned to Mars after a 20-year absence. Mars Pathfinder and Mars Sojourner Rover returned images and other data from the surface for nearly three months. The lander's many images included a spectacular panoramic shot, *above,* known as the "Monster Pan."

The Sojourner Rover, which roamed free after the landing, took about 550 pictures of its own, including a color-enhanced sunset over hills in the distance, *right,* and a dramatic portrait of Pathfinder and the now-deflated airbags that cushioned its landing, *below.*

Littered with a jumbled variety of rocks, Pathfinder's landing site was apparently an ancient flood plain. Sojourner's chemical analyses of rocks in the area supported the theory that Mars once had abundant flowing water on its surface. Such conditions would have been suitable for the emergence of primitive life forms.

Images taken by Pathfinder's cameras tracked Sojourner as it rolled across the Martian surface, *above,* and approached nearby rocks, such as "Yogi," *above right,* and made measurements of their chemical makeup. Cameras on Sojourner were used to produce a high-resolution image of "Yogi," *right,* that clearly shows the large rock's dual nature. One explanation for Yogi's mixed coloration is that wind-blown, reddish dust has accumulated on the gray surface.

Viking 1 and 2. Both Viking probes consisted of an orbiter and a lander. The orbiters took thousands of images of the surface. Many showed what appeared to be ancient riverbeds, with teardrop-shaped islands formed by powerful rivers or streams. Other images showed chaotic, rocky terrain that may have been caused by catastrophic flooding. In the meantime, the landers made weather observations and conducted chemical experiments on the soil, some of which were designed to detect the presence of microbes.

Although the Viking probes failed to detect any trace of life or water, they found many indications that water had flowed on Mars in the distant past. This meant that Mars might once have been hospitable to life, a possibility that intrigued NASA enough to spur the development of another Mars probe mission. Sixteen years passed, however, before that probe, Mars Observer, was ready to go. Mars Observer, launched in September 1992, was a $1-billion spacecraft packed with instruments to study the atmosphere, chemical composition, weather, mineral content, and other properties of Mars. It had the potential to gather more data than all previous Mars missions combined. But in August 1993, three days before Mars Observer was due to reach orbit, controllers on Earth lost contact with it. Whether the probe went sailing on past Mars and into the depths of interplanetary space or simply blew up may never be known. An investigative committee decided that the most likely explanation was that an undetected fuel leak had damaged the probe or rendered it inoperable by causing it to spin out of control.

After the loss of Mars Observer, NASA adopted a new strategy for future probe missions. Instead of launching complex, expensive probes designed to perform many tasks, the agency would develop smaller, simpler spacecraft that carried fewer instruments. Mars Global Surveyor, for example, was built from spare parts and instruments from Mars Observer and was about half of Mars Observer's size. By reducing its costs per mission in such ways, NASA was able to develop a greater number of probe missions. Mars Pathfinder and Mars Global Surveyor, the first probes of this new generation, were launched in 1996.

Hints of life in a Martian meteorite?

In August 1996, a few months before Pathfinder and Surveyor were launched, a team of scientists reported tantalizing evidence of past life on Mars. The evidence did not come from a probe, however, but from a meteorite found in Antarctica in 1984. Scientists from NASA's Johnson Space Center in Houston, Texas, and Stanford University in California said the meteorite, known as ALH84001, contained tiny amounts of biochemical substances. These substances indicated that life may have existed on Mars more than 3.6 billion years ago.

Most experts believe that ALH84001 was blasted off Mars millions of years ago by a cataclysmic event, most likely an impact by a small asteroid. ALH84001 probably drifted in space for millions of years before crossing Earth's path and falling to Antarctica about 13,000 years ago.

The researchers found that ALH84001 contained *carbonates,* compounds formed when water and carbon dioxide chemically erode certain kinds of rock and redeposit the minerals elsewhere to form new rock. The presence of carbonates in ALH84001 suggested that the meteorite may once have been under water—perhaps at the bottom of an ancient Martian sea. The team also detected other kinds of carbon-based molecules in ALH84001 that could have resulted from biological processes. For example, they found polycyclic aromatic hydrocarbons (PAH's), carbon-based compounds that can be created by the decay of dead microorganisms. In association with PAH's, team members found iron sulfides and magnetite—common compounds that sometimes result from biological activity. On top of all this, researchers using an electron microscope observed several tiny structures in the rock that appeared to be microscopic fossils of a type of primitive bacteria. These "nanofossils" found within ALH84001 closely resembled the fossils of bacteria sometimes found in Earth rocks, but they were much smaller than the tiniest known life forms on Earth.

These discoveries sparked great interest among scientists, journalists, and the public in 1996. However, later reports from other research groups that studied ALH84001 cast doubt on the findings. Some scientists said, for example, that the PAH's could have been the result of nonliving chemical processes or biological contamination by Earth organisms after the fragment landed on Antarctica.

Pathfinder, Sojourner, and Surveyor

As the debate over ALH84001 continued, Pathfinder and Surveyor began their study of Mars. During nearly three months of operation, Pathfinder took about 16,000 pictures, including spectacular panoramas of the surrounding terrain. Sojourner Rover examined the chemical makeup of the soil and nearby rocks, traveling a total of about 100 meters (325 feet), and took some 550 photographs.

Like its Viking predecessors, Pathfinder studied the atmosphere near the surface, making more than 8 million individual measurements. It revealed that *dust devils,* small whirlwinds that lift soil particles into the air, appear to be common on Mars. The lander also documented abrupt changes in air temperature and pressure at certain times during the day. These findings led investigators to conclude that airborne dust is the chief absorber of solar energy in the Martian atmosphere. Thus, it is dust that drives the weather of Mars, filling the role that a denser atmosphere and water vapor play on Earth.

Observations by Pathfinder and Sojourner also added weight to the theory that there once had been open water on the Martian surface. Pathfinder images of the terrain near the landing site showed numerous rocks, pebbles, and boulders that appeared to have been shaped by flowing water, sandstorms, or other natural processes. Some of these rocks appeared to be *conglomerates,* rocks formed by the cementing together of dissimilar mineral fragments and sand particles in the presence of water. Pathfinder images also showed distant cliffs with a layered structure—possible evidence that the rock was built up by the

The view from Mars orbit

Early observations from orbit by Mars Global Surveyor revealed the remnants of a planet-wide magnetic field and confirmed for the first time that water once did flow across the surface of Mars.

Surveyor began gathering valuable information even before it achieved orbit. Its images revealed surface features, including a stunning, highly detailed view of a canyon's edge, *right,* in greater detail than had ever been possible before. Surveyor images provided many clues about the nature of the Martian climate, both at present and in the distant past. For example, scientists now believe that Nanedi Vallis, *below,* a wide, sharp-edged canyon, was cut by a flowing river about 1 billion years ago.

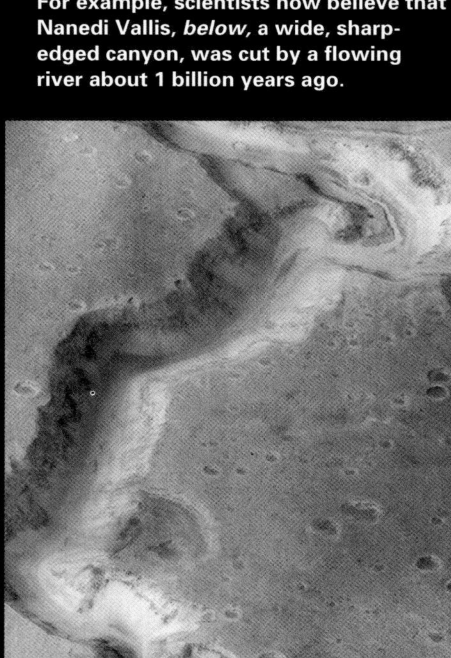

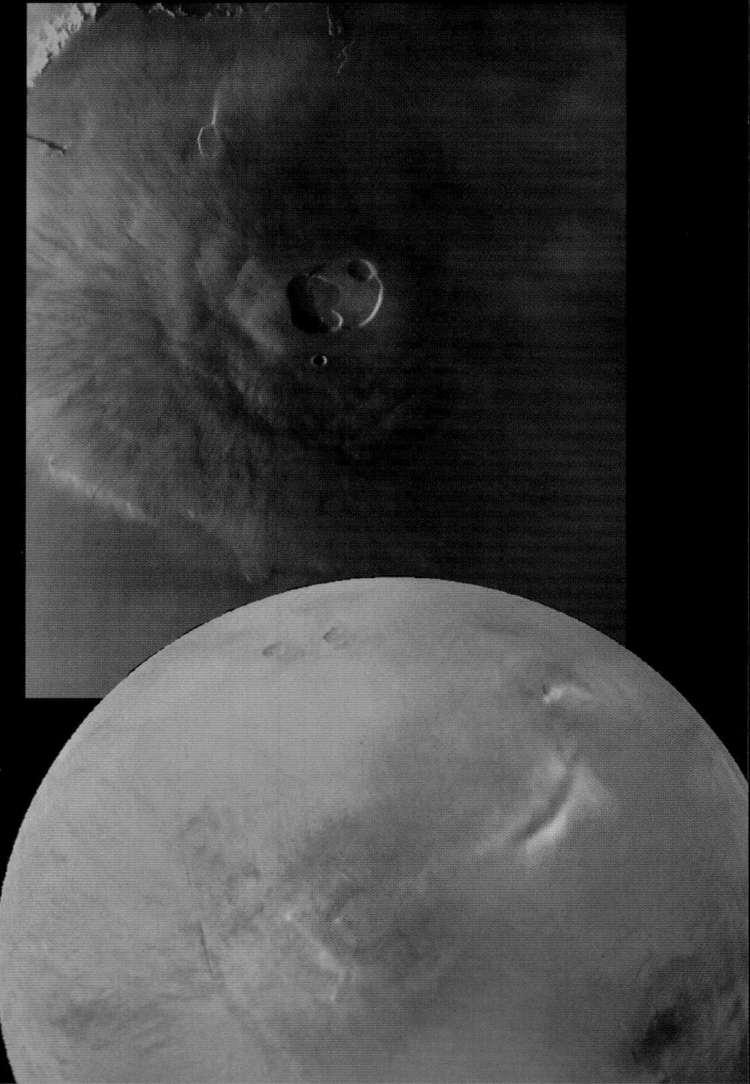

Other images by Surveyor showed clouds of condensed water vapor over Olympus Mons, a huge volcanic peak, *above right,* and over Tharsis, an enormous bulge on the Martian surface about 4,000 kilometers (2,485 miles) across and 7 kilometers (4 miles) high, *right.* The finding reinforced interest in determining how much water may remain in the frozen crust of Mars.

deposit of sediments. But if they are sediments, scientists could not determine whether they were laid down over time by water in an ancient lake bed or by dust storms. Sojourner's chemical analysis of the soil near the landing site revealed traces of what may be carbonates and sulfates, which are commonly found in dry lake beds on Earth. But again, the evidence was not conclusive. Scientists hoped that Mars Global Surveyor, equipped to analyze the mineral content of the Martian surface from orbit, would gather more definite evidence.

To conserve fuel, the Surveyor orbiter was designed to use the friction of Mars's atmosphere to help it settle into an orbit about 380 kilometers (235 miles) above the planet. This method of slowing a spacecraft is known as aerobraking. During the aerobraking process, however, technicians discovered that one of Surveyor's solar panels, which generate power for the probe, was in danger of failing due to the stress of the maneuver. To preserve the vital solar panel, mission controllers decided to slow the orbiter more gradually than planned, delaying completion of aerobraking until March 1999 instead of March 1998.

Despite the delay, Surveyor made many useful observations almost immediately. The most significant of these came in January 1998, when NASA scientists announced that Surveyor had beamed back an image with the first conclusive evidence that a river had once flowed on Mars. The picture showed a feature that NASA officials named Nanedi Vallis. This wide, sharp-edged canyon contains terraces and channels strongly suggesting that the canyon was cut by a river. Scientists estimated that water had flowed through the canyon about 1 billion years ago.

Other early findings by Surveyor, though perhaps not as dramatic, were also of great interest to scientists. An instrument aboard the orbiter measured the heights of surface features by bouncing laser pulses off them and timing how long it took the reflected pulse to return to the orbiter. The probe found that the Northern Plains, which make up a large part of the northern hemisphere, are extremely flat, with nothing much taller than a small hill across a distance of about 4,800 kilometers (3,000 miles). One explanation for such a huge expanse of flat terrain is that the region was once covered by an immense lake or sea.

Magnetic fields on Mars

Another instrument on Surveyor confirmed that, unlike Earth, Mars does not have a global magnetic field. Instead, there are several weaker, localized magnetic fields, each aligned in a different direction. Earth's magnetic field, which scientists think is generated by the circulation of molten iron and nickel in the outer core, behaves as if a huge bar magnet is buried deep within the planet. Because Earth's magnetic field is aligned roughly with the planet's axis of rotation, a compass on Earth always points northward. Mars's magnetism, on the other hand, behaves as though individual magnets are buried at locations around the planet. Therefore, a compass on Mars would point in different directions, depending on where it was on the surface.

Scientists think it is likely that Mars did once have a global magnetic field similar to Earth's. Today's localized magnetic fields on Mars may represent masses of the crust that solidified at different times during a period when Mars still had a global magnetic field. The orientation of this field probably changed over time, as has happened on Earth. If so, adjacent regions of crust could have developed considerably different magnetic-field orientations because the global field shifted after one of those sections of crust hardened. The new direction of the global field was then frozen into the neighboring area of crust when it solidified. Or perhaps geological processes shifted regions of the crust over hundreds of millions of years, thus mixing up the magnetic-field orientations of different crustal areas. It is also possible that a combination of these scenarios led to the current state of magnetic fields on Mars. Since then, the circulating molten core that probably generated Mars's magnetic field has apparently cooled and solidified. The magnetic concentrations that remain are like fossils that hint at conditions that prevailed on the early Mars.

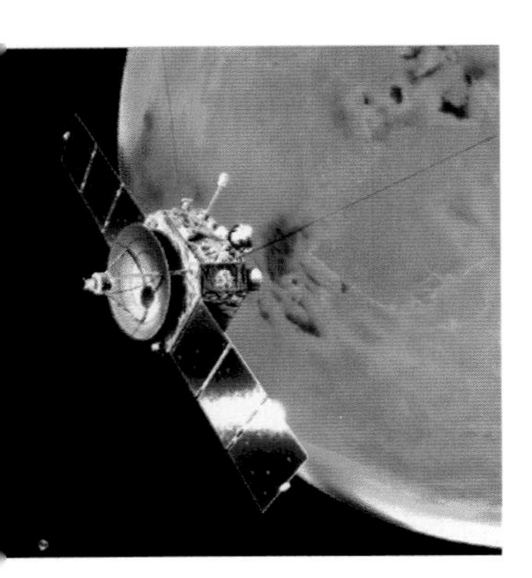

Upcoming missions
Planet-B, *above,* a Japanese Mars probe, was scheduled for an August 1998 launch. An orbiter, Planet-B was designed primarily to study the interaction between Mars and the *solar wind* (a stream of electrically charged particles emitted by the sun). The Japanese probe was just one of three unmanned missions to the red planet being prepared in 1998, including two from the United States. Two other U.S. spacecraft were scheduled for launch in 2001.

Future missions to Mars

Plans called for Mars Global Surveyor's primary mission to continue into early 2000. Afterward, the spacecraft was to serve as a communications relay for the next unmanned mission to Mars, the two-probe Mars Surveyor 98 mission. These probes were designed to study *volatiles* (substances, like carbon dioxide, that readily change from a solid state directly to a gas) and the history of the Martian climate. The two spacecraft, the Mars 98 Climate Orbiter and the Mars 98 Polar Lander, were scheduled to be launched in December 1998 and January 1999, respectively, and to arrive in late September and early December 1999. The orbiter was to measure such factors as atmospheric temperature, dust, water vapor, and cloud composition. The lander was to touch down near the south polar cap. Among its tasks, it was to deploy two probes to penetrate the ground in search of water ice below the surface.

Another Mars mission slated for a 1998 launch was Planet-B, a probe developed by Japan's Institute of Space and Astronautical Science. Planet-B was an orbiter designed primarily to study the interaction between Mars's atmosphere and the *solar wind,* a steady flow of electrically charged particles from the sun. It was scheduled for an August 1998 launch and arrival in October 1999.

The mineral content of the Martian surface was to be studied by the Mars Surveyor 2001 mission, the next pair of U.S. Mars probes. Surveyor 2001, which NASA planned to launch in the spring of 2001, is an orbiter and a lander-rover combination. The space agency's plans called for the orbiter to study the distribution of minerals on Mars, while the lander and rover conduct experiments on the ground. The 2001 rover was to be similar in design to Sojourner, but it would be larger and would travel much farther from the landing site—perhaps 10 to 20 kilometers (6 to 12 miles). It would analyze the soil and rock for clues about Mars's geological history and for evidence of ancient life.

NASA was also making plans for Mars missions beyond 2001. A third orbiter-lander pair was likely to be launched in 2003. In 2005, a mission to bring Martian rock and soil to Earth for study was possible.

The greatest uncertainty in 1998 involved the possibility of sending human explorers to Mars. Even advocates of a manned Mars mission conceded that such a project would be possible only after many major obstacles were overcome. For instance, engineers would need to prove that a manned flight between planets can be made safely. One danger for astronauts traveling so far beyond Earth would be occasional bombardment by subatomic particles released by *solar flares* (explosions on the sun). People on a long trip to Mars and back would need to be shielded from this harmful radiation. In addition, travelers living in a weightless environment for an extended period might be vulnerable to such damaging physiological effects as the loss of bone mass.

Another major roadblock is cost. In the late 1980's, the estimated price tag for a manned flight to Mars was $450 billion. A number of methods were proposed in the 1990's to cut this figure to around $25 billion. For example, it might be possible to extract fuel—such as oxygen or methane—from the Martian atmosphere for the return to Earth. This would save money by greatly reducing the amount of fuel—and thus mass—that a spacecraft would have to carry to Mars. However, researchers would need to identify actual fuel sources on

Visiting Mars
Sometime around the mid-2010's, astronauts from Earth may set foot on Mars. A future version of TransHab, *below,* an inflatable module now being developed for use on an orbiting space station, might be transported to Mars to provide living quarters for the period of several months that the astronauts would remain on the Martian surface.

Mars beforehand and devise a reliable method for refining the fuel.

Yet another major technological challenge was to develop safe and efficient living quarters for the Mars voyagers, both on the spaceship and at the landing site. One promising solution, an inflatable module called TransHab, was being developed for possible use on the International Space Station, which was to be built in orbit beginning in 1998. Because it is inflatable, TransHab takes up much less space before use than a comparably sized rigid module. For travel beyond Earth orbit, a TransHab module might incorporate a layer of water that would help shield the astronauts from the dangerous solar particles. However, such a water shield would add substantial mass to the spacecraft and reduce available space. Upon reaching Mars, another version of TransHab might provide living and working space on the surface.

Finally, given the great expense and complexity involved in a manned Mars mission, it would most likely require the participation of several nations. Many political, diplomatic, and economic issues would have to be resolved before such a mammoth project could be undertaken.

Terraforming: the final step?

Despite the many hurdles, a number of experts predicted in 1998 that a manned flight to Mars was likely within 20 or 30 years, and they envisioned a permanent colony on the red planet in the more distant future. The long-term settlement of Mars, however, may require that the Martian environment be altered to resemble that of Earth, with a breathable atmosphere, water to grow crops and sustain life, and a warmer climate. The process of creating an Earthlike environment on another planet is called terraforming. Although the concept of terraforming has appeared in science fiction stories since the 1940's, it has only recently become the subject of serious scientific discussion.

Terraforming Mars
One day, some futurists believe, human colonists might be able to restore climatic conditions on Mars to what they probably were billions of years ago. Thickening the depleted atmosphere may begin a chain reaction, raising the average surface temperature and restoring the planet's long-evaporated seas.

Terraforming advocates regard Mars as the most likely candidate to be transformed into a more Earthlike planet. The present average surface temperature on Mars is –63 °C (–81 °F), compared with 15 °C (59 °F) on Earth, and the atmosphere, composed of 95 percent carbon dioxide, is much thinner than Earth's. Because evidence suggests that water once flowed on Mars, however, the normal average temperature of the planet must once have been at least above 0 °C (32 °F)—the freezing point of water. To account for this temperature difference between then and now, some scientists theorize that the atmosphere of ancient Mars was much thicker than it is today. Mars's atmosphere is presently so thin that it traps little heat from the sun. A denser atmosphere, rich in carbon dioxide, would have prevented some of the heat from the sun from radiating back into space. This phenomenon, which also helps keep Earth warm, is known as the greenhouse effect.

Scientists believe that, over billions of years, most of the carbon dioxide in Mars's atmosphere became trapped in the rocky Martian surface layer and in ice at the poles. Terraforming advocates argue that releasing carbon dioxide from these sources back into the atmosphere might trigger a chain reaction: The carbon dioxide would trap more solar heat, which would gradually lead to increased global temperatures. This, in turn, would begin to *sublime* (change from a solid to a gas) Mars's polar ice caps, releasing still more carbon dioxide, small amounts of oxygen, and perhaps melting water ice in the caps as well. One possible method for liberating the trapped carbon dioxide on Mars involves sprinkling dark dust (which could be mined on Mars) on the dry-ice caps at the poles. Because dark colors absorb more sunlight than light colors, the dust would absorb light that is normally reflected by the ice. This would heat the dry ice, releasing carbon dioxide.

But whatever method is used, experts estimate that releasing enough carbon dioxide to complete the first stage of terraforming would take at least 100 Earth years. By that time, the average temperature might be a more bearable –8 °C (18 °F) and the atmosphere might be dense enough to allow people to work outdoors without space suits. The air would still not be breathable, however, so the next stage might involve introducing plants to Mars to produce oxygen.

But Martian terraforming, and even the first permanent colony on Mars, will have to wait for another day. In the meantime, an armada of unmanned probes will serve as our advance scouts, gathering information crucial to making a manned mission successful. But in some distant future, the existence of intelligent beings on Mars, as envisioned by Percival Lowell and many later scientists and science fiction writers, may become a reality. These Martians, however, will not be an alien race. They will be people from Earth.

For further reading:

Clarke, Arthur C. *The Snows of Olympus.* W. W. Norton & Company, 1995.

Sheehan, William. *The Planet Mars: A History of Observation and Discovery.* The University of Arizona Press, 1996.

Questions for thought and discussion

Suppose you are a U.S. legislator in the year 2010, preparing to vote on a bill for funding a manned mission to Mars. The proposed bill calls for $300 billion in spending. The exploration of space has led to significant scientific and technological advances, especially in recent years. In addition, private investment in space-based industries has created many jobs, in both the industrialized and developing countries of the world. However, social problems like hunger and poverty are still common in many parts of the world, including the United States. Lobbyists for humanitarian organizations and agricultural research institutions ask you to vote against the Mars mission legislation. They argue that the funding would be better used to finance research on developing new food sources.

Questions: How would you vote and why? What arguments would you use to support your decision?

Species: long-toed salamander
Range: southeastern Alaska to northern California
Status: declining in some areas

Species: golden toad
Range: Costa Rica
Status: thought to be extinct

Species: mountain yellow-legged frog
Range: California
Status: declining throughout range

Species: northern leopard frog
Range: much of Canada, northern and western U.S.
Status: declining in some areas

By Michael J. Lannoo

Scientists are trying to learn why many frogs, salamanders, and other amphibians are declining in number and developing deformities.

Amphibian Mystery

Glossary

Amphibian: A vertebrate (animal with a backbone) that lacks such skin coverings as scales, feathers, or hair, and typically lives part of its life on land and part in water.

Aquaculture: The raising of aquatic organisms, such as fish, shellfish, and water plants; commonly known as fish farming.

Endocrine disrupters: Synthetic chemical compounds, such as certain pesticides, that have chemical properties similar to human and animal hormones and are alleged to sometimes disrupt the balance of those natural hormones.

Metamorphosis: The physical transformation of an animal from one stage of development to another, such as from a tadpole to an adult frog.

Mutation: A change in the structure of genetic material, which in some cases can lead to disease or abnormalities.

Ozone layer: A layer of the upper atmosphere consisting of molecules composed of three oxygen atoms that filters out much of the sun's ultraviolet radiation.

Population: A group of interacting individuals of the same species living in a particular region.

Wetland: An area of shallow water, such as a swamp or marsh, that typically dries up periodically.

The author:
Michael J. Lannoo is an associate professor at the Indiana University School of Medicine in Muncie. He is also a researcher with the Declining Amphibian Populations Task Force.

The golden toads of the Monteverde Cloud Forest Preserve, a rain forest in the mountains of Costa Rica, were numerous and easy to find as recently as 1987. The yellow-orange males and the brownish females would be seen during breeding season after they emerged from their burrows. In 1988, however, scientists reported a huge drop in the number of these toads—they saw only 10 of them. The next year, they could find only one. In December 1997, after no golden toads had been seen for several years, biologists said that the species was almost certainly extinct. Moreover, they said, the toad's disappearance could not be attributed to natural causes. Although animal numbers fluctuate naturally from year to year, the golden toad's precipitous decline led biologists to suspect that human activities were involved. Whatever the cause of its disappearance, the golden toad was not the only amphibian species in the Monteverde preserve to be affected. Six other species of frogs or toads in the forest also declined in numbers during the same period.

The frogs and toads of the Monteverde rain forest, like other amphibians, are characterized by a life cycle that is partially *terrestrial* (on land) and partially *aquatic* (in water). A typical amphibian lives on the land until it is time to breed and lay eggs, when it returns to the water. Eggs hatch into aquatic *larvae* (immature organisms), which live in the water until undergoing *metamorphosis* (changing into adults). Not all amphibians, however, go through this cycle. Many species live entirely on land or in water. Amphibians are also characterized by a lack of protective body covering, such as scales, feathers, or fur. The skin of amphibians is exposed and easily penetrated by substances in the water and air, allowing them to breathe through their skin as well as with their lungs.

Scientists have been reporting since the late 1980's that certain *populations* (regional groups within a species) of amphibians are in decline. And in the mid-1990's, there was another worrisome development. Many people began seeing amphibians with various kinds of malformations, including extra limbs and abnormal sex organs. Scientists are concerned about these problems because amphibians are thought to be good indicators of environmental health. The ability of amphibians to readily absorb substances from the water and air makes them more sensitive than most other animals to environmental conditions. Amphibians may thus be providing an early warning of a deteriorating environment. If so, researchers caution, whatever is harming amphibians might also pose a threat to human health.

Some biologists, however, see little cause for concern. They contend that many amphibian population declines are just natural fluctuations. Some scientists also question whether malformations are truly on the rise among amphibians. They note that deformities in frogs had been reported as far back as the 1800's.

Researchers were engaged in many studies in the 1990's to learn what—if anything—was happening to amphibians. By 1998, these studies had uncovered a number of important clues that helped investigators better understand this amphibian mystery.

Frogs, toads, and their kin

There are more than 4,000 species of amphibians in three main groups: frogs and toads (3,500 species), salamanders and newts (350 species), and caecilians (*see SIL ee uns*) (160 species). Frogs and toads have short tailless bodies with long hind legs. They range in length from 1 centimeter (0.4 inch) to 35 centimeters (14 inches). Toads differ from frogs in having rougher skin and spending most of their life on land. Most frogs and toads lay their eggs in water. Tadpoles (the aquatic stage of frogs and toads) develop from these eggs and live in the water until growing legs and becoming land-based adults. Many frogs, however, continue to spend much of their time in or around water.

Salamanders and newts have long bodies with tails. They range in length from 3 centimeters (1.2 inches) to 160 centimeters (63 inches). Although most salamanders and newts have two pairs of limbs, some aquatic forms have no hind limbs and only small forelimbs.

Caecilians are limbless, wormlike amphibians found throughout the tropics. They range in length from 7 centimeters (2.8 inches) to 150 centimeters (59 inches). Some species lay eggs, while others give birth to their offspring.

The scientific community first began to learn about amphibian population declines at the First World Congress of Herpetology [the study of reptiles and amphibians], held in Canterbury, England, in 1989. David Wake, director of the Museum of Vertebrate Zoology at the University of California at Berkeley, presented evidence of dwindling frog populations in central and northern California. In 1990, an inter-

What are amphibians?

There are three major types of amphibians—caecilians, *top right;* salamanders and newts, *center right;* and frogs and toads; *bottom right.* Most amphibians spend part of their lives in water and part on land. For example, the life cycle of a typical frog, *below,* begins when tadpoles hatch out of underwater eggs and cling to vegetation until capable of swimming. As the tadpoles mature, they develop limbs and lose their tail. When a frog's development is complete, it can live on land.

national group of about 40 biologists gathered in Irvine, California, to discuss research findings. As the scientists exchanged information, it became apparent to them that amphibian disappearances were a global phenomenon. The most disturbing development was that amphibians seemed to be disappearing not only from areas known to be disturbed by human activity but also from areas thought to be pristine wilderness.

In 1991, the Species Survival Commission, a division of the World Conservation Union (an international organization made up of more than 500 environmental groups), established the Declining Amphibian Populations Task Force (DAPTF). The DAPTF, based at the Open University in Milton Keynes, England, recruited more than 1,200 scientists, including me. Our research goals were to determine: (1) if amphibian numbers are, in fact, declining; (2) if so, why; and (3) what can be done to halt these losses?

Evidence for population declines and malformations

Proving that amphibian populations are truly dwindling has been challenging for scientists. A major problem with documenting amphibian declines is that, in most regions and for most species, there are no reliable data on past population sizes to serve as a basis for comparison with recent observations. Because populations fluctuate naturally—due to such factors as drought, variations in the availability of food, and changing numbers of predators—an apparently worrisome decline could be just a short-term phenomenon. In addition, it is difficult to confirm the suspected extinction of a species. Just because a species is not found by a group of researchers at a certain time and in a certain place does not necessarily mean that the species no longer exists there.

Because of these difficulties, scientists participating in the DAPTF focus their studies on areas where historical data on amphibians are available. And they make observations over several years under a variety of conditions. A number of such studies have found unmistakable declines in amphibian populations. For example, in the mid-1990's, my research team (affiliated with Iowa Lakeside Laboratory near Milford) found a hundredfold to a thousandfold decline in amphibian numbers in the Prairie Pothole Region of northwest Iowa, compared with the early 1900's. Of the seven native species of amphibians in the region, we believe that one, Blanchard's cricket frog, has become extinct. A second species, a type of salamander called the mudpuppy, has not been seen in the region in about 30 years. It's hard to say if this species is extinct however, because its silent, *nocturnal* (active at night) nature makes it difficult to track down.

Other research teams have made similar findings. A group at the University of Alberta in Canada discovered that the northern leopard frog, once common in Alberta, had virtually disappeared since 1979. In the Canadian province of Quebec, researchers documented a dramatic decline in the number of chorus frogs and chorus-frog habitats throughout the St. Lawrence River Valley since 1988. And Australian

Declining numbers and deformities
The natterjack toad, *above,* is one of many amphibian species whose numbers are declining. Although this toad was once common throughout much of Great Britain, by the 1990's it could only be found in scattered sites there. In the United States and Canada, many deformed frogs have been seen since the mid-1990's. *Right,* a researcher displays a frog that is sprouting an extra pair of hind legs.

scientists found that 14 species of frogs in the rain forests of eastern Australia had either disappeared or greatly declined in number since the late 1970's. These are just a few examples among many.

In contrast to these reports of declines, other studies have found that some amphibian populations are stable. For example, researchers studying amphibians in the Canadian province of New Brunswick over several years found no evidence that any species there have declined in number. And studies since the early 1980's at the Savannah River Ecology Laboratory, affiliated with the University of Georgia, have discovered that amphibian populations on the Georgia-South Carolina border undergo considerable fluctuations. But the researchers have found no indications that any of the populations in the area are in permanent decline.

Some studies have obtained contrasting results by looking at different populations of the same species. For example, in North America, studies indicate that northern populations of Blanchard's cricket frogs have declined, while southern populations have remained robust. Such studies underscore the fact that amphibians in different regions confront different environmental conditions. In addition, within particular regions, some species of amphibians may be in decline while others are stable. Although researchers have documented cricket frog declines in the upper Midwest, many other amphibian species that share the cricket frog's wetland habitat in the region appear to be doing well. These sorts of findings indicate that different amphibian species respond in their own individual ways to the same kinds of environmental influences.

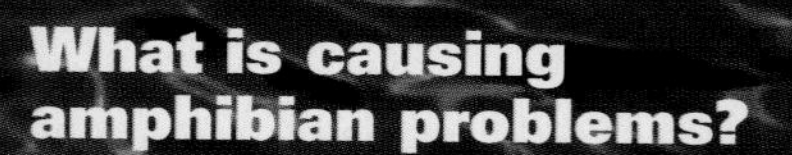

What is causing amphibian problems?

Scientists in 1998 were trying to determine the causes of amphibian population declines and malformations. Most researchers believed that a combination of factors was responsible for these problems and that much of the blame could be attributed to humans.

Destruction of habitats

Scientists believe that the destruction of such amphibian habitats as wetlands and rain forests is the largest single cause of amphibian population declines. Two aerial photos, *right,* of the same plot of land near the Everglades in Florida show the conversion of wetlands (mottled, reddish areas in these infrared photos) to roads and residential subdivisions between 1984, *top,* and 1990, *bottom.* In Brazil, a rain forest is burned, *below,* to make room for farmland.

Ultraviolet radiation

A scientist records the status of salamander eggs shielded by protective filters from the ultraviolet (UV) rays of the sun in an Oregon mountain lake in 1997, *below.* Researchers found that eggs that were shielded developed normally, but eggs that were left exposed to UV rays contained embryos with odd bumps and other deformities, *left.* Other studies have linked an increase in the level of UV radiation to certain atmospheric pollutants.

Disease

The fungus *Saprolegnia, above,* has caused disease in a growing number of Oregon amphibians during the 1990's. Scientists believe that an increase in UV radiation may be making amphibians more susceptible to diseases.

Pesticides

A crop duster releases chemical pesticides over a farm field, *above.* Water and wind sometimes carry pesticides from farms to amphibian habitats, where they can cause malformations and other problems.

Introduction of new species

Newts mate in a California stream, *right.* Research in 1997 indicated that California newts were declining in number because nonnative crayfish were forcing them out of the water and onto land, where they cannot mate. The crayfish had been introduced into the newts' streams by humans.

While some researchers are studying trends in amphibian populations, other investigators are focusing on amphibian malformations. The phenomenon of amphibians with various kinds of physical abnormalities came to national attention in 1995, when a group of Minnesota students discovered hundreds of deformed frogs during a field trip to a wetland. Soon, other reports of malformed amphibians were being reported from many parts of the United States and Canada. Malformations have also been reported from Japan. Observed deformities have included misshapen, extra, or missing limbs; missing or shrunken eyes; and abnormally small sex organs. The relationship between amphibian population declines and amphibian malformations —if any—was still unclear in 1998. Observers have seen malformations in amphibian populations that appeared to be stable, and they have noted a decline in numbers among many amphibian populations that have no malformations.

An analysis of the various studies done in the 1980's and 1990's has led most biologists affiliated with the DAPTF to conclude that amphibians in many parts of the world are suffering population declines and malformations. Researchers in 1998 were trying to determine the causes of these problems. They were investigating causes both originating in nature and stemming from human activities.

Some possible causes of amphibian problems

One natural factor that scientists think is probably involved in some amphibian population declines is drought. Because most amphibians lay their eggs in water, droughts can have a devastating effect on them. Food shortages and greater numbers of predators can also decimate amphibian populations. However, biologists believe that when a population experiences a steady decline over many years, other factors must be involved.

One additional possibility of natural origin is disease. Since the late 1980's, many amphibians in Oregon have been sickened by a fungus known as *Saprolegnia.* Researchers in 1997 reported that an unknown type of protozoan or fungus was killing frogs in Panama's Fortuna Forest Reserve, and frogs in Australia were found to be dying from a similar cause. Also in 1997, investigators in the United States implicated a bacterium called *Aeromonas hydrophila* in the disappearance of several populations of western toads in Colorado.

Parasitic organisms have been implicated in certain amphibian malformations. Flukes—a type of parasitic worm—burrow into tadpoles and form small cysts, which some investigators believe can cause deformities as the tadpole develops.

Although natural causes may contribute to amphibian problems, human disruptions of the environment undoubtably play a more important role. In fact, most biologists believe that the destruction and alteration of wetlands is the single biggest cause of amphibian population declines. In September 1997, the U.S. Fish and Wildlife Service released a report estimating that more than 405,000 hectares (1 million

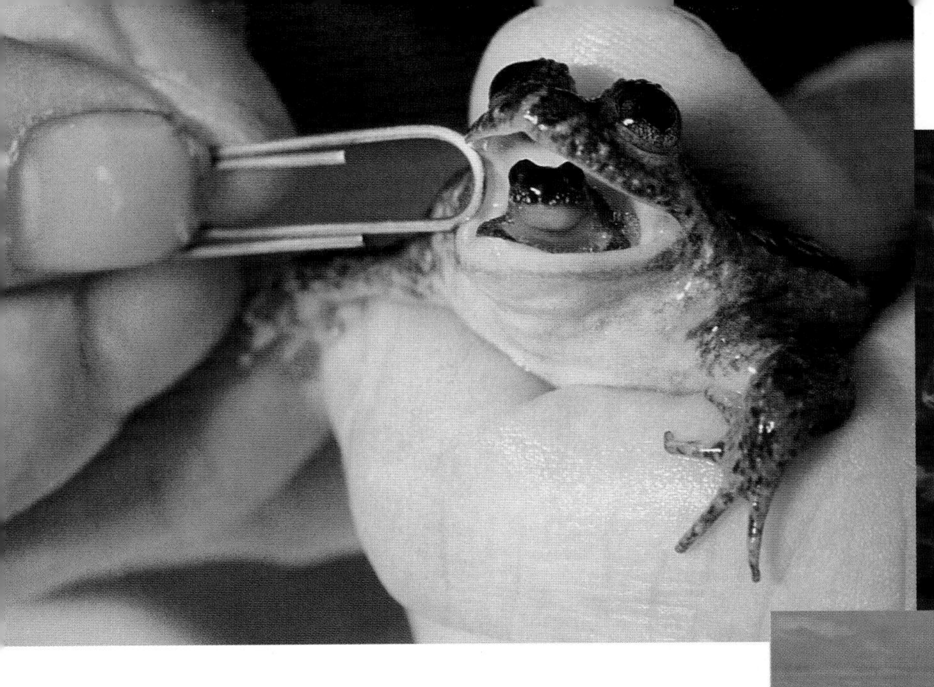

Implications for people

The extinction of amphibian species and the occurrence of deformities among other species have serious implications for people. The loss of a species may mean the loss of a valuable drug to treat a disease. And an amphibian suffering from a malformation may be an early warning of problems that could eventually affect humans.

acres) of wetlands—which include various kinds of watery terrain, such as marshes and swamps—vanished in the United States between 1985 and 1995. This loss of wetlands has harmed many animals besides amphibians, including numerous species of birds and mammals. Construction has been a major cause of the problem. The building of roads, housing developments, and commercial strips fragments the landscape, reducing wetlands and cutting them off from one another.

The use of surviving wetlands for *aquaculture* (fish farming) has also imperiled wildlife. Since the 1980's, an increasing number of natural wetlands in North America have been converted into aquacultural basins for the raising of both game fish and bait fish. In Wisconsin alone in 1997, some 7,000 wetlands were made available to commercial interests for the creation of fish farms. During the conversion of a wetland to an aquacultural basin, toxic chemicals are sometimes added to the water to kill existing fish in order to make room for fish of commercial value. These chemicals may also kill amphibians. In addition, when wetlands are too shallow for aquaculture, they may be dredged to make them deeper. Amphibians, which generally prefer shallow water, may not be able to survive in such an altered habitat.

Nonnative animals released in aquacultural basins, streams, and other habitats can have devastating effects on the ecology of an area by preying on amphibian eggs and larvae. Salmon and trout added to Sierra Nevada mountain streams in Yosemite National Park, for example, have been implicated in the decline of the park's mountain yellow-legged frogs. Even bullfrogs, which were introduced to western North America from eastern North America around 1900, have caused the decline of many native western am-

Top, a gastric brooding frog's mouth is opened to reveal the frog's offspring in the 1970's. A study of how this frog, which is now thought to be extinct, switched off digestive enzymes to brood its eggs in its stomach might have led to new ulcer treatments. *Above,* a sunbather soaks in ultraviolet rays. Studies linking an increase in UV rays to amphibian health problems indicate that people should be more concerned about the risks of contracting skin cancer and eye disease from too much sun.

phibians by competing with them for food and preying on them. Nonnative animals may also spread diseases to native amphibians.

Another factor that scientists believe is responsible for amphibian problems is the use of agricultural pesticides, which may be transported by water or wind into amphibian habitats. Pesticides can produce fatal genetic mutations and developmental malformations in animals and can weaken their immune systems. Biologists suspect that agricultural pesticides have played a role in the collapse of the amphibian populations in the Monteverde rain forest and in Alberta and Yosemite National Park.

Even pesticides that have been banned from use may be killing amphibians. Studies in the 1990's found that toxic residues from the breakdown of the insecticide dichloro-diphenyl-trichloroethane (DDT) were present in the bodies of frogs in Point Pelee National Park, in southern Ontario, Canada. Although DDT, which was once widely used to control mosquitoes in the park, was banned in the United States and Canada in the 1970's, its toxic breakdown products are still present in the environment. In addition, DDT itself is carried to North America by winds and migratory birds from countries where it is still used.

Scientists have found that certain pesticides, fertilizers, and other synthetic chemicals mimic female hormones. These synthetic chemicals, sometimes called endocrine disrupters, have been linked to defects of sex organs in amphibians and other animals in various locations in the United States and Canada, including the Minnesota region where amphibian deformities were first observed in 1995. Such animals may be unable to breed.

Another environmental threat that may be harming amphibians is *acid rain* (precipitation containing sulfuric or nitric acid), formed when certain kinds of pollutants in the atmosphere combine with other chemicals. The pollutants come mainly from the burning of fossil fuels. Excessively acidic water and soil can kill amphibians, affect their behavior, or cause them to suffer developmental disorders. In England, population declines among natterjack toads have been attributed, in part, to acid rain.

Other factors that may be harming amphibians

Many studies indicate that amphibian declines are most prevalent in high-altitude areas, a finding that points to increases in ultraviolet radiation from the sun as a possible cause of some amphibian deaths and malformations. This intense form of light energy, which can cause genetic changes and other physical damage, is strongest at high altitudes, where there is less atmosphere to filter damaging solar rays. Ultraviolet radiation may have grown stronger in these areas since the 1970's due to the thinning of the ozone layer, a protective blanket of oxygen molecules that absorbs ultraviolet rays. Scientists believe that the ozone layer has been eroded by certain reactive chemicals—principally chlorofluorocarbons—that have drifted into the upper atmosphere.

In December 1997, Oregon State University researchers reported that ultraviolet radiation was killing *embryos* (an early developmental stage) of the long-toed salamander in lakes of the Cascade Mountains. The scientists said that more than 90 percent of the embryos they studied that had been exposed to direct sunlight died before hatching or hatched with deformities. In contrast, almost all of the embryos that the researchers had shielded from sunlight with ultraviolet filters developed normally.

One last factor that should not be overlooked is the collection of amphibians—mostly frogs—for biological-supply companies and the food industry. The biological-supply trade provides dissection specimens for school classrooms. Such collecting is not well regulated and has reduced the number of frogs in many areas. Hunting frogs for their legs, a gourmet delicacy, has also depleted frog populations.

The various factors affecting amphibians do not occur in isolation. Scientists believe that several negative influences probably work in combination to push an amphibian population into decline. For example, pesticides and ultraviolet radiation could weaken amphibian immune systems, making them more vulnerable to bacterial infections and parasitic diseases.

Effects on both nature and people

A die-off of amphibians can have a profound impact on a wetland area. Amphibians are often the most abundant *vertebrates* (animals with a backbone) in wetlands. Therefore, their disappearance can lead to a proliferation of insects, their main prey. In addition, reptiles, birds, and mammals that feed on adult amphibians or their larvae are liable to face a serious food shortage.

The problems threatening amphibians may have implications for human life. Some researchers believe that humans are being affected by the same hormone-mimicking chemicals linked to amphibian sexual abnormalities. A few disputed studies in the 1990's found that men were producing fewer sperm than in previous decades, and some scientists speculated that the decline might be due to endocrine disrupters. Other studies in the 1990's linked lower IQ scores in children to these chemicals, but that research, too, was controversial. A more accepted scientific conclusion was that the higher levels of ultraviolet radiation that seemed to be affecting amphibians may be causing a greater incidence of human skin cancer and eye disease.

The loss of amphibian species may in itself have human implications, particularly in the development of new drugs. In 1973, for example, a frog called the gastric brooding frog was discovered in Australia. This frog was of interest to scientists because it was somehow able to switch off the production of digestive chemicals in order to brood its eggs in its stomach. Researchers hoped that learning how the frog did this might lead to new treatments for stomach ulcers. Unfortunately, the gastric brooding frog has not been seen in the wild since 1979 and is thought to be extinct.

Seeking solutions
Researchers affiliated with the author study amphibian habitats in Iowa. Their studies have found that the ideal amphibian habitat in the upper Midwest consists of a series of wetlands containing no fish and connected by stretches of undisturbed landscape. The study, restoration, and preservation of amphibian habitats help ensure that amphibian populations will be adequately protected.

Steps toward protection

The loss of such a potentially valuable amphibian species—as well as those species whose primary value was in preserving life in wetland communities—emphasizes the importance of protecting these vulnerable creatures. The most important step we can take is to preserve amphibian habitats. In the upper Midwest, my colleagues and I have found that the ideal amphibian habitat is a series of wetlands containing no fish and connected by stretches of undisturbed landscape. In other regions of the United States and the world, ideal amphibian habitats include old-growth forests, diverse prairies, open savannas, flowing streams, and vegetation-fringed lakes.

Another important step that could protect amphibians, and people as well, is a further reduction of chemical pollutants in the environment. By the mid-1990's, a number of nations had taken action to reduce the use of chemicals responsible for acid rain and ozone depletion. However, some other chemicals suspected of harming amphibians—including endocrine disrupters—were still being widely used in the United States and other countries.

I and other scientists affiliated with the DAPTF acknowledge that more research is needed before the extent and causes of amphibian disappearances and deformities are fully understood. Nonetheless, we believe that the findings to date clearly show that there is indeed a problem and that humans are partly to blame for it. Habitat destruction, pesticide use, and ozone depletion seem to be magnifying such natural factors as disease and drought. Perhaps the possibility that amphibians are sounding an early warning of environmental damage with potential human consequences may prompt people to practice better stewardship of the Earth as we enter a new millennium.

For further reading:

Blaustein, Andrew R. and Wake, David B. "The Puzzle of Declining Amphibian Populations." *Scientific American,* April 1995, pp. 52-57.

Gannon, Robert. "Frogs in Peril." Popular Science, December 1997, pp. 84-88.

Halliday, Timothy R. and Heyer, W. Ronald. "The Case of the Vanishing Frogs." *Technology Review,* May/June 1997, pp. 56-63.

Lannoo, Michael J. *Okoboji Wetlands: A Lesson in Natural History.* University of Iowa Press, 1996.

Lannoo, Michael J. (editor). *Status and Conservation of Midwestern Amphibians.* University of Iowa Press, 1998.

Phillips, Kathryn. *Tracking the Vanishing Frogs: An Ecological Mystery.* St. Martin's Press, 1994.

For additional information:

The Froggy Page (An information clearinghouse with many links to scientific and popular Web sites about amphibians.)— frog.simplenet.com/froggy

Questions for thought and discussion

Scientists believe that the destruction of wetlands is the largest single factor contributing to the decline of amphibian populations. Suppose you lived near a marsh that you enjoyed visiting in order to see the many frogs, birds, and other animals that lived there. One day you read in the newspaper that a company has decided that the marsh would make an ideal site for a development it is planning. The company wants to fill in the marsh and build an industrial park that would offer many good, high-paying jobs for your community. You yourself are interested in one of the jobs that the company mentions.

Questions: Would you favor or oppose the construction of the industrial park? What would be the advantages and disadvantages of the project versus the preservation of the marsh? Can you think of a compromise that would permit the existence of both the industrial park and the marsh? Explain how such a compromise might be accomplished.

Art conservator Pinin Brambilla Barcilon cleans a portion of Leonardo da Vinci's famous painting *The Last Supper* in Milan, Italy. Various scientific methods were used to assess the damage to the mural and guide the painstaking restoration.

New Tools of the Art Conservator

Art conservators increasingly rely on sophisticated scientific methods to rescue precious works of art.

By Gordon Graff

In 1998, after almost 20 years of labor, a team of artisans at the monastery of Santa Maria delle Grazie in Milan, Italy, reached the final stages of a slow and painstaking project. They had restored much of what was left of Leonardo da Vinci's badly deteriorated wallpainting *The Last Supper,* which the Italian Renaissance master had completed exactly 500 years earlier, in 1498. Armed with magnifying glasses, scalpels, solvents, and cotton swabs, the specialists, led by conservator Pinin Brambilla Barcilon, had slowly cleaned away centuries of dirt, as well as encrusted layers of paint and glue applied to the renowned mural by generations of well-meaning restorers. They also brushed in neutral shades of watercolors where the original paint had flaked off.

Science played a role in this prolonged effort to prevent Leonardo's haunting image of Christ and his disciples from crumbling to dust. The restorers used microscopic and chemical analyses of the painting's surface to guide them. They photographed the painting in visible light to record its appearance at different stages of the restoration and in infrared light to peer beneath its surface. They also converted their photographs into computer images so they could easily track their progress.

Such scientific methods are becoming more common as art conservators race against time to rescue priceless works of art from the ravages of aging and the elements. The use of scientific techniques in the art world is nothing new, of course. Over the years, for example, art historians have often used X rays to reveal hidden paintings and alterations beneath well-known works of the old masters. Museum curators have also relied on chemical analyses of paint to determine whether pieces in their collections are originals or clever fakes. But today, specialists are increasingly harnessing the power of advanced chemical and physical tools to preserve works of art that are threatened with decay or to restore those that have already begun to deteriorate.

Terms and concepts

Consolidant: A chemical solution used in the conservation of stone artwork that strengthens the bond between the particles of stone.

Fresco: A style of wallpainting in which paints are applied to a wet lime-based plaster.

Infrared light: Light waves that lie just beyond the red end of the visible-light spectrum.

Inorganic compounds: Chemical compounds, such as pigments made with metals, that do not contain carbon.

Organic compounds: Chemical substances based on carbon.

Oil paints: Paints made by adding pigments to linseed oil or other kinds of oil.

Pigment: A substance, often a metal or plant material, that gives color to paints and dyes.

X rays: Electromagnetic waves that are much shorter and higher in energy than those of visible light and are able to penetrate into most substances.

Ultraviolet light: Light waves that lie just beyond the violet end of the visible-light spectrum.

The modern assault on artwork

Conservation efforts are necessary because most works of art are under constant assault by an array of destructive forces. Natural agents such as water, sunlight, oxygen, and carbon dioxide have always caused statues, monuments, and paintings to deteriorate over time. But since the beginning of the 1900's, the problem has been much worse, with smoke, soot, dust, acid rain, and corrosive gases greatly accelerating the processes of deterioration.

Deterioration of stone monuments has become an international concern. The Great Sphinx of Giza in Egypt, the ancient Maya pyramid at Chichen Itza in Mexico's Yucatan Peninsula, Wells Cathedral in England, and the Acropolis in Athens are all deteriorating rapidly, particularly from the ravages of modern pollutants. Even the Lincoln Memorial in Washington, D.C., which was completed in 1922, shows signs of decay.

Chemical agents from industrial pollution are the biggest culprits in the decay of outdoor monuments and statues. For example, sulfur dioxide, produced by the burning of fossil fuels such as coal or oil, combines with oxygen and water in the atmosphere to form sulfuric acid. This corrosive compound returns to the Earth in "acid rain." On bronze or copper statues, such as the copper-clad Statue of Liberty in New York City, sulfuric acid attacks the green *patina* (surface coating) that forms on these materials and helps protect them from the elements. Sulfuric acid also reacts with calcium carbonate, the principal mineral in limestone and marble, to form calcium sulfate dihydrate, or gypsum, a white crystal. Rain can wash away the gypsum, creating unsightly pits or destroying details carved in the stone.

Paintings and sculptures in museums, churches, and other indoor areas are also subject to damage by airborne pollutants. Many frescoes, which were originally painted on a wet lime-based plaster on interior walls, are deteriorating from exposure to sulfur dioxide. The gas also produces brittleness and discoloration in works of art made with various natural materials, such as paper, cotton, silk, and leather. Ozone, a three-atom form of oxygen created by the interaction of auto exhausts with sunlight, discolors paint pigments and dyes. In addition, human activities in places where art is displayed frequently contribute to its de-

The author:
Gordon Graff is a free-lance science writer.

Challenges for art conservators

Artwork deteriorates over time because of the aging of artists' materials and exposure to natural elements, such as sunlight and moisture. But in the 1900's, pollutants increased the urgency for conservation efforts.

A 1908 photograph, *far left,* shows the details of a sandstone sculpture created in 1702 for the Herten Castle in Recklinghausen, Germany. By the 1970's, acid rain and other airborne pollutants had destroyed the face and other features of the sculpture, *left.*

terioration. For example, the smoke from candles and incense used in religious services over the centuries gradually caused Michelangelo's epic paintings in the Sistine Chapel of the Vatican in Rome to darken.

Additional destructive agents

Even in the absence of pollutants, fairly normal conditions can cause severe damage to artwork. Excess moisture in the air can encourage the growth of molds on wooden objects such as old furniture, warp canvases by causing their fibers to swell, and create an environment favorable to destructive insects. And sunlight filtering into a room can gradually fade the *pigments* (colored substances) in paintings.

The quality of the original materials or the technical skill of the artist can also affect the durability of art. For example, an oil painting naturally develops a *craquelure* (*KRAK kloor*), a series of fine lines on the surface, as the oil ages. But improperly prepared or applied oil paints can develop deep cracks that make the layers of paint unstable or cause them to separate from the canvas. And varnishes that artists use to protect oil paintings often

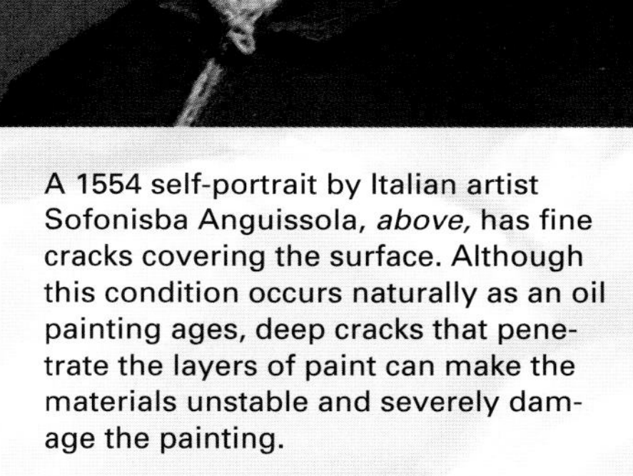

A 1554 self-portrait by Italian artist Sofonisba Anguissola, *above,* has fine cracks covering the surface. Although this condition occurs naturally as an oil painting ages, deep cracks that penetrate the layers of paint can make the materials unstable and severely damage the painting.

yellow over a long period of time and obscure the colors of the paints.

Unfortunately, past attempts to preserve or restore art sometimes did more harm than good. Waterproof coatings intended to protect stone statues trapped moisture in the stone's pores, causing the stone to crack when the water froze. Glues that were intended to secure loose fragments of paint absorbed dirt, which darkened the paintings. Many crumbling or fading frescoes were protected with paraffin wax, beeswax, various oils, or shellac—materials that could radically alter the appearance of the underlying colors. And occasionally, conservators tried to restore a decaying masterpiece by repainting it—as was done with *The Last Supper* twice in the 1700's and once in the 1800's, thereby masking Leonardo's colors and brushstrokes.

Modern restoration methods are more cautious than earlier techniques, though they are not without their critics. Michelangelo's Sistine Chapel frescoes, which were cleaned with distilled water and detergents in a project that ended in 1993, are a case in point. The restoration was received favorably by most people, but some critics claimed that the restorers had transformed Michelangelo's subtle, shadowy tones into flat, garish colors that were not in keeping with the master's intentions. While such controversy is probably unavoidable, conservators usually try to minimize it by consulting art historians and experts in the chemistry of art materials before starting a restoration.

A close look at the damage

The first step in bringing a piece of art back to at least some semblance of its former glory is deciding exactly what needs to be done. The restoration of a painting, for instance, usually begins with a careful visual inspection. Conservators may study the painting's surface with a *stereomicroscope,* a low-power microscope that gives a three-dimensional view. This close-up view enables them to analyze the artist's brushstrokes and to detect damaged or fragile areas of the painting that require special attention.

For an even more detailed view, conservators sometimes scrape tiny flakes from a painting, usually from a corner or another inconspicuous place. Using high-power microscopes, they can observe the separate layers of paint in the samples. They are also able to magnify the image of a sample enough to make out individual pigment particles. This technique can reveal how an artist mixed different pigments to achieve a particular effect.

Conservators can also get a new view of a painting by photographing it under wavelengths of light that are shorter or longer than visible light. These include X rays, ultraviolet rays, and infrared rays. X-ray photography, one of the oldest scientific art-inspection tools, is useful for peering beneath the surface of a painting. An X-ray image is created by passing a beam of X rays through a painting onto a piece of photographic film. Different pigments absorb the rays in varying degrees. Pigments containing lead, for example, absorb the rays completely and produce a white image on the film. Other pigments, because they absorb fewer rays, appear as varying shades of gray. Pigments that do not absorb the rays at all appear black. This effect can reveal hidden images that have long been obscured by grime or overlying layers of paint.

Ultraviolet light causes pigments to *fluoresce* (glow) in wavelengths of light that can be recorded on film. Different pigments vary in their fluorescence, even when they have the same color under normal lighting. For example, a yellow pigment made from cadmium sulfide and a yellow pigment made from a synthetic material might look the same in visible light but would appear different in an ultraviolet photograph. Consequently, an ultraviolet image can sometimes allow a conservator to distinguish between the work of the original artist and touch-ups done by a later restorer.

Art restorers can also see below the surface of a painting by photographing it in infrared light, which passes through the first layer of paint and then is reflected back. A photographic image of the reflected light, known as a reflectogram, enables conservators to get a detailed picture of an underlying painting or the original artist's preparatory drawings.

Another imaging method, called neutron autoradiography, creates a series of images that allow the conservator to look at only a few pigments at a time. To use this procedure, specialists bombard the canvas with a stream of *neutrons* (subatomic particles with no electrical charge inside the nuclei of atoms) that are generated by a nuclear reactor. The neutrons impart energy to the atoms of the *inorganic* pigments (those lacking carbon). The atoms then reemit that energy as other forms of radiation, mostly *beta particles* (electrons) and high-energy gamma rays. The time it takes for the atoms to shed the excess energy varies, depending on the pigment. For example, it takes about 2½ hours for manganese to return to a normal energy state and three days for gold. Researchers record the atom's radioactive emissions on several pieces of film at various times. Because only a few pigments appear on each piece of film, the process essentially separates details of a painting for careful inspection.

Chemical analysis of paintings

In addition to creating a variety of images of a painting, conservators like to conduct a complete analysis of the pigments, varnishes, oils, and other materials in a painting. This information can guide the selection of solvents and detergents used to remove grime, glue, and other unwanted coatings. Also, because certain pigments can be linked to particular historical periods, understanding the chemical composition of pigments helps conservators determine what parts of a painting do not belong to the original work.

One method of chemical analysis is called infrared spectroscopy. Conservators place a tiny scraping from a painting in an instrument that bombards the sample with various wavelengths of infrared rays. Sensors in the instrument record on a graph the degree to which

Taking a deeper look at art

Various analytical tools help art conservators peer beneath the surface of paintings or see details that do not appear in normal lighting. This kind of analysis can help conservators evaluate the condition of a piece and reconstruct its history.

Young Lady in a Tricorn Hat by Italian artist Giovanni Battista Tiepolo, *above left,* depicts a young woman holding a closed fan. An X-ray photograph of the painting, *above right,* reveals that the artist had originally painted an open fan. X rays allow conservators to look beneath the surface of paintings because some pigments in paint absorb the rays more than others. Lead-based pigments absorb the rays completely and appear white in an X-ray image.

The Abduction of the Sabine Women by Italian artist Luca Giordano, *left,* was photographed under visible light. When photographed under ultraviolet light, *below,* the materials in the painting appear to have a different color. The varnish, which normally appears colorless, now looks yellow.

The ultraviolet light reveals the work of a conservator. Synthetic paints used to touch up portions of the painting were applied on top of the varnish. Those materials, which do not *fluoresce* (glow) in ultraviolet light, appear as dark splotches, revealing the places where the new paints were added.

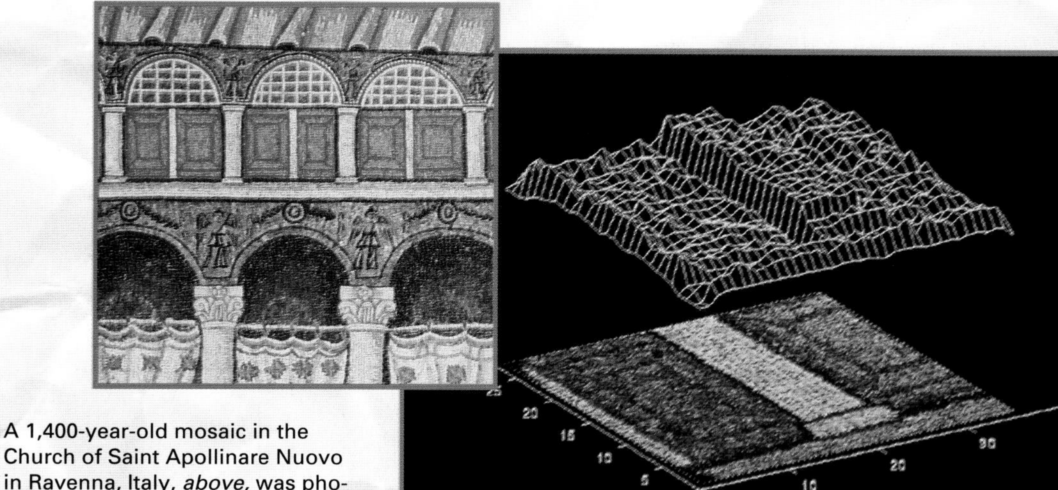

A 1,400-year-old mosaic in the Church of Saint Apollinare Nuovo in Ravenna, Italy, *above,* was photographed with a special camera to create computerized images to assist conservation efforts. The "bumpy" image, *above right,* reveals the degree of reflectance of each tile under certain wavelengths of light. The variations in reflectance—the differing "heights" of the tiles—depend on what the tiles are made of. Consequently, conservators can distinguish the original tiles from those added later. By using a computer program that combines different images of reflected infrared light, *right,* conservators could reveal the condition of the mosaic below the surface. The final image in the bottom right corner reveals white splotches, areas where tiles have become detached from damaged plaster.

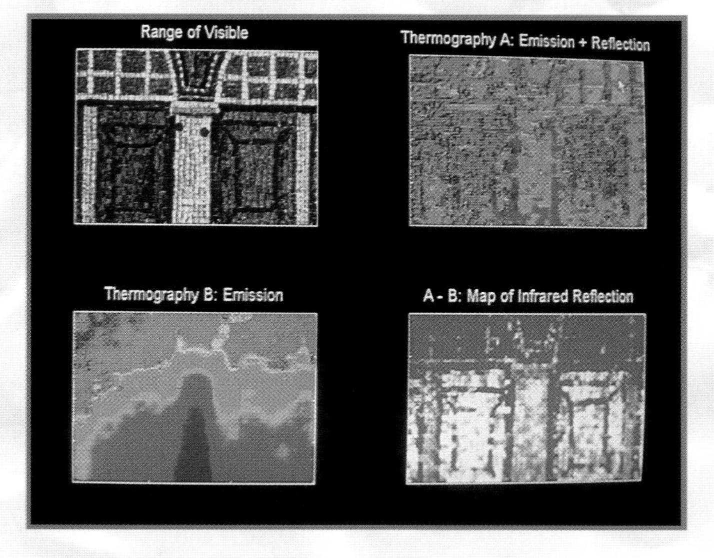

organic (carbon-containing) materials absorb the rays. Because the degree of absorption of various wavelengths of infrared rays is unique to many chemical compounds, infrared spectroscopy can graphically display which chemicals are present in a paint sample. This chemical analysis was crucial in helping conservators identify the composition of the soot, grime, glues, and waxes covering the Sistine Chapel ceiling. Specialists can also determine the various chemical constituents of a painting by probing the surface of a painting with a procedure called X-ray fluorescence spectroscopy. This technique identifies inorganic materials on the basis of how their atoms absorb and reemit radiation from X rays.

Using a combination of these physical and chemical inspections, a conservator can reconstruct a painting's history, its original appearance, how it has deteriorated, and the alterations made by previous restorers. With that information in hand, the restoration can proceed.

With many paintings, conservators start by stripping away layers of glue or varnish using organic solvents and cotton swabs. They may remove dirt with mixtures of distilled water and mild detergents. Often they must use scalpels to painstakingly scrape away layers of paint covering the original images.

After that work has been completed, the conservators must decide how closely they want to return the artwork to its original condition. Extensive repaintings to replace missing or damaged paint are less common today than they once were. Conservators often fill in gaps with neutral hues when they do not know which colors are missing or they determine that repainting would violate the spirit of the original work. When restorers do apply new paints to oil paintings, they often use nonoil paints—watercolors or pigmented acrylic resins—that are easily removed. Future conservators who need to do further restoration work would then be able to remove the touch-ups and begin once again with only the original material.

Restoration of statues and monuments

As with paintings, the restoration of sculptures, statues, and monuments also begins with an assessment of their condition. This is usually done by making a visual inspection of their surfaces and a chemical analysis of surface grime or corrosion. Conservators also study the history of the works, including old photographs and drawings.

Next, restorers must select an appropriate cleaning method and determine how extensively they should apply it. For metal and stone artwork, many effective cleaning methods can also cause damage. Consequently, in a few cases, conservators have decided not to clean outdoor statues and monuments, because surface dirt was not causing any damage and even provided protection from pollutants. In most cases, however, restorers conduct a thorough cleaning to remove harmful grime and to improve the appearance of a piece.

In the past, abrasive cleaning methods, such as sandblasting, were commonly used on marble, limestone, and granite. But these processes removed tiny bits of the stone, leaving the surface scarred and pitted. Likewise, chemical cleansers, though frequently used on both metal and stone, can strip the patina from metal surfaces and leave stone looking discolored or stained. A gentler approach is the use of *poultices,* mixtures of clay and a solvent, that are applied to the surface of stone to absorb salts and deep-seated stains. This process, however, does not effectively remove heavy layers of grime.

To avoid some of the problems associated with cleaning stone and metal artwork, conservators conduct tests to determine the best solvents to use, the proper concentration of the solution, and the optimum length of time to leave it on the surface. Also, they often use a combination of treatments to prevent damage. For example, conservators in 1993 cleaned grime, paint, and corrosion from the bronze Statue of Freedom, which sits atop the Capitol in Washington, D.C.,

Studying the pigments in paintings

Conservators use a variety of specialized equipment to study the coloring materials used in works of art and to determine the condition of those materials.

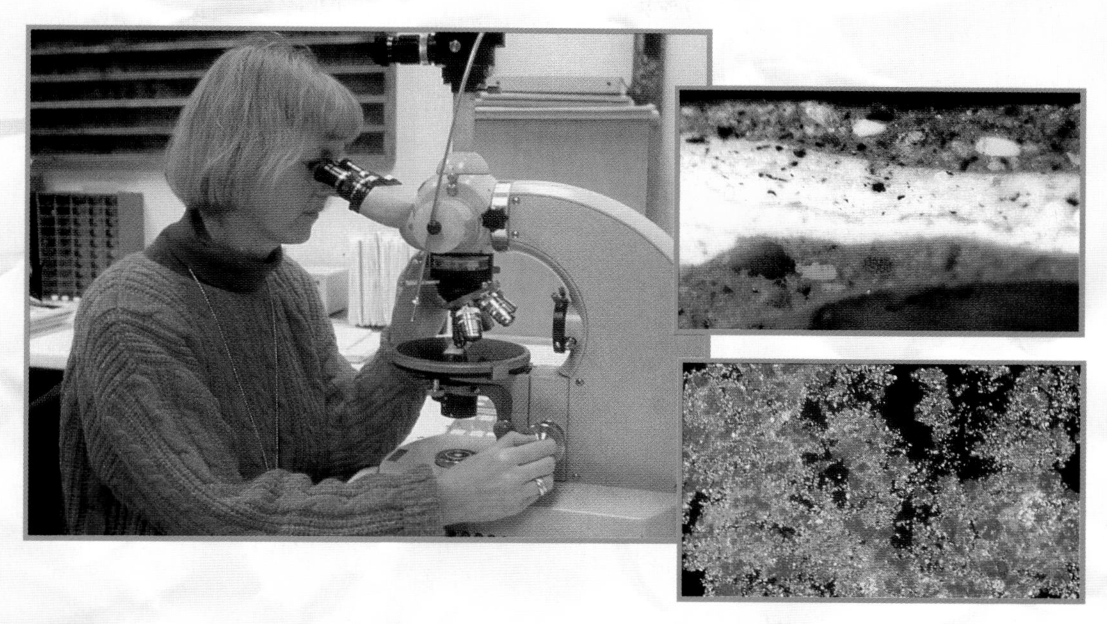

An art expert uses an instrument called a portable tristimulus colorimeter to measure the color of a cave painting at the Mogao Grottoes near Dunhuang, China. The colorimeter translates each color into a numerical code. By making repeated measurements of the painting over time, specialists can determine if the colors are fading. Such information helps conservators determine what methods are necessary to preserve the art.

A microscopist examines a tiny sample of a painting with a high-powered instrument called a polarizing light microscope, *above left.* The magnified image, *top,* shows a cross section of the layers of materials in a painting. The tiny particles are pigments that are bound together with an oil. The multicolored particles, *above,* are a mixture of three pigments—chrome yellow, vermilion, and lead white—which the artist used to make a bright orange color. These details help conservators understand a particular artist's selection of pigments and his or her painting techniques.

with high-pressure water hoses, which stripped away some of the patina. Then they applied a solution that caused a chemical reaction on the surface of the bronze to re-create the brownish patina and another solution to prevent further deterioration.

Many stone sculptures need more than a cleaning. The surfaces of many statues have eroded so much that features such as faces have completely disappeared. Because such artwork was carved from single pieces of stone, most conservators agree that trying to replace a statue's eroded features by some artificial means does not truly reflect the artist's skill and cannot re-create the original effect. Therefore, conservation efforts focus mainly on preventing any further deterioration of the stone.

In such cases, conservators sometimes apply *consolidants,* solutions that restore chemical bonds between the stone's particles and thus slow the disintegration caused by pollutants. One consolidant is a water solution of calcium hydroxide, also known as limewater. When it dries, it reacts with carbon dioxide in the air to form rocklike calcium carbonate that binds loose particles of limestone. Limewater has been used extensively in the ongoing effort to preserve the masonry and statues of Wells Cathedral in England. Unfortunately, this newly reinforced surface is still subject to the corrosive effects of pollutants.

Another group of consolidants consists of chemicals called *alkoxysilanes* (*al kok see SIGH laynz*). When coated on stone, these compounds of silicon, oxygen, carbon, and hydrogen form *polymers* (long-chain molecules) that resemble silica, a mineral found in granite and sandstone. As the polymer forms, it binds free particles into a tough and durable coating that resembles the chemical structure of quartz and glass. But alkoxysilanes have been of limited use, because they do not adhere well to the calcium carbonate in marble and limestone.

The search for better conservation technology

Even though art conservators have an extensive arsenal of physical and chemical tools at their disposal, they are constantly experimenting with new ones. These emerging techniques include both new methods of analyzing art and assessing damage and better ways of preserving works of art.

One analytical technique that came to the fore in the 1990's for studying the materials in a painting was gas chromatography/mass spectroscopy (GC-MS), a laboratory tool long used by chemists to determine the identities of unknown organic compounds. Scientists heat a tiny sample from a painting until it vaporizes. The resulting gases pass through a long glass column that separates the gases into different chemical compounds based on how fast they travel through the column. The separated compounds then flow into a machine that bombards them with electrons, which break up the molecules of each chemical into *ions* (electrically charged atoms or molecules). Electric and magnetic fields are then used to separate the ions based on their masses and the strength of their charges. Electronic instruments plot the results of the test, enabling specialists to determine the type and amount of organic chemicals in the sample.

Another technique, called digital imaging, has also aided art conservators. This procedure involves taking pictures of a work of art under different wavelengths of light using a camera equipped with an electronic light sensor called a charge-coupled device (CCD). The camera breaks up each image into tiny dots, called pixels, and records the light intensity of each pixel in *digital* form (a series of 0's and 1's). Art conservators in Italy, led by image-processing specialist Leonardo Seccia of the University of Bologna, have used digital imaging during the restoration of 1,400-year-old mosaics in Ravenna, Italy. This system has provided an effective way to choose the appropriate restoration techniques and to monitor the mosaics for future damage.

In one application of the CCD camera, the conservation team captured reflected-light digital images of the colored, glassy mosaic tiles with wavelengths of visible light from the near-ultraviolet to the near-infrared. With the help of computers, they were able to reconstruct the degree of reflectance from each individual tile in the mosaic. Differences between tiles—indicating the various materials used to make them—allowed the conservators to distinguish the original tiles from those that were added in restorations.

Seccia's team also produced digital images using various wavelengths of infrared light. With this technique, the researchers were able to identify places in the mosaics where the underlying plaster was damaged or where the tiles had become detached. Images made with other wavelengths revealed cracks, discolorations, and fading that were too subtle to be visible to the human eye.

Identifying—and stopping—problems in the making

Art conservators in 1998 were also developing analytical methods to spot problems before they arise so damage can be avoided. At the Carnegie Mellon Research Institute in Pittsburgh, Pennsylvania, for example, conservation scientist Paul Whitmore and his colleagues experimented with a measuring device that can rapidly determine how fast pigments will fade under the display lighting in museums. Fading can be a serious problem with paintings. Ironically, it is more common with the works of modern artists, who often employ unusual—and unstable—pigments to achieve special effects.

The testing apparatus, known as a micro-lightfastness detector, uses an *optical fiber* (a slender, light-carrying glass filament) to direct a beam of intense white light from a xenon lamp onto a tiny and inconspicuous part of a painting. A detector records the color spectrum of light reflected from the illuminated spot. Changes in the color spectrum during a 5- to 10-minute test indicate the amount of fading and how display lighting over a much longer period of time will likely affect the painting. Essentially, the apparatus artificially speeds up the natural fading processes, but the area affected is so small it can be detected only with the aid of a microscope. Other methods of monitoring pig-

Recovering images from the past

To restore works of art as closely as possible to their original appearance and to prevent further damage, conservators rely both on new scientific techniques and on the reliable methods that have been used for years.

A conservator at the Art Institute of Chicago, *above,* uses a swab and solvents to clean a painting from the museum's collection. Old, yellowed varnish altered the appearance of *Lady Reading the Letters of Heloise and Abelard* by French artist Jean-Baptiste Greuze, *left.* After cleaning the lower left corner of the painting, conservators revealed the original—or near original—colors.

Textile specialists at the Minneapolis Institute of Arts, *right,* weave new threads into a damaged tapestry. New fibers, *far right,* are used to replace missing portions of a tapestry and stabilize the existing threads. Although these methods have changed little over the years, scientific research has enhanced these efforts. A team of conservation scientists, led by David Howell of the Hampton Court Palace in England, presented research in September 1997 that calculated the stress placed on fibers when tapestries are hung for display. The researchers used computer models to evaluate how much weight different kinds of fibers and tapestry structures can support.

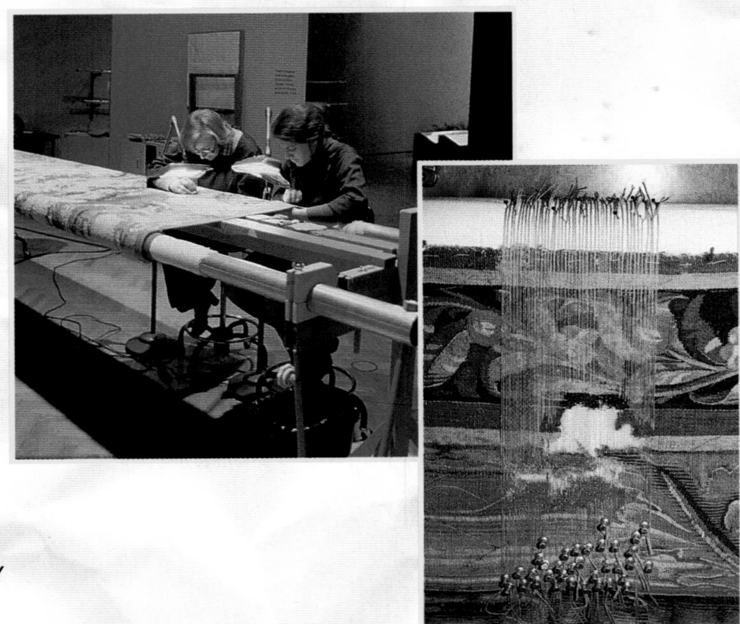

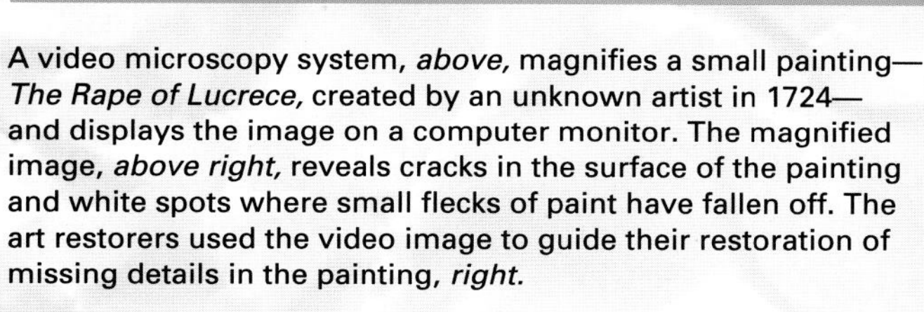

A video microscopy system, *above,* magnifies a small painting—*The Rape of Lucrece,* created by an unknown artist in 1724—and displays the image on a computer monitor. The magnified image, *above right,* reveals cracks in the surface of the painting and white spots where small flecks of paint have fallen off. The art restorers used the video image to guide their restoration of missing details in the painting, *right.*

A conservator uses a laser instrument to clean a statue in the Cremona Cathedral in Italy. The particles of dirt and grime on the statue absorb the energy of the laser beam, heat up, and expand until they fly off the surface. The laser cleaning method is safer than the use of traditional chemical cleaners, which can leave stains or discolorations on stone.

ment fading require repeated measurements over a period of years, during which time considerable damage can occur to a painting.

Science is also providing new chemical and physical methods for treating artwork that has already begun to deteriorate. For instance, a team of researchers at Sandia National Laboratories in Albuquerque, New Mexico, and the Metropolitan Museum of Art in New York City has developed an improved alkoxysilane treatment that binds well to the calcium carbonate in limestone and marble. The investigators formulated a solution that—when applied to the stone before it is coated with alkoxysilane—acts as a chemical link between the stone and the consolidant. The active ingredient in the solution is a molecule with an organic side and an inorganic, silicon-based side. The organic portion of the molecule bonds strongly to carbonates in limestone or marble, while the inorganic portion bonds just as strongly to alkoxysilane. The researchers reported in 1997 that limestone treated with the dual coatings weathered only about one-tenth as fast as uncoated limestone when both stone samples were exposed to very dilute acids similar to acid rain.

Space-age solutions for age-old problems

Several new high-technology techniques for cleaning paintings and sculptures were also being used and developed in the late 1900's. One procedure, pioneered by scientists at the National Aeronautics and Space Administration (NASA), involved the use of atomic oxygen to strip off old, yellowed layers of varnish on paintings or to remove soot deposited on paintings by fires. Atomic oxygen, which is found in the Earth's upper atmosphere, consists of single atoms of oxygen, rather than the two-atom molecules we breathe. NASA had always been concerned about atomic oxygen because it is highly reactive with organic materials and can erode coatings on satellites in low Earth orbits. But scientists at the NASA Lewis Research Center in Cleveland reasoned that artificially generated atomic oxygen—created by splitting ordinary oxygen gas molecules with electric discharges—could be used to strip away unwanted coatings on artwork.

In a collaborative effort with the Cleveland Museum of Art, NASA researchers placed varnish- and soot-coated test patches cut from paintings—not works from the museum's collection—in a vacuum-sealed test chamber. The single oxygen atoms sent into the chamber reacted with the hydrocarbons in the soot and varnish, forming by-products such as carbon monoxide and water vapor, which were pumped out of the chamber. The atoms stripped away the layers on top of the painting, but not the paints, which were made from inorganic substances.

In 1998, the atomic oxygen technique was not ready for general use, and the researchers were trying to determine if the method is safe and practical for cleaning valuable works of art. They know it could be risky. For example, if the process were used on paintings with organic pigments, it would have to be monitored very closely and shut off before the atomic oxygen reached the paints and caused irreversible damage. Nonetheless, the procedure could provide some benefits over other cleaning methods. The organic solvents now used to remove varnish and soot from a painting can cause the underlying paint to swell, and the rubbing required to clean an image can wear away details of the painting.

Among the latest physical treatments for restoring art, particularly stone sculptures, is laser cleaning. This process removes dirt and grime by blasting it with very brief pulses of powerful laser beams, while leaving the underlying stone untouched. Lasers used to clean artwork belong to a type called neodymium-YAG, which emit light in the infrared region of the spectrum. When the energy from the laser beams is absorbed by dirt particles, they heat up and expand violently for a fraction of a second. The sudden expansion causes the particles to fly off the stone's surface. Art conservators have extended the laser cleaning technique to sandstone, terra cotta, aluminum, and ivory. And researchers are experimenting with lasers to clean paintings, stained glass, and parchments.

Conservators are even using electron beams in their battle to rescue threatened artifacts. They use the beams to artificially age new silk fibers before they are woven into the fabric of ancient scrolls to replace missing sections. Such aging is necessary because new fibers are usually much stronger than the original fibers and, if untreated, can exert excessive stress on the fragile fabrics, causing further deterioration. In a restoration project in the early 1990's for the Freer Gallery of Art in Washington, D.C., restorers in Japan used silk threads aged by electron beams to repair thousands of holes in the 600-year-old Fugen scroll, a priceless Japanese painting on silk.

Despite the effectiveness of modern conservation methods, science will never be able to completely halt the visible aging of paintings, sculptures, ceramics, manuscripts, and other works of art. Indeed, many art lovers cherish the aura of great antiquity that surrounds masterpieces from past centuries. But when time, weather, or pollution threaten to obliterate irreplaceable works of art forever, modern technology can help ensure that these cultural treasures will be saved for the enjoyment of future generations.

For further reading:

Oddy, Andrew, ed. *The Art of the Conservator.* Smithsonian Institution Press, 1992.

Science for Conservators: An Introduction to Materials. Vol. 1. Conservation Science Teaching Series. The Conservation Unit of the Museums & Galleries Commission with Routledge, 1992.

Shulman, Ken. *Anatomy of a Restoration: The Brancacci Chapel.* Walker and Company, 1991.

For additional information:

The Getty Conservation Institute Web site: www.getty.edu/gci.

National Center for Preservation Technology and Training, *NCPTT Notes:* www.ncptt.nps.gov/notes.

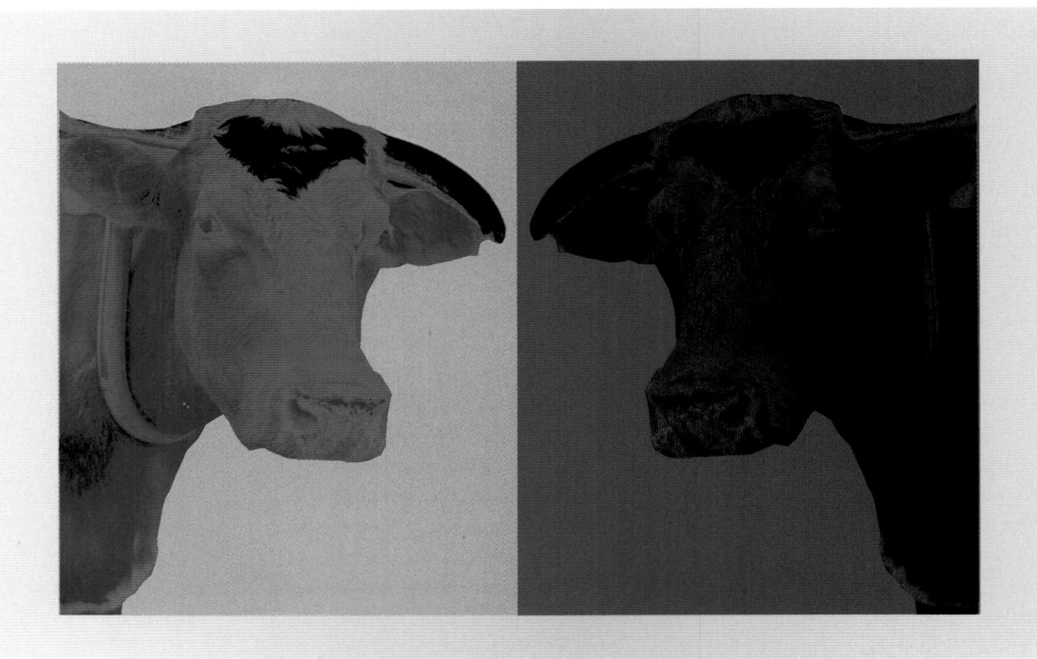

Cloning Isn't Science Fiction Anymore

by Robert H. Tamarin

An achievement previously thought impossible may offer great promise in medicine, agriculture, and industry, but it also raises many serious issues—including concerns about human cloning.

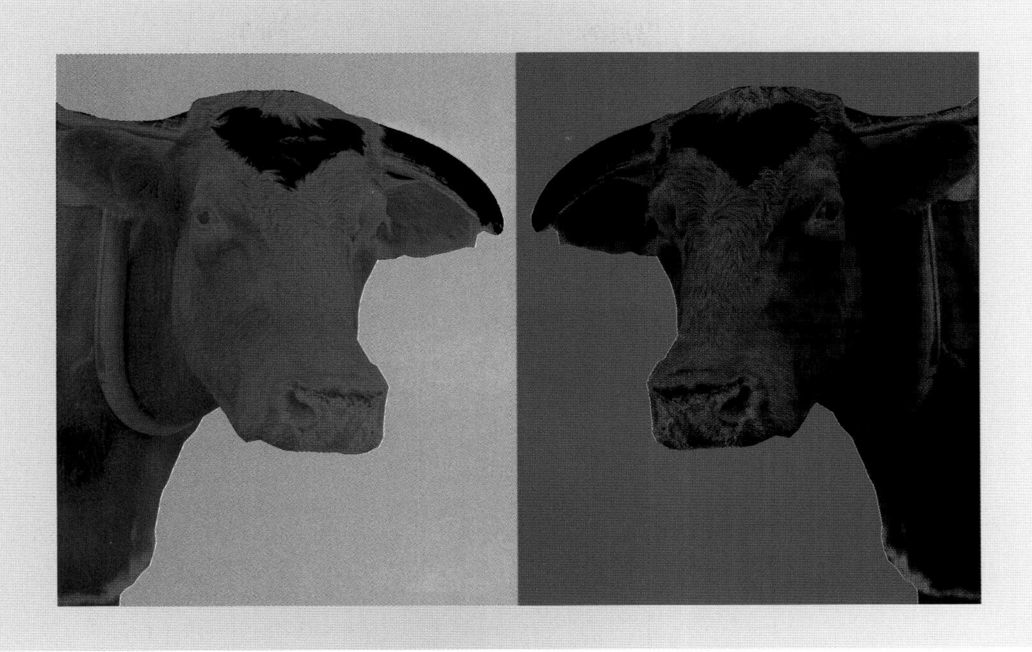

In February 1997, the world was startled by the news that scientists in Scotland had produced a *clone*—an exact genetic duplicate—of a sheep. This achievement marked the first time that researchers had cloned an adult *vertebrate* (animal with a backbone). The scientists, from the Roslin Institute and PPL Therapeutics PLC, a biotechnology company—both near Edinburgh, Scotland—reported the birth of Dolly, a lamb produced from an udder cell taken from a 6-year-old female sheep. Even the scientific community was surprised by this accomplishment, which most geneticists had thought was impossible.

While scientists and others were absorbing the implications of this news, other cloning breakthroughs were made. Oregon researchers announced in March that they had used an older cloning technique to produce two monkeys, marking the first time that primates, the group of animals to which humans belong, had been cloned. Then in July, the same scientists who created Dolly announced that they had produced other lambs through cloning. These animals, the researchers reported, carried a human gene that had been added to their own genetic material. Cloning grabbed headlines again in August, when Wisconsin scientists said that they had produced a calf using a more efficient cloning technique than that used by the scientists in Scotland. During 1998, cloning developments continued to make the news.

Researchers said the cloning of animals, especially those that have been genetically modified in certain ways, could have a number of medical, agricultural, and industrial applications. For example, cloning could result in the mass production of genetically modified cattle that secrete valuable drugs into their milk. But the cloning of animals indicated that it may also be possible to clone humans. Much of the public expressed revulsion toward the prospect of human cloning, and some politicians vowed to outlaw it. Its proponents, however, saw human cloning as a way to help people, such as by allowing infertile couples to have children.

What is a clone?

What exactly is a clone? A clone is an exact genetic copy of a gene, cell, or whole organism. Cloning commonly occurs in nature. For example, when a cell in a person's body divides, the two resulting cells are clones of the first cell and contain genes that are exact duplicates of the genes in the original cell. Organisms that reproduce *asexually* (without sex) produce clones as offspring. Examples are certain corals and fungi, which develop buds that detach from the parent and grow into individuals that are genetically identical to the parent. Strawberry plants sprout *runners* (modified stems), which give rise to separate plants that are genetically indistinguishable from the original plant. Offspring produced through asexual methods obtain all of their genes from a single parent.

With sexual reproduction, on the other hand, an egg cell from the female is fertilized by a sperm cell from the male. Because each sex cell contains half the number of *chromosomes* (the structures that carry an organism's genes) needed to create a new individual, fertilization results in a cell with a full assortment of genes—one set of chromosomes coming from each parent. That cell then begins dividing and develops into a complete organism. With half its genes derived from its father and half from its mother, the offspring possesses physical traits of both parents.

Cloning occurs naturally even in sexually reproducing organisms, including humans. In such organisms, the splitting of a fertilized egg to form identical twins is a natural form of cloning. Identical twins are clones of each other.

Early scientific attempts at cloning

Scientists have long been intrigued by the possibility of artificially cloning animals. In fact, people have known since ancient times that some *invertebrates* (animals without backbones), such as earthworms and starfish, can be cloned simply by dividing them into two pieces. Each piece regrows into a complete organism. The cloning of vertebrates, however, was much more difficult. The first leap forward in the cloning of these more complex organisms came in the 1950's with work done on frogs.

Glossary

Chromosomes: Threadlike structures in cells that carry an organism's genes.

Clone: An exact genetic copy of a gene, cell, or whole organism.

Embryo: The developmental stage of an organism prior to birth.

Embryo splitting: A cloning technique in which an embryo is divided into individual cells or groups of cells, which then develop into genetically identical organisms.

Fetus: An embryo in its later stages of development.

Gene: An inherited molecular unit that carries coded information, typically for the production of a particular protein.

Genetic engineering: The manipulation of an organism's genetic material, including the insertion of genes from other species.

Nuclear transplantation: A cloning technique in which the nucleus is removed from an egg cell and replaced with the nucleus of another cell; the resulting cell develops into an embryo genetically identical to the organism from which the transplanted nucleus was obtained.

Nucleus: The structure within a cell containing most of an organism's genetic material.

Totipotency: The ability of a cell's genetic material to direct the development of a complete organism.

The author:
Robert H. Tamarin is a professor of genetics and dean of sciences at the University of Massachusetts in Lowell.

Sexual reproduction and the development of an embryo

Sexual reproduction involves the union of two sex cells—a sperm cell from the male and an egg cell from the female. The union leads to the formation of a collection of cells called an embryo, which develops slowly into a fully formed organism. During the embryo's development, its cells become specialized. That means that under normal circumstances, cells destined to be part of the organism's skin, for example, can give rise only to more cells of that type. Although specialized cells contain all of the genes needed to create an entire organism, most of the genes are "turned off."

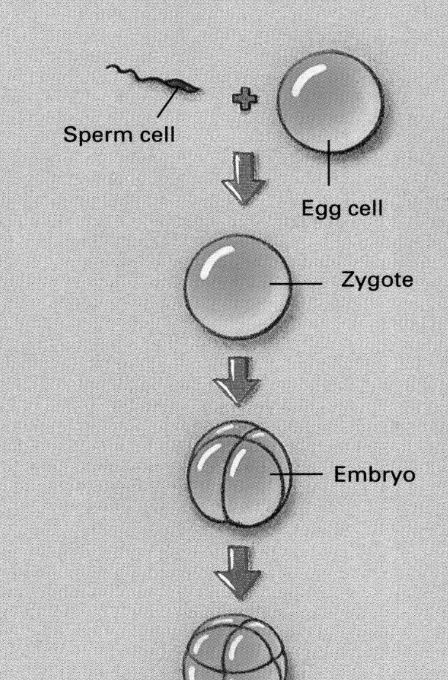

1. A sperm cell and an egg cell—in this case, from frogs—combine to form a cell called a zygote.

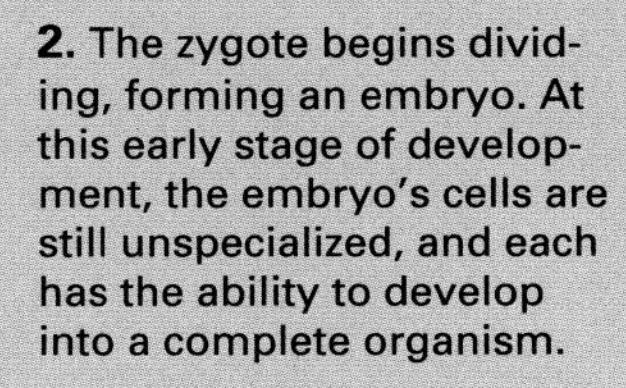

2. The zygote begins dividing, forming an embryo. At this early stage of development, the embryo's cells are still unspecialized, and each has the ability to develop into a complete organism.

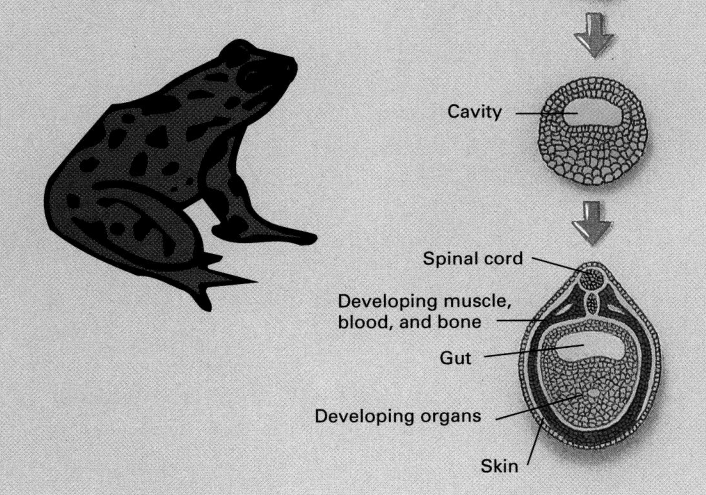

3. Cell specialization begins when a cavity forms within the embryo. This cavity will eventually develop into the embryo's gut.

4. As the embryo continues to develop, its cells become increasingly specialized as they begin to form the organism's various parts, such as the spinal cord and skin. By this point, many of the embryo's cells have lost their ability to develop into complete organisms, and eventually all the body cells will lose that ability.

Beginning in 1952, Robert Briggs and Thomas King, developmental biologists at the Institute for Cancer Research (now the Fox Chase Cancer Center) in Philadelphia, developed a cloning method called nuclear transplantation, or nuclear transfer, which was first proposed in 1938 by the German scientist Hans Spemann. In this method, the *nucleus*—the cellular structure that contains most of the genetic material and that controls growth and development—is removed from an egg cell of an organism, a procedure known as enucleation. The nucleus from a body cell of another organism of the same species is then placed into the enucleated egg cell. Nurtured by the nutrients in the remaining part of the egg cell, an *embryo* (an organism prior to birth) begins growing. Because the embryo's genes came from the body cell's nucleus, the embryo is genetically identical to the organism from which the body cell was obtained.

In their experiments, Briggs and King used body cells from frog embryos. From these cells, they were able to produce several tadpoles.

Briggs and King used embryos consisting of only a few thousand cells as the source for body cells and nuclei, because at that stage of development an embryo's cells are still relatively unspecialized. As an embryo develops into a completely formed organism consisting of billions of cells, its cells become increasingly specialized. Some cells become skin cells, for example, while others become blood cells. Skin cells can normally make only more skin cells. Likewise, blood cells can normally make only blood cells. By contrast, each of the unspecialized cells of an early embryo is capable of producing an entire body. At the time of Briggs's and King's experiment, researchers were not sure whether specialization occurs because different cells get different assortments of genes or because genes that are not needed in a particular kind of cell become inactive.

Additional research on nuclear transplantation was conducted in the 1960's and 1970's by John Gurdon, a molecular biologist at Oxford University in England. In 1966, Gurdon produced adult frogs using nuclei from tadpole intestine cells. This experiment proved that even cells that have undergone a great amount of specialization remain *totipotent*—capable, under certain circumstances, of directing the development of a complete organism. Totipotency implied that all of a fully developed organism's body cells contain a complete set of genes and that specialization occurs because certain genes are active in some cells and inactive in other cells.

Despite the demonstrated totipotency of body cells, scientists were repeatedly frustrated in their attempts to use nuclear transplantation with nuclei taken from the cells of adult vertebrates. In the rare cases in which offspring resulted from such experiments, the young never survived to adulthood.

Embryo splitting

A different and simpler cloning procedure, called embryo splitting, or artificial twinning, was developed in the 1980's and was adopted by livestock breeders. In this procedure, an early embryo is simply split into individual cells or groups of cells, as happens naturally with twins, triplets, and other multiple births. Each cell or collection of cells develops into a new embryo, which is then placed into the womb of a host mother animal, who carries it to a full term. Although this technique permits the production of multiple clones, the clones are derived from an embryo whose physical characteristics are not completely known rather than from an adult animal with known characteristics—a serious limitation for practical applications of the procedure. By the early 1990's, embryo splitting and nuclear transplantation using cells from embryos had been used to clone a number of animals, including mice, cows, pigs, rabbits, and sheep.

Two or more offspring from a single fertilized egg

Nature has its own cloning methods, such as when a zygote, or fertilized egg, divides spontaneously to create identical twins or triplets. Because they came from the same fertilized egg, each of the offspring has exactly the same set of genes. Scientists have learned to duplicate the natural cloning process artificially with animals by dividing early embryos into separate cells or groups of cells in the laboratory. This procedure, called embryo splitting or artificial twinning, has been used to create identical livestock animals.

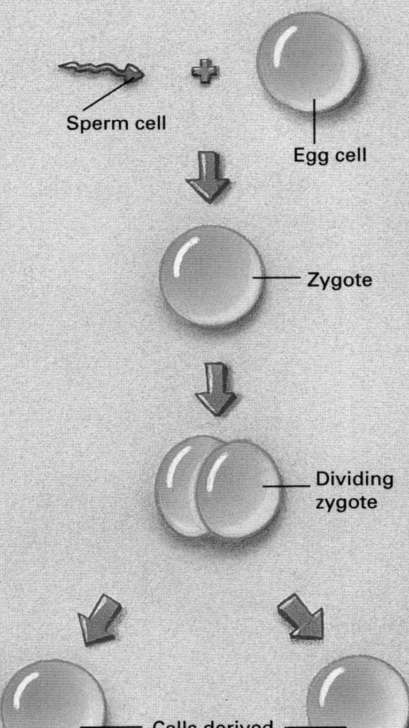

1. A sperm cell combines with an egg cell to form a zygote.

2. The zygote divides into two cells.

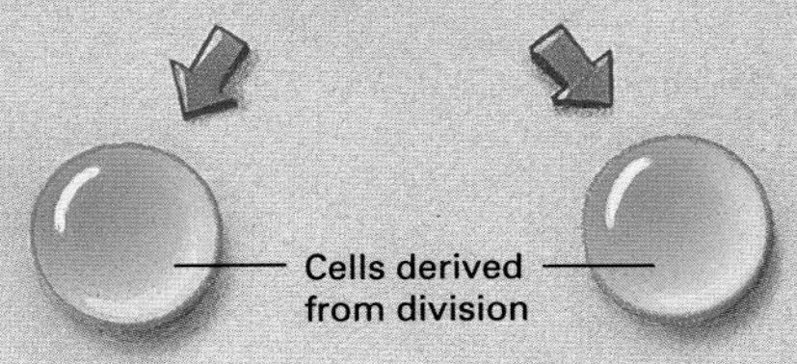

3. The cells split apart from each other.

4. The two cells develop into identical embryos, which grow into natural identical twins, such as the ones at left, who are clones of each other. (When embryos are split artificially, the resulting embryos are placed into the wombs of surrogate mothers to complete their development.)

The big breakthroughs of 1997

The failure of nuclear transplantation to work with cells from adult animals led most scientists to conclude that the cells of mature organisms are simply too specialized to be cloned. In 1996, however, Roslin Institute researchers led by embryologist Ian Wilmut found a way to do the seemingly impossible.

Wilmut and his colleagues took mammary-gland cells from an adult sheep and placed them in a solution that essentially starved them of nutrients and caused them to stop growing for a few days. Then, with a spark of electricity, they fused each mammary cell with an enucleated egg cell. The resulting cells were allowed to grow into embryos, which were then transplanted into surrogate mother *ewes* (female sheep) to

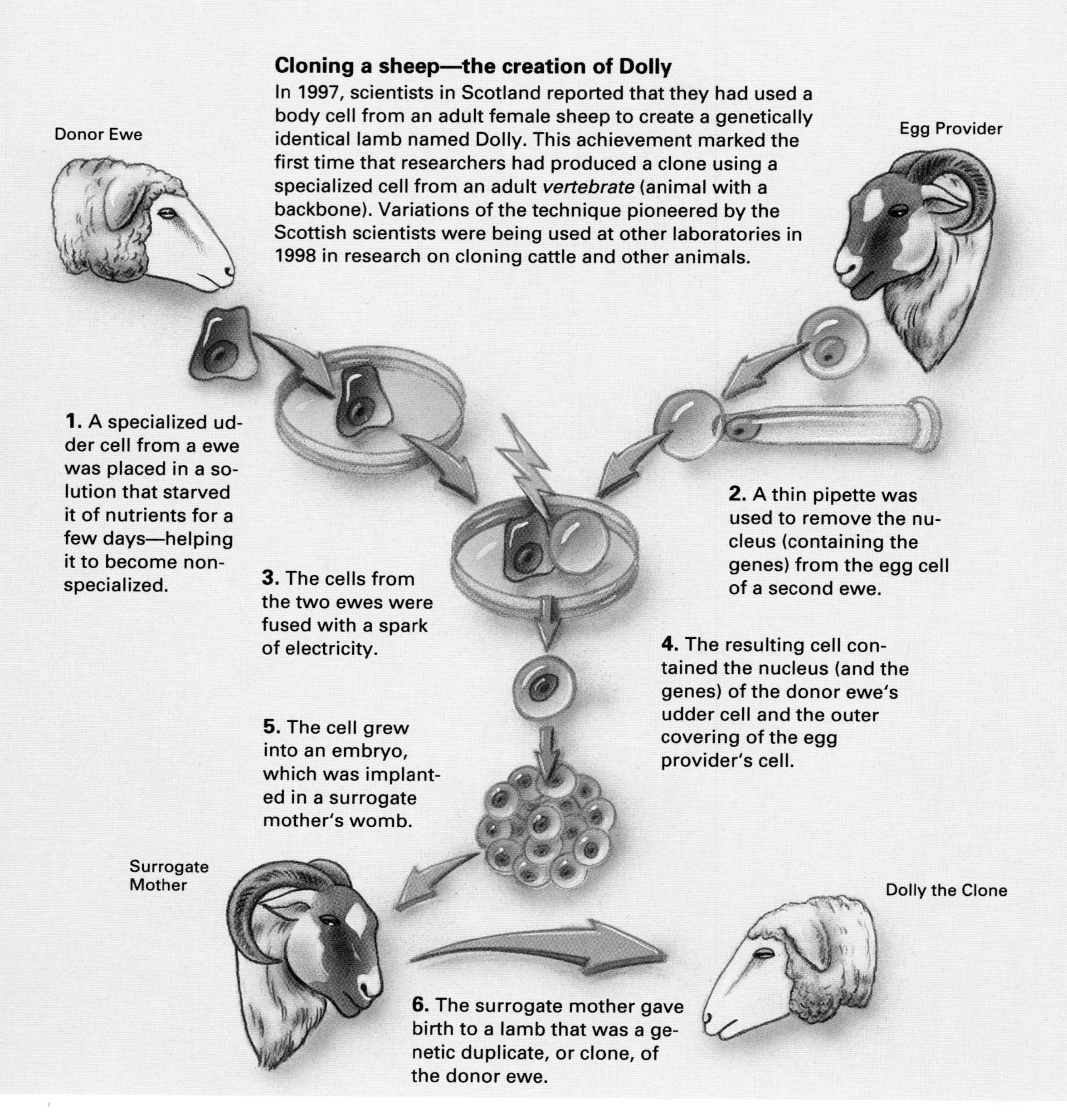

complete their development. Nearly 300 attempts at this technique resulted in failure for the scientists. Some eggs did not accept mammary cell nuclei, embryos that were produced died, and lambs that were born were abnormal and died. But one lamb, apparently healthy, survived the procedure: Dolly, who was born in July 1996.

The most important technical point of this research was that the nucleus of a specialized body cell from an adult animal could be reprogrammed to direct the development of a new organism—that is, the cell could be restored to totipotency. The cell-starvation technique used by the scientists was based on the theory that cells that are not growing become easier to reprogram. The programming of a nucleus

Dolly and Gene
A very young Dolly stands next to her surrogate mother, *top,* the ewe that gave birth to her in July 1996. Dolly, like the ewe that she is a clone of, is a Finn Dorset sheep. Her surrogate mother, however, was of a different breed—Scottish Blackface. In August 1997, Wisconsin scientists claimed that they had developed a more efficient cloning procedure than was used with Dolly. The first animal they produced with their procedure was a calf, named Gene, *right.* Although Gene was produced from a cell taken from a bull fetus, the scientists said that they were also experimenting with cloning techniques using cells from adult cattle.

is an interactive process between the *cytoplasm* (the cellular substance surrounding the nucleus) and the genes in the nucleus. The cytoplasm sends to the nucleus signals that determine which genes are turned on or off and therefore which proteins—the end product of genes—are produced by the cell. As the cell becomes specialized and loses its totipotency, it becomes unable to produce many proteins.

In nuclear transplantation, the longer the cytoplasm has to work on the transplanted nucleus before protein production begins, the more likely it is that the nucleus can be reprogrammed to direct the normal development of an embryo. Contributing to the success of the Roslin scientists was the fact that protein production begins relatively late in the development of sheep embryos, compared with, for example, mouse and human embryos. Therefore, there was a lot of time for the nucleus to be reprogrammed.

An apparent advance on the Roslin technique was reported in August 1997, when scientists at ABS Global Incorporated in DeForest, Wisconsin, a biotechnology company specializing in livestock reproductive services, said that they had produced a calf using a cloning procedure more efficient than that used with Dolly. The Wisconsin scientists began by performing a nuclear transplantation to produce an embryo from a body cell of a 30-day-old bull *fetus* (a more fully developed embryo). Then they added another step. They took one of the cells from the embryo they had produced and performed a second nuclear transplantation. The embryo resulting from the second transplantation was then placed in a surrogate mother. The double nuclear transplantation helped make the nucleus even more susceptible to reprogramming and increased the success rate of the procedure. The scientists said that only 15 attempts were required before success was achieved with the birth of the calf, named Gene.

Although ABS Global researchers used a fetus as the source of the nucleus used to produce Gene, they said they were also experimenting with nuclei from the cells of adult cattle. The scientists hoped that their procedure would eventually enable animal breeders to produce an unlimited number of identical animals with superior traits.

Transgenic animals

In July 1997, Ian Wilmut and his colleagues announced that they had combined their cloning procedure with genetic engineering techniques to produce lambs with a human gene for a particular blood protein in their cells. First, Wilmut's team inserted the human gene into a sheep fetus cell. This cell was then allowed to divide into many cells, each of which was then used in a nuclear transplantation to produce an embryo. Of three identical female lambs born with the human gene, two—named Polly and Molly—survived and were introduced to the press in December.

Animals engineered to carry genes from other species are called transgenic animals. The human gene that Polly and Molly were born with causes their bodies to produce factor IX, a human blood-clotting protein useful for treating hemophilia, a disorder in which blood clotting does not occur normally. The protein is secreted into the animals' milk and can be extracted to create a drug for treating hemophilia.

In January 1998, scientists at the University of Massachusetts and Advanced Cell Technology, Incorporated, a biotechnology firm in Worcester, Massachusetts, announced the birth of three transgenic calves created through a method similar to that used with Polly and Molly. Although these calves carried only an experimental gene that had no effect on the animals, the scientists said that other calves they planned on creating were to carry a gene to produce human serum albumin, a protein needed by people who have lost large amounts of blood.

Transgenic animals can be made to produce a wide variety of proteins that could be sold as drugs, as well as other proteins, called enzymes, that could be used to speed up industrial chemical reactions. Although the creation of transgenic animals began in the 1980's, the application of the technology has been limited. Cloning was expected to make it possible for transgenic animals to be mass produced. Large numbers of transgenic animals could produce vast quantities of needed drugs and other useful substances more efficiently and at much lower cost than is possible with bioengineering methods. As of mid-1998, most genetically engineered proteins were being manufactured in *bioreactors,* large steel vessels in which billions of genetically modified microorganisms produce proteins that are then extracted and purified.

Researchers involved in cloning envision a number of other practical applications for their work, including the creation of genetically modified animals that could provide organs for human organ transplants; the mass production of faster-growing and leaner livestock; and the perpetuation of endangered species.

Are human clones next?

The same newly developed procedures used to clone sheep and cattle could theoretically be used to clone humans. However, human cloning would probably be more difficult than sheep or cattle cloning, because the cells of human embryos start producing proteins at a relatively early stage. Thus, there would not be as much time for the egg cytoplasm to reprogram a transplanted nucleus.

Nonetheless, most geneticists believe that this problem can be overcome, and they foresee a number of practical applications for the cloning of humans. Infertile couples who do not wish to adopt, for instance, could use cloning to have children who are biologically related to them. Cloning could also be used to produce offspring free of certain diseases. For example, a number of disorders, including some affecting the eyes, brain, and muscles, are (at least partially) caused by flawed genes located in the mitochondria, energy-producing structures in the cytoplasm. If a woman were to carry a gene for one of these disorders, she could conceive a healthy child by having the nucleus of one of her body cells inserted into an enucleated egg cell from a woman who does not have anything wrong with her mitochondrial genes. The resulting embryo could then be implanted into the woman who donated the nucleus, and she would carry the baby to term.

Nevertheless, much of the public finds the prospect of cloning people extremely disturbing. In February 1997, soon after the announcement of Dolly's birth, a Gallup Poll indicated that almost 90 percent of the people in the United States thought that human cloning was "morally wrong," and a Time/CNN poll found that almost 75 percent thought that it was "against God's will." United States President Bill Clinton responded to public concerns in 1997 by issuing a moratorium on the use of federal funds for human cloning research and proposing that Congress make human cloning illegal for at least five years. Many

Cloning leads to "pharming"

Pharmaceutical companies make many drugs using bioreactors—large, expensive vessels containing microorganisms that have been genetically altered to secrete medically useful proteins. Scientists hoped that cloning would enable companies to make drugs less expensively by producing herds of animals that have been genetically altered to secrete human proteins in their milk. The proteins would be extracted from the milk to be used as drugs. The production of drugs using genetically modified animals is called "pharming."

Scientists responsible for creating Dolly, including Ian Wilmut, *above,* at right, pose in December 1997 with Polly and Molly, two genetically modified sheep they also produced through cloning. These animals have a human gene that causes their bodies to make a protein useful for treating hemophilia, a disorder in which the blood does not clot normally.

George and Charlie, two genetically altered calves created through cloning by Massachusetts scientists, greet each other in their pen in January 1998, *above.* The scientists said that the male calves were a first step toward the mass production of female cows that, like Polly and Molly, will secrete medically useful human proteins in their milk.

Above, a technician adjusts the controls on a bioreactor to maintain optimum conditions, such as the ideal temperature and pressure, for the growth of microorganisms.

states moved to prohibit human cloning, though only California had succeeded in passing a ban as of mid-1998. A number of other states outlawed experiments with human embryos, effectively prohibiting human cloning. In January 1998, 19 European nations signed a treaty in which they agreed to enact laws banning human cloning.

Some public concerns arose from the possibility that cloning could be combined with genetic engineering to create human beings who have been genetically altered for some purpose. Those concerns were heightened in October 1997, when researchers at the University of Bath in England announced that they had used genetic-engineering techniques to create tadpoles without heads. In January 1998, University of Texas researchers said that they had done the same thing with mice. Some biologists suggested that, in the future, headless human clones might be produced. Such *nonsentient* (nonfeeling and nonthinking) clones, they said, could be used as a source of organs for organ transplants. Other biologists contended, however, that getting a headless human to survive long enough to be a usable organ donor would be virtually impossible. But technical issues aside, most people regarded the creation of headless clones as a horror.

Politicians and members of the public were not the only people who had reservations about human cloning. Many scientists were also uncomfortable with the idea. In September 1997, the Federation of American Societies for Experimental Biology, the largest professional association of biologists in the United States, announced its adoption of a five-year voluntary moratorium on attempts to clone humans. A number of other scientific and medical associations adopted similar moratoriums.

Some proposed practical applications of cloning

- The mass production of animals engineered to carry human genes for the production of certain proteins that could be used as drugs; the proteins would be extracted from the animals' milk and used to treat human diseases.
- The mass production of animals with genetically modified organs that could be safely transplanted into humans.
- The mass production of livestock that have been genetically modified to possess certain desirable traits.
- The perpetuation of endangered species.
- The production of offspring by infertile couples.
- The production of offspring free of a potentially disease-causing genetic flaw carried by one member of a couple; the individual without the defect could be cloned.

Unanswered questions

Because no cloning technique had been perfected as of mid-1998, scientists expected that any attempt to clone a human would—just as in the work that led to Dolly—result in the death of many embryos and newborns before success was achieved. In addition, even if an infant clone survived, there was no guarantee that it would develop normally. The genetic material in body cells accumulates subtle molecular changes as an organism ages. Because the cell used to create Dolly came from a 6-year-old animal, Dolly's chromosomes had certain characteristics normally found only in older animals. This finding led some scientists to wonder whether Dolly, though appearing normal, might have inherited genetic damage that would eventually show up as premature aging or some other disorder.

An indication that Dolly could at least breed normally came in April 1998, when she gave birth to a lamb that Roslin scientists said was healthy. The lamb was not a clone, but rather the offspring of a natural mating between Dolly and a male sheep.

While some biologists remained concerned about how Dolly would age, others raised doubts as to whether she was, in fact, a clone of an adult sheep. These biologists noted that, as of mid-1998, no other researchers had succeeded in cloning an adult vertebrate—leading them to question the validity of Wilmut's research.

Because of the many unresolved issues, many scientists said that human cloning should be postponed until such questions were answered. Nonetheless, some people wanted attempts at human cloning to begin as soon as possible. They included a Chicago scientist who announced his intention to open a group of clinics to clone humans. In response to that proposal, the U.S. Food and Drug Administration (FDA) said in January 1998 that any such project would require FDA approval—which the agency indicated it might not grant.

Changing attitudes?

Some observers believed that negative attitudes toward human cloning would eventually change. Lori Andrews, a reproductive law specialist at the Chicago-Kent College of Law, noted in January 1998 that attitudes toward previously developed reproductive technologies, such as *in-vitro fertilization* (test-tube babies), had shifted from "horrified negation" to "very slow but steady acceptance." She observed that a number of scientists who had once expressed strong opposition to the idea of ever cloning humans had come to believe that human cloning should be allowed under certain circumstances after the technical problems were solved. In a sign that politicians were viewing cloning less negatively, the U.S. Senate in February 1998 rejected a bill that would have outlawed nuclear transplantation using human cells.

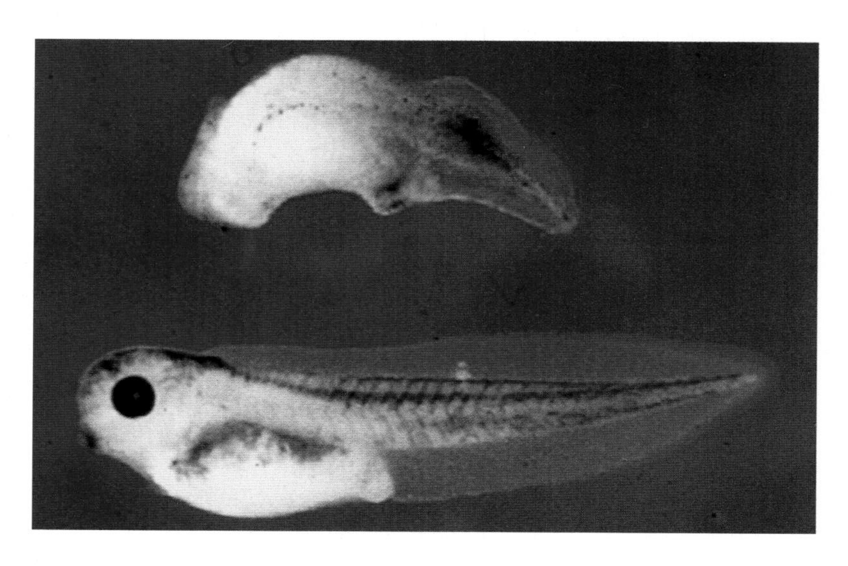

A frog tadpole that was genetically manipulated to lack a head, top, is shown with a normal tadpole. Scientists in England created headless tadpoles in 1997 as part of research into how genes control development. Some biologists suggested that headless humans might be cloned someday as a source of organs for organ transplants, but such proposals were highly controversial.

As of mid-1998, no laboratory was known to be attempting human cloning. However, researchers at a number of institutions were doing work on nuclear transplantation, animal cloning, and other techniques that could eventually be applied to humans. Some laboratories were interested in cloning human embryos for research purposes, though none were known to be actually doing so.

Although a great deal of controversy during 1997 and 1998 revolved around the idea of cloning humans, most cloning research was concerned strictly with producing animals for medical, agricultural, and industrial purposes. Many researchers believed that significant advances in such areas as drug and food production were promised by cloning and that these advances would provide much benefit to society. The net effect on society of the scientific breakthrough announced in February 1997, however, remained to be seen.

For further reading:

Kolata, Gina Bari. *Clone: The Road to Dolly and the Path Ahead.* William Morrow & Company, 1998.

Pence, Gregory. *Who's Afraid of Human Cloning?* Rowman & Littlefield Publishers, 1998.

Silver, Lee M. *Remaking Eden: Cloning and Beyond in a Brave New World.* Avon Books, 1997.

For additional information:

www.phrma.org/genomics/documents/cloning.html#general (Information on scientific, ethical, and legal aspects of cloning; links to other cloning Web sites.)

Questions for thought and discussion

1. Why was the cloning of an adult vertebrate thought to be impossible before the creation of Dolly? How were scientists finally able to achieve that goal?
2. What are some uses for animal cloning?
3. Do you think that there are ethical problems involved in cloning animals? If so, what are they?
4. What are some of the remaining technical problems involved in cloning that must be solved before it can be applied to humans?
5. Do you think that there are ethical problems involved in cloning humans? If so, what are they?
6. Under what circumstances do you think human cloning should be allowed? Why?
7. Do you think the government has a right to ban cloning? Explain.
8. If you were the parent of a child produced through cloning, how much would you tell that child about how he or she was produced?
9. What special problems do you think a clone might have in life, compared with other people?
10. Would you ever have yourself cloned? Why or why not?

The Hunt for the Giant Squid

By Clyde F. E. Roper

Incredible as giant squids may seem, marine biologists know that they really exist. Now, the search is on to find these elusive creatures in their own deep-ocean domain.

It is a scene famous in literature and film: An enormous squid, its large eyes glaring with malice, attacks a submarine and forces it to the surface. The crew rushes onto the deck, fighting desperately to free the sub from the writhing arms of the monstrous beast. Terrified sailors are squeezed, crushed, and drowned as the frightful creature slithers aboard and threatens total destruction.

This thrilling episode was most famously portrayed in the 1954 feature film *20,000 Leagues Under the Sea,* adapted from Jules Verne's classic novel. Similar accounts have been a part of seafaring legend for centuries, but we know today that most of them had little, if any, basis in fact. The book and movie were fiction, of course, but—incredible as it may seem—giant squids are real.

But even though we have absolute proof that giant squids do exist, we know very little for certain about their existence: Where do they live? What do they eat? How fast do they swim? How long do they live? Are they really as dangerous as the legends suggest? In 1996 and again in 1997, my colleagues and I set out to answer some of these basic questions by finding and observing a giant squid in the wild. Although both attempts were unsuccessful, we resolved to try again in 1999.

At the time Verne published his novel in 1870, the giant squid was much more a creature of mythology than biology. The only evidence

Opposite page:
The crew of Captain Nemo's submarine, the *Nautilus,* battles a giant squid in an illustration from Jules Verne's 1870 novel *Twenty Thousand Leagues under the Sea.*

for the existence of this monster of the deep was a handful of alleged sightings by ocean voyagers—including one by a French naval vessel in Verne's own time—and body parts found on remote shores or in the sea. No complete specimen of a giant squid had ever been examined by scientists.

By the late 1990's, scientists had studied a number of giant squids. But even this most recent knowledge is of limited value, because every specimen found so far was either dead or dying. Complicating matters further, the squids that scientists are able to examine are nearly always damaged. Moreover, no giant squid has ever been observed alive in its natural habitat. So the nature of these mysterious creatures remains largely unknown.

Since intact specimens of giant squids are hard to come by, the study of these great animals—the largest *invertebrates* (animals without backbones) ever known to exist—is much like detective work. Each new specimen can potentially add to scientists' knowledge about the anatomy and geographical distribution of giant squids. The process is slow and painstaking, however. It would be so much easier if we could simply observe a living giant squid in its natural setting.

Terms and concepts

Architeuthis: The animal genus to which giant squids belong; the name is Greek for *ruling squid.*

Autonomous underwater vehicle (AUV): An unmanned submersible vehicle that roams through the sea guided by a preprogrammed computer.

Bioluminescence: Biochemically generated light, similar to that of a firefly, that is produced by many ocean creatures.

Cephalopods: A group of marine animals that includes squids and their relatives—octopuses, cuttlefishes, and the chambered nautilus.

Kraken: A huge, many-armed sea monster in old Norwegian legends; in the 1850's, it was identified as a giant squid.

Some facts about squids

A giant squid is not just a small squid that grew to be very large. Giant squids belong to their own family, *Architeuthidae* (*ahr kuh TOO thih dee*), with a single genus, *Architeuthis.* Architeuthidae is one of 25 different families of squids that together contain about 650 species. Squids, along with their zoological cousins—octopuses, cuttlefishes, and the chambered nautilus—are classified as *cephalopods,* from the Greek words for "head" and "foot." Marine biologists chose this name because the feet and arms of these animals, which evolved from ancient relatives, grow out from their head. Adult squids range in length from about 2 to 3 centimeters (0.8 to 1.2 inches) up to the giant squid, the largest of which measured to date was nearly 18 meters (60 feet) from the tip of its tail to the ends of its two long feeding tentacles.

All squids have six basic external structures: (1) an elongated, cylindrical body known as the *mantle,* which contains the internal organs and bears two fins near the tail end; (2) a short head with two large eyes; (3) eight arms studded with suckers or hooks; (4) a mouth, surrounded by the arms, with a hard beak resembling a parrot's beak; (5) two long, cordlike tentacles extending from the base of the arms, each ending in a club covered with suckers; and (6) a funnel protruding from the mantle, beneath the head. Many variations on this basic layout exist because squids have evolved to occupy almost every habitat and ecosystem in the seas of the world. Some species of squid are long and slender, with whiplike tentacles many times longer than the mantle. Others are so short and stout—with body-length fins, stubby arms, and no tentacles—that they look like big-eyed flying saucers.

Squids have been called the jet-propelled torpedoes of the sea. They swim by sucking water into a chamber in the mantle, locking the man-

The author:
Clyde F. E. Roper is a zoologist in the Department of Invertebrate Zoology at the National Museum of Natural History at the Smithsonian Institution and an authority on giant squids.

tle opening, and squeezing a powerful jet of water out through the funnel. They maneuver by pointing the funnel in different directions. Some of the fastest squids can attain speeds of 55 kilometers (35 miles) per hour. These squids use speed to escape fast-swimming predators such as tunas, billfishes, and dolphins. In turn, they are active predators, preying upon small fishes—perhaps even the juveniles of the very species that prey on them! The rest of their diet consists of shrimp and other kinds of squid.

Squids are fast-growing animals—most species that scientists have been able to study in detail become fully mature and reproduce in a year or less. A two-year-old squid would be considered ancient. Females and males of most squid species gather in schools to mate. Males have a modified arm, the *hectocotylus,* that they use to deposit *spermatophores* (long, cylindrical packets of sperm) into the female. The female lays hundreds of thousands of eggs at a time, which are fertilized and then released into the sea. The eggs hatch in a few weeks or months, depending on the species and the environmental conditions. The young of most squid species begin life entirely on their own, without any sort of parental care.

A terrifying sea monster is identified

Although some kinds of squids are quite common and are a popular seafood in many parts of the world, the giant squid has always been a reclusive creature. It keeps mostly to the cold, dark recesses of the deep ocean. Nonetheless, throughout seafaring history there have been occasional encounters with giant squids, events that filled sailors with dread and gave rise to hair-raising myths. An old Norwegian legend, for example, told of the *kraken,* a many-armed sea monster so large that it could drag entire ships to the ocean bottom. Similar stories and myths—some dating back thousands of years—told of ships attacked and sunk, their crews spilled into the water and devoured. While most people in the superstitious ancient world must have been terrified by such tales, they had no way of knowing if the accounts were factual or contained even a grain of truth.

The mystery persisted until the 1800's. In 1857, an eminent Danish biologist, Jappetus Steenstrup, deduced from ancient drawings and stories that the fearsome kraken was actually a squid—a very big squid. He placed it in a new squid genus, which he named *Architeuthis*—Greek for *ruling squid.* Steenstrup's conclusion was verified in 1873, when two dead "krakens" were found within one month of each other in Newfoundland, Canada. Professor Addison E. Verrill of Yale University, an authority on marine animals, examined and described the creatures, confirming that they were essentially supersized relatives of the common squid.

Although *Architeuthis* entered marine-biology textbooks as a real group of animals, knowledge about giant squids accumulated slowly. Everything scientists presently know about the giant squid has been gathered from three primary sources: (1) carcasses or body parts

The anatomy of a giant squid

Giant squids are elusive creatures that have only been seen dead (or near death) and always in a damaged condition. Therefore, many details of their anatomy remain a mystery. However, all squids share some basic physical characteristics. The largest giant squid ever measured was nearly 18 meters (60 feet) from the tip of its tail to the ends of its tentacles.

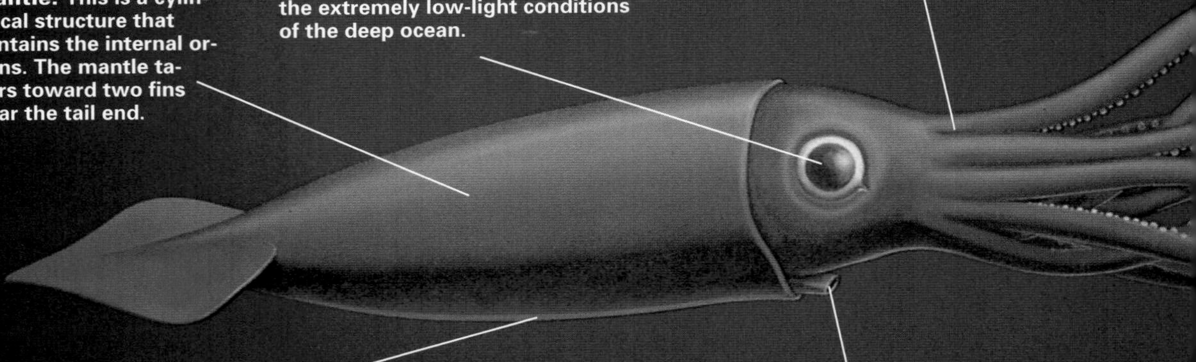

Mantle. This is a cylindrical structure that contains the internal organs. The mantle tapers toward two fins near the tail end.

Eyes. The eyes of a giant squid are the largest in the animal kingdom. They are about 25 centimeters (10 inches) in diameter—about the size of a human head. Such large eyes may help the giant squid to see in the extremely low-light conditions of the deep ocean.

Mouth. Centered amid the arms, the mouth has a beak remarkably similar to that of a parrot. The sharp, powerful beak is used to chop up its prey into more easily eaten pieces.

Ink sac (internal). The giant squid has an ink sac, but how it functions is not certain. The giant squid's ink may glow due to bioluminescent bacteria, as has been observed with some other species of deep-sea squids. Glowing ink can be seen in the dark, and thus it may serve to distract potential predators, just as black ink in sunlit waters does.

Funnel. The funnel is a short, flexible nozzle that protrudes from the mantle, beneath the head. The squid pumps a powerful jet of water out through the funnel and maneuvers by pointing the funnel in different directions. Some of the fastest squids can reach speeds of 55 kilometers (35 miles) per hour. Giant squids, like other kinds of squids, most likely swim tail first most of the time.

washed onto shore or floating at sea; (2) remains found in the stomachs of sperm whales and sucker scars on the whales' skin; and (3) specimens captured in fishing nets.

More than 100 specimens of *Architeuthis* have been found on beaches and coastlines. These discoveries are strictly a matter of chance because no one can predict where or when a giant squid will die and drift ashore. It is also pure luck to find remains that are reasonably intact. Typically, the tentacles and some of the arms are missing, having been broken off by the action of waves or eaten by scavengers. Because a lifeless squid gets tossed and rolled along a shoreline by surf before coming to rest, its skin normally is scraped off. Often, the animal's beak is removed by souvenir hunters.

Many giant squids that reach shore are examined where they lie and then left there. They are so big and cumbersome that they can be preserved only in a large tank, which many labs and museums cannot af-

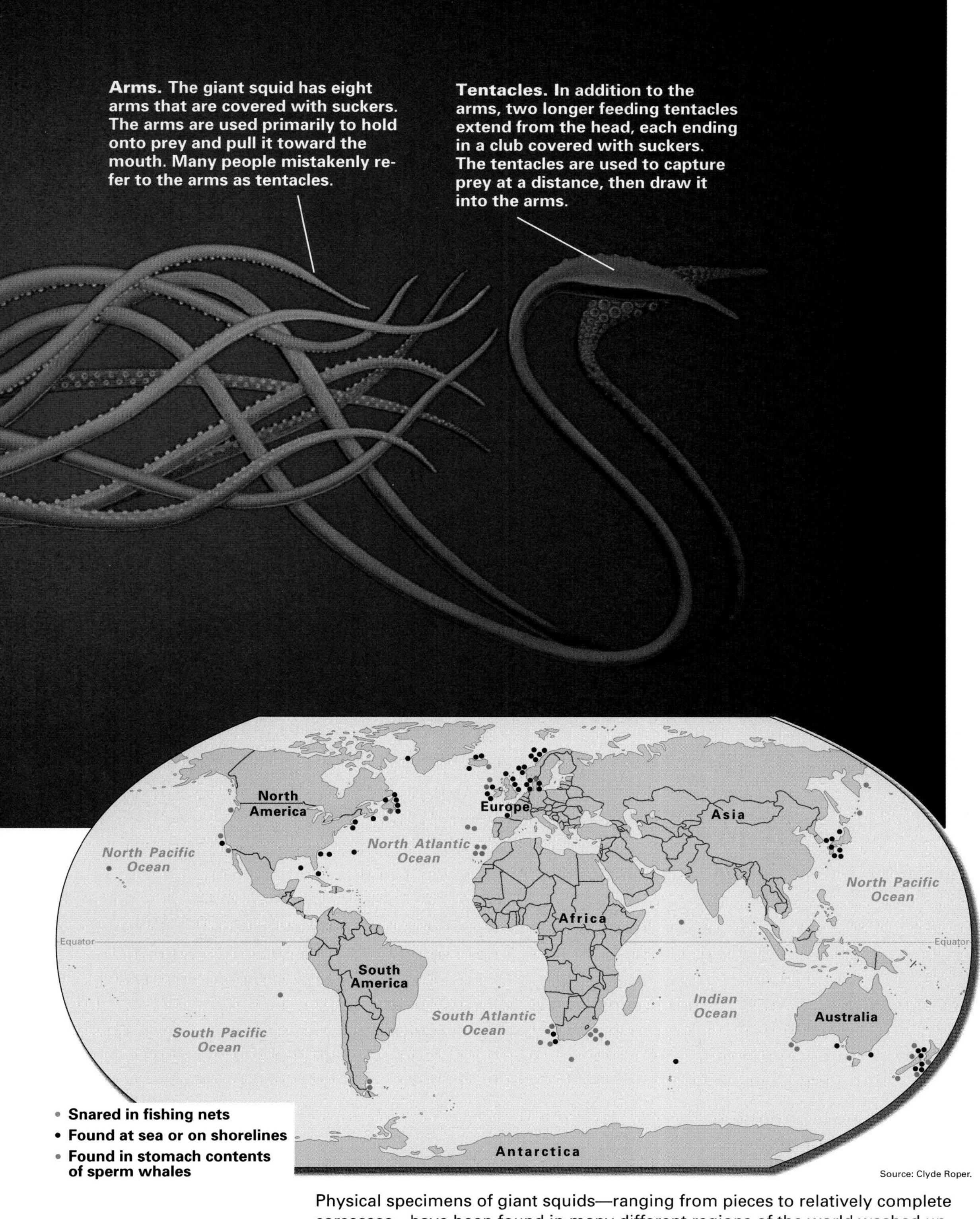

Physical specimens of giant squids—ranging from pieces to relatively complete carcasses—have been found in many different regions of the world washed up on shorelines, floating in the open sea, snared in deep-sea fishing nets, or in the stomach contents of sperm whales.

ford to maintain. Likewise, squid carcasses and body parts discovered floating at sea are usually too big and too damaged to be preserved for study. Still, many *Architeuthis* specimens have been saved by researchers. Because the specimens are incomplete, however, and because they are widely scattered at museums and marine laboratories around the world, it is difficult to know if they all belong to one species or to several. The same identifying features are seldom present on every specimen, but whether this is due to damage or to differences in genetics is not known.

The evidence from sperm whales and fishing nets

Much of our knowledge about giant squids has come from sperm whales. Whale hunters of the 1800's often reported seeing large hunks of animal flesh floating on the ocean surface in areas populated by sperm whales. Occasionally, even whole "kraken" carcasses were sighted. These accounts fit nicely with other stories about sperm whales rising to the surface with huge beasts in their mouths—monsters with many long, writhing arms that encircled the head of the whales in a struggle to the death. Once the kraken had been identified by Jappetus Steenstrup as a giant squid, scientists concluded that sperm whales must prey on giant squids.

To confirm this theory, scientists began to examine stomach contents of captured sperm whales. They discovered that the whales feed mostly on squids and in some parts of the world they eat squids exclusively. Most of the remains taken from sperm whale stomachs were of small to medium-sized squids. The most common remains found were squid beaks, because these hard structures are indigestible and remain in the digestive tract for up to a week before being expelled. A whale's stomach might contain thousands of beaks. In each jumble of beaks, a few were usually much larger than the rest—evidence that sperm whales do indeed prey on giant squids.

The bodies of sperm whales bear striking evidence of the whales' battles with giant squids. The skin on whales' jaws and heads often is heavily scarred by circular marks up to 5 centimeters (2 inches) in diameter. These once-mysterious marks are now known to be healed wounds made by the suckers of large squids as they struggled and fought to escape the whale. Reports of sucker scars 10 to 30 centimeters (4 to 12 inches) in diameter led some imaginative writers to estimate that the squids making such huge marks were 35 to 75 meters (115 to 250 feet) long. However, my colleagues and I believe these large scars were made when the whales were young and expanded as the whales grew. In some cases, similar marks may be caused by a type of parasitic ringworm that lives only on sperm whales and grows as a large, circular patch. So the evidence still supports the conclusion that the giant squid reaches a maximum length of 18 or perhaps 20 meters (65 feet).

Occasionally, giant squids are hauled up in fishing nets. These animals, though dead or dying, are sometimes in excellent condition.

The *kraken,* a fearsome beast famous in old Norwegian sea legends, was a many-armed sea monster so huge that it could drag entire ships to the bottom of the ocean. In 1857, a Danish biologist deduced that the kraken was actually a giant squid—a conclusion that was verified in 1873.

Thus, they yield more complete and accurate data than can be obtained from specimens that have been drifting at sea for days.

Since the early 1980's, Ellen Forch, a colleague of mine from New Zealand, has examined 16 *Architeuthis* specimens caught in nets. Her research is one of the most thorough studies of giant squid anatomy and relationships to date. As a result of her work, we now know that physical characteristics among individuals of *Architeuthis*—no matter whether there is one species or several—can be quite diverse. However, because populations of *Architeuthis* are so widespread, many mysteries remain regarding their biology and even their anatomy.

My associates and I solved one nagging puzzle about giant squids: why so many of them are found beached or floating far from deep water. Wouldn't a dead or dying animal in the deep sea be eaten by hungry scavengers before it floated to the surface? Do giant squids migrate to the surface or into shallow water to die? The solution to this mystery came in a completely unexpected way in 1974, while a colleague, Malcolm Clarke, and I were visiting a fellow squid specialist, Chung-Cheng Lu, in Newfoundland, Canada. Lu and his wife had prepared a

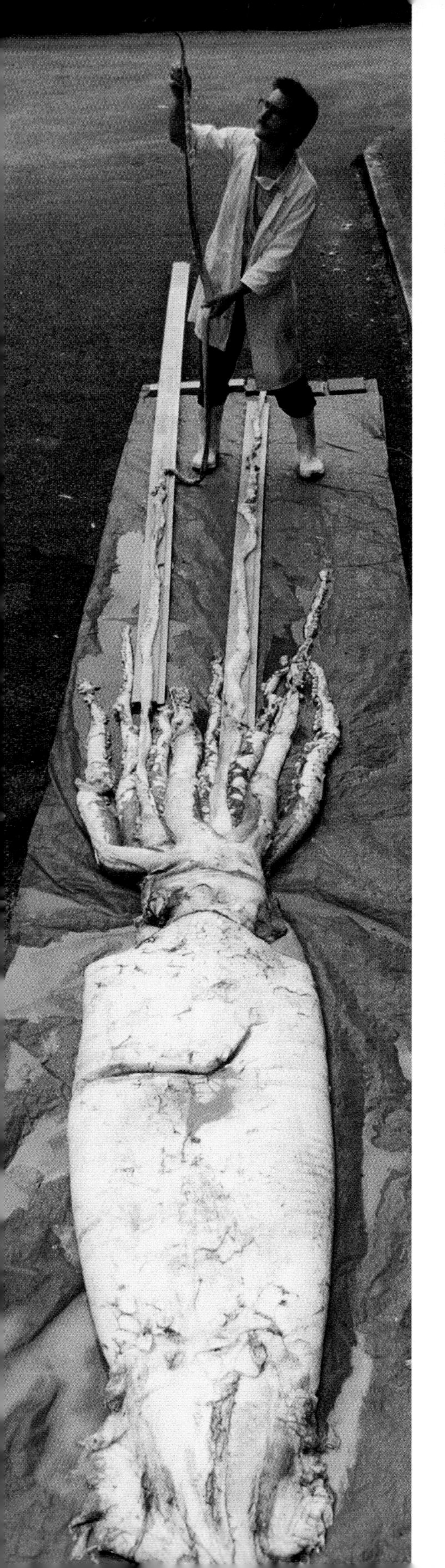

A researcher examines one of three dead giant squids captured in fishing nets off the eastern coast of New Zealand in late 1995 and early 1996. This specimen, measuring 6 meters (20 feet) long, was hauled in from a depth of 300 meters (1,000 feet).

dinner for us featuring the local squid, cooked in 12 different ways. During the meal, I wondered aloud what a giant squid tasted like. Lu said that he just happened to have a piece of giant squid in his freezer. It was taken from an animal that had washed ashore a few months earlier. We prepared the sample in the simplest way and then tasted it. It was awful! Very bitter. But that bitterness gave us an idea, so the next morning we went to Lu's laboratory and tested the remaining uncooked piece of squid.

We learned that in place of the sodium chloride (salt) found in the tissues of many animals, the flesh of the giant squid contains ammonium chloride, which is lighter than seawater. This compound makes the squid buoyant, which prevents it from sinking when it is not swimming. This natural buoyancy causes a squid to float to the surface when it dies. From there, surface currents and waves drive the animal ashore. So it turns out that without the presence of ammonium chloride in *Architeuthis,* we would not have stranded giant squid specimens to study.

Speculation about the nature of giant squids

Over many years, the careful analysis of even the smallest pieces of physical evidence has given squid researchers a general picture of where and how giant squids live. We know, for instance, that *Architeuthis* is found in all the oceans of the world. Clues from their anatomy and their run-ins with sperm whales and deep-sea fishing boats indicate that they live at a depth of 300 to 1,200 meters (980 to 3,900 feet). They grow to full length and a weight of over 400 kilograms (880 pounds) in a very short time, certainly less than five years. Because they start life as hatchlings only 2 to 3 millimeters (0.08 to 0.12 inch) in length, they must be real eating machines. Their prey consists of deep-sea fishes and other species of squids known to inhabit the deep ocean. Females tend to grow larger than males; in fact, some males less than 3 meters (10 feet) long have been found to be fully mature and capable of mating. Mature females produce millions of tiny eggs, but we

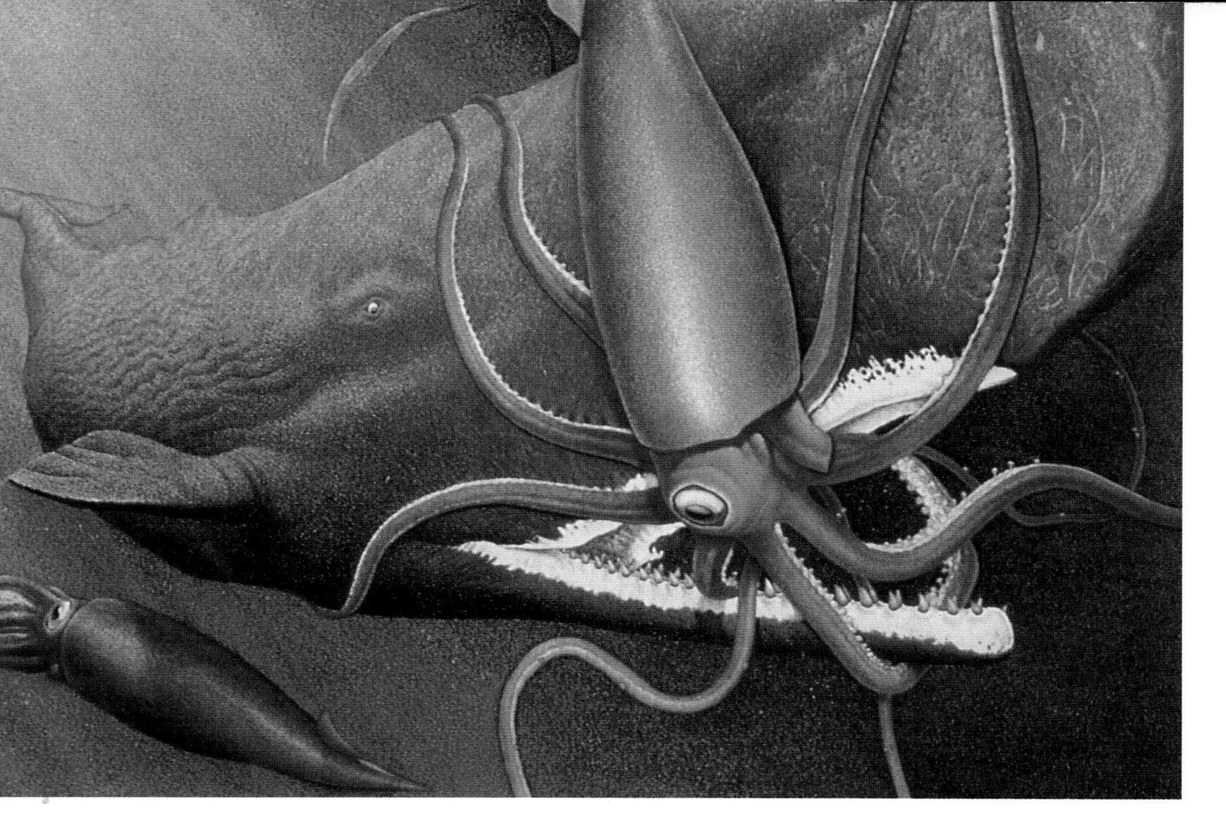

Giant squids are a favorite prey of sperm whales. The remains of many giant squids have been found in the stomachs of captured sperm whales. But the squids apparently battle fiercely for their lives when attacked by a whale. The skin around the jaws and head of sperm whales often is heavily scarred by circular sucker marks left by giant squids as they fought to escape being eaten.

know little for certain about the mating habits of giant squids or the early development of their young.

We can speculate about some other aspects of their lives with some confidence. Their anatomy provides clues. For example, the giant squid's eyes are the largest in the animal kingdom—each is about the diameter of a human head. Such huge eyes must have evolved for gathering light in dimly lit waters, so *Architeuthis* may be able to see in the near blackness of its deep-ocean lair. It also may be able to detect the *bioluminescence* (biochemically generated light) emitted by other deep-sea creatures. In addition, giant squids have a dark-red skin, another indication that they live in deep water. Red light does not penetrate to great depths—its energy is rapidly absorbed in the upper 20 meters (65 feet) or so of water. For this reason, a red animal in deep water is as invisible as a black land animal on a moonless night. Being invisible would be a great help to the giant squid in both hunting prey and avoiding the jaws of sperm whales.

Does the giant squid have other ways of escaping whales? We know that shallow-water squids eject clouds of black ink as a defensive decoy to help them confuse and escape from potential predators. The giant squid also has an ink sac, but is it functional? If it is, what good would it do to squirt out black ink in the darkness of the deep sea? No one would notice! For this reason, I believe that the giant squid's ink glows with bioluminescent bacteria, as my associates and I have observed with some other species of deep-sea squids. Glowing ink can be seen in the dark, and thus it functions as a decoy in the same way as black ink in lighted waters.

Giant squids may need evasive tactics to get away from sperm whales because they probably are not especially powerful or fast swimmers. The muscles of their mantles, though quite thick, are not as solid or tightly packed as those of other squids known to be strong swimmers. Furthermore, the muscles of the giant squid's mantle, head, and arms contain significant amounts of ammonium, giving the muscles a soft, spongy texture—another indication of weaker swimming ability. Giant squids are most likely passive, slow-moving animals that wander through their domain hunting for equally passive prey.

But that's mostly a theory, and one that is not shared by some of my colleagues, who believe that *Architeuthis* is probably a speed demon. This kind of uncertainty emphasizes why the giant squid is often called "one of the last great mysteries of the sea." It's a mystery I would love to clear up by observing giant squids in their natural surroundings. It has been a dream of mine for many years.

The search begins

My recent efforts to encounter a live giant squid were sparked in 1994, when the National Museum of Natural History in Washington, D.C.—where I have worked since 1966—opened an exhibit on giant squids. My colleagues and I worked diligently to make the presentation as informative, accurate, and interesting as we could. The research that went into creating the exhibit renewed my ambition to seek the giant squid and gave me several new insights into where and how to look. I concluded that the best places to look for *Architeuthis* were areas where large numbers of sperm whales are known to feed. It occurred to me that the whales might serve as "hound dogs," leading me into the deep-sea habitat of the giant squid. My intent was not to capture a giant squid but only to find one and film it doing what it normally does. Just a few minutes of video footage and remote measurements of its environment could add immeasurably to our understanding of *Architeuthis*. For example, we could record the actual depth and location of its habitat, how it moves, its ways of escaping sperm whales, and—with extraordinarily good luck—its hunting and mating behavior.

Opportunity knocked one day in September 1994, when producers from National Geographic Television approached me about my plans for an expedition to search for the giant squid. The National Geographic Society offered to finance part of the expedition if I would consent to let their team participate. I agreed at once and scouted the two places I considered to be the most likely "hot spots" for finding giant squids: the Azores Islands in the central North Atlantic and Kaikoura Canyon, an underwater gorge in the South Pacific off the South Island of New Zealand. We decided to try the Azores first. That phase of the expedition began in July 1996.

Our objective in the Azores was to film the hunting strategies of sperm whales using an innovative camera system developed by Greg Marshall, a National Geographic technologist and marine biologist. Marshall had built a miniaturized video camera system called the "crit-

During a 1996 expedition to the Azores Islands, a member of the research team approaches a sperm whale—swimming just below the surface—to attach a "crittercam" to its head with a large suction cup. The crittercam is a small video camera specially designed to be carried underwater by marine animals. Footage of sperm whales obtained with the camera was featured in the February 1998 National Geographic television special "Sea Monsters: Search for the Giant Squid." Unfortunately, the crittercam failed to capture any footage of giant squids.

tercam," which was designed to be attached to large marine animals in order to study and film their behavior in their natural habitat. The crittercam typically is attached to an animal's back with a 30-centimeter (12-inch) suction cup. In addition to its video capabilities, the device can make sound recordings and collect data on depth, temperature, swimming speed, and other variables. After a preset length of time, the crittercam detaches itself and floats to the surface. A homing beacon on the crittercam enables researchers to find the device and retrieve it. The apparatus had performed successfully with sea turtles, dolphins, and even sharks, but would it work on the animal we wanted to use—a huge sperm whale?

Aboard the *Silvery Light,* a restored English fishing vessel built in 1884, we spent day after day in the waters off the Azores looking for a sperm whale to serve as our camera platform. When we finally recruited a suitable candidate—an 11-meter (36-foot) female—Greg Marshall and a colleague paddled out to the whale in a small, inflatable sea kayak and attached the crittercam to the top of her head. We held our breaths as the whale slipped below the surface of the water and disappeared. About an hour later, the crittercam bobbed back to the surface and we recovered it.

During the rest of the cruise we attached the crittercam to the heads of seven other sperm whales. After each recovery of the crittercam, we watched the footage of its latest journey with wonder and excitement. We were able to capture the first images ever made of sperm whales during a dive. We also got video and sound recordings of sperm whales socializing at great depths and of baby whales swimming unexpectedly

The *Odyssey II*

Off the coast of New Zealand in 1997, Clyde Roper and his team scouted the waters for giant squids using the *Odyssey II.* This small unmanned submersible scanned its surroundings with a nose-mounted video camera and a variety of sensors.

Bob Grieve of the Massachusetts Institute of Technology (MIT), a member of the *Odyssey II* team, makes some last-minute adjustments before launching the submersible on one of its many runs.

Jim Bellingham of MIT, *left,* designer of the *Odyssey II,* Brad Moran of MIT, and Roper monitor the *Odyssey II's* progress. Each dive lasted two to three hours and covered several kilometers at depths of 300 to 750 meters (980 to 2,460 feet).

Its mission completed, the *Odyssey II* returns to the surface and is hauled back aboard the research vessel for the trip back to port. After each dive, team members spent hours reviewing the little sub's videotapes for noteworthy images. But again, there were no pictures of giant squids.

far below the surface. We also obtained a number of scientifically important glimpses of these magnificent creatures.

What we didn't see was a giant squid. In fact, we saw no squids at all. So it was on to New Zealand.

In New Zealand, Roper and another squid expert, Malcolm R. Clarke, photograph sucker marks made by a giant squid on the head of a stranded sperm whale. The relatively recent scars on the dead whale's skin confirmed that the waters off New Zealand are in fact giant squid territory.

Another attempt in New Zealand

The second phase of our hunt for the giant squid kicked off in January 1997. We chartered a New Zealand fishing boat, the *Tanekaha,* as our research vessel. Home base for the expedition was the marine station in the small town of Kaikoura on the northeast coast of South Island. Each year, tourists and researchers come to Kaikoura from all over the world to observe and study the abundant ocean life, especially the sperm whales and other marine mammals. We were optimistic about seeing a giant squid in this region because numerous specimens had been found there over the years in the usual places—on beaches, in fishing nets, and floating at sea. Our plan was to use three kinds of search tools: the crittercam, an unmanned submersible robot, and several submerged cameras.

Our use of the crittercam ran into delays, however, because we needed a permit from the New Zealand Department of Conservation to approach the whales and attach the device. By the time permission was granted, the weather had begun to deteriorate, making it impossible to safely approach the whales.

With the crittercam sidelined, our only mobile camera would be aboard a small unmanned submersible, called an autonomous underwater vehicle (AUV). This remarkable little submarine, named the *Odyssey II,* was designed and operated by Jim Bellingham of the Undersea Vehicles Laboratory of the Massachusetts Institute of Technology. It is equipped with a nose-mounted camera, a powerful computer, a navigation system with scanning sonars, and a variety of sensors. Just over 2 meters (6.6 feet) long and weighing 160 kilograms (350 pounds), the *Odyssey II* can dive to a depth of 6,000 meters (19,700 feet). Guided by its computer, which is programmed before each launch, the AUV roams through the sea completely on its own. The preprogrammed instructions tell the vehicle how deep to dive, which directions to go, how fast to travel, and when to come home.

Each of the *Odyssey II's* dives lasted two to three hours and covered several kilometers at depths of 300 to 750 meters (980 to 2,460 feet). We always knew precisely where the AUV was because it frequently reported its position back to us. Still, we were always a bit tense as we waited for the sub's return until its nose finally burst through the surface. Then we had to wait, with growing anticipation, until we got back to the laboratory before we could view the videotapes. Studying the tapes took three to four hours every night. Often, we would have to replay a segment several times to identify an animal that had suddenly whizzed past the camera. We also frequently stopped the tape to discuss images of particular interest: What kind of shrimp was that? How big was that fish? Was *that* thing a squid?

Over the two-week period of research operations with the *Odyssey II,* the sub performed superbly. We obtained many hours of interesting, informative videotape and data. But again, we failed to capture an image of a giant squid.

We were no more successful with a system of submerged video cameras dubbed "ropecams." The ropecam was designed by Emory Kristoff, a deep-sea photographer at National Geographic. It consisted of a camera mounted in a steel frame, which was lowered to the sea floor on a long rope attached to a floating marker. A battery pack provided power to run the camera for a few minutes at 15-minute intervals for up to eight hours. From a pole attached to the frame, big chunks of tasty fish dangled in the camera's field of view to attract deep-sea creatures. In addition, a drum periodically released a cloud of fish juice that we hoped would have the same effect on marine animals that the smell of a sizzling steak has on most people.

Over 50 deployments of the ropecam at depths of 500 to 1,500 meters (1,640 to 4,900 feet) produced some exciting footage of sharks, eels, shrimps, and crabs, and—much to my delight—squids! These were not *Architeuthis,* however, but New Zealand arrow squids, creatures about 1 meter (3 feet) long. The squids flashed in to attack the bait again and again. Several small sharks, also about 1 meter long, joined the free-for-all, tearing at the bait. In the midst of all this activity, we saw an arrow squid attack a shark. But the squid let go immediately and jetted away just as fast as it had come in. The shark raced off too. I suspect this was an accidental attack, stimulated by the frenzied feeding activity.

Thus, the expedition ended without our obtaining any images of giant squids. Nonetheless, we gathered a considerable amount of information about the marine life and sea-floor terrain of Kaikoura Canyon, which should be helpful to marine scientists working in the canyon area in the future.

A sad sight—but solid evidence of giant squids

There was one other noteworthy find that we made during our research in New Zealand. One afternoon after we returned to the lab from a day's work on the *Tanekaha,* we heard on the radio that five sperm whales had become stranded on a remote beach in Golden Bay, at the north end of South Island. We quickly arranged a charter flight on a small plane and reached the scene before dark. What a somber sight, those five huge animals lying dead side by side, far from the open sea. No one knows for certain why apparently healthy whales become stranded (or strand, as marine biologists say), and it is unusual for so many sperm whales to get trapped on a beach together. We speculated that these whales, all males 11 to 13 meters (36 to 43 feet) long, were cruising through Cook Strait, which divides North and South Islands. Instead of continuing directly through, they must have turned into the bay, which has a gently sloping, silty bottom. They probably became disoriented in the shallow water and became stranded after the receding tide left them stuck on a sand bar.

At least one useful piece of information resulted from this sad event: Each whale bore fresh circular scars on its skin, marks left during the past few weeks by the suckers of giant squids. In death, these magnificent whales confirmed that we really were in giant squid territory.

Another chapter in the search for the giant squid was planned for 1999. This trip would be part of a major expedition back to New Zealand involving scientists from several nations and a variety of scientific specialties. This time, we planned to use a manned research submersible, the Johnson Sea Link (JSL), which is operated by the Harbor Branch Oceanographic Institution in Fort Pierce, Florida. The JSL can carry up to four people to a depth of 1,000 meters (3,300 feet) and can remain submerged for more than eight hours. It is equipped with sophisticated sensors, lights, cameras and video recorders, and bioluminescence recorders—everything a scientific team needs to probe the cold, black depths of Kaikoura Canyon and other interesting sites in New Zealand waters. No deep-diving submersible has ever worked in the ocean around New Zealand before, so we felt that this expedition might produce many interesting and useful discoveries. Among our anticipated research projects were a biological survey of deep-sea fishes and invertebrates around New Zealand and an evaluation of the area's marine invertebrates and algae for compounds useful as drugs and medicines. And of course we planned to look long and hard for *Architeuthis.*

Our chances for finding and filming a giant squid on this expedition looked better than ever before. Not only had we learned much from our previous expeditions, but being able to have human observers in a submersible would significantly enhance the prospects for success. I could think of no more exciting way to close out the 20th century than to be part of such an important oceanographic expedition of research and exploration.

For further reading:

Conniff, Richard. "Clyde Roper Can't Wait to be Attacked by the Giant Squid." *Smithsonian Magazine,* May 1996, pp. 126-128.

Fisher, Arthur. "The Hunt for the Giant Squid." *Popular Science,* March 1997, pp. 74-77.

Roper, Clyde, and Boss, Kenneth J. "The Giant Squid." *Scientific American,* April 1982, pp. 96-100.

Mr. Nanotechnology

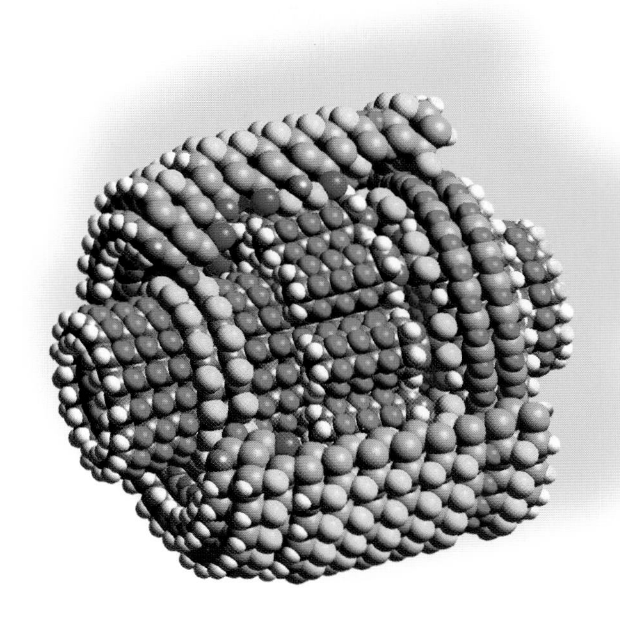

An engineering visionary discusses the possibility of manipulating matter at the smallest levels and how that ability could shape the world's future.

An interview with K. Eric Drexler, conducted by David Dreier

In the exotic field of nanotechnology, K. Eric Drexler has been one of the most visible figures and one of the most controversial. Nanotechnology, which is still more dream than reality, is the precise manipulation of individual atoms and molecules to create larger structures. Since the late 1970's, Drexler has argued that the absolute control of matter at the smallest scales is an achievable goal and would lead to molecular manufacturing—producing consumer goods and many other things with the aid of multitudes of submicroscopic "assemblers." Many scientists dismissed Drexler as more of a nanotechnology "guru" than a serious researcher, particularly because he wasn't afraid to make predictions about the ultimate possibilities of nanotechnology. They called molecular manufacturing a pipe dream. By the mid-1990's, however, nanotechnology was becoming a mainstream pursuit involving a growing number of researchers in the United States, Europe, and Japan. In 1994, the White House science adviser urged U.S. government support for nanotechnology, commenting that molecular manufacturing "could fuel a powerful economic engine providing new sources of jobs and wealth." Although the problems to be overcome in developing this incredible new technology were formidable, Drexler's vision was looking less like a fantasy and more like an attainable goal. Drexler himself was increasingly being viewed as a man ahead of his time.

Opposite page: Drexler is backed by a blow-up of a computerized model of a proposed nanodevice of his design, a type of gear. Each small sphere represents an individual atom. *Above:* A model of a molecular planetary gear—a device to vary the rotational speed transmitted from a central drive shaft—is cut away to reveal its internal structure.

Science Year: What all is encompassed by the term *nanotechnology*?

Drexler: It refers, in general, to engineering things on a nanoscale. By nanoscale, I mean measured in units called nanometers, or billionths of a meter—dimensions much smaller than a human cell. So the term nanotechnology is sometimes applied, for example, to building microchips with extremely small features and a whole host of other technologies that involve very tiny pieces. When I use the term, though, I'm usually referring to a technology that, as it matures, would enable us to make almost anything with molecular precision by manipulating individual atoms and molecules.

Science Year: And that's why you usually prefer the term molecular manufacturing?

Drexler: Yes. Nanotechnology was the first term I came up with, though I later found out that it had first been coined by a researcher in Japan. I thought it was a really nice word that could be applied to a variety of engineering applications, and so did a lot of other people—all kinds of researchers started using the term. Molecular manufacturing was the second term I invented, and I think it better communicates the central idea of making useful things with the submicroscopic machines I call assemblers.

Science Year: How did you become interested in the possibility of engineering things on such a tiny scale?

Drexler: I started on this path when I was an undergraduate student at MIT [Massachusetts Institute of Technology]. I was doing some reading in molecular biology, and I learned that cells have structures called ribosomes, which molecular biologists often refer to as molecular machines. Ribosomes read instructions from DNA [the molecule that carries the genetic code] to produce the many proteins needed by an organism. And when I found out about this, I started asking myself, what if we could find a way to design and build tiny biological systems like that—to make custom-designed proteins? Could those be used as a starting point to create other kinds of tiny structures? And that's what led to the idea of molecular manufacturing.

Science Year: So you thought that if nature can make things one atom or molecule at a time, engineers can do it?

Drexler: Basically. I reasoned that if nature can do it, it must at least be possible for engineers to do it. I couldn't see any scientific reason preventing it. And after I started working on these ideas, I learned about a talk that the American physicist Richard Feynman gave back in 1959, in which he said that the manipulation of atoms and molecules to build things wouldn't violate any physical principles. Feynman went on to win the Nobel Prize, so I guess he knew a thing or two about what's possible according to physical law. And it's pretty clear that, in fact, we can indeed move atoms and molecules around to suit our own purposes. Researchers have already used tiny probes on the tips of scanning tunneling microscopes and atomic force microscopes [two types of scanning probe microscopes] to put groups of atoms into precise arrangements. Admittedly, that's a long way from molecular manufacturing, but it's a start.

Welcome to the Nano World

The realm of nanotechnology is the world of atoms and molecules. It is a very, very small world. The prefix *nano* means billionth and refers in this case to nanometers—billionths of a meter. The average atom is about four-tenths of a nanometer in diameter, and the smallest speck of matter that can be seen under an ordinary microscope contains more than 10 billion atoms. Atoms combine to form molecules, which in turn form larger structures.

Nanotechnology, which was still very much in a developmental stage in 1998, aims at building things by connecting individual atoms and molecules—much like stacking bricks, though in a very precise way. This will be done, say K. Eric Drexler and other proponents of nanotechnology, by submicroscopic machines called assemblers. These nanomachines will be guided by computers, including ones with dimensions also measured in nanometers. As of 1998, however, both assemblers and nanocomputers were still just engineering dreams.

Nanotechnology, whenever it becomes a reality, offers the promise of producing a vast array of consumer goods. But some of the first fruits of nanotechnology may be tiny electronic devices, such as ultrasmall sensors.

Other products of nanotechnology will be used in science and medicine. For example, researchers predict the creation of microscopic robotic devices that will patrol the human body and fight disease.

The possibilities of nanotechnology are truly staggering, and no one can predict with certainty where the technology will lead.

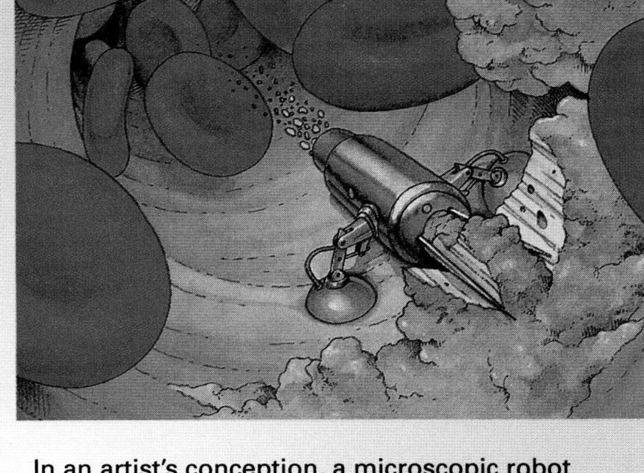

In an artist's conception, a microscopic robot cleans deposits from a blood vessel. Such devices, and many consumer products, may someday be made with nanotechnology.

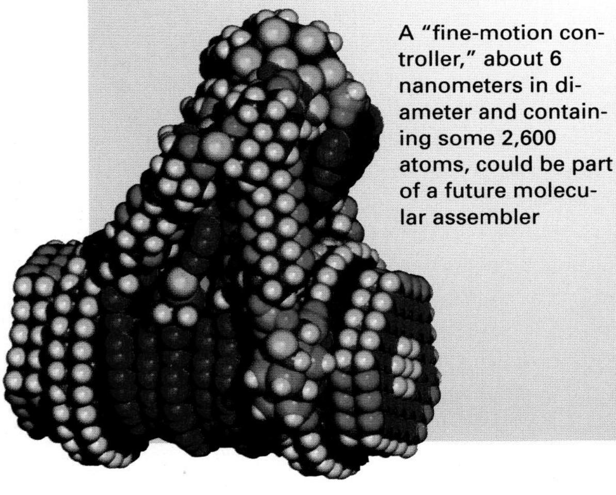

A "fine-motion controller," about 6 nanometers in diameter and containing some 2,600 atoms, could be part of a future molecular assembler

Science Year: Despite what Feynman said, some chemists and physicists have argued that constructing things atom by atom is beyond human capability.

Drexler: Oh, yes, I know. They've said atoms are too difficult to work with because they're these "fuzzy," unpredictable things; you can't pin down their location. Then you see the arrangements of atoms made with scanning probe microscopes, and atoms don't seem so mysterious. I mean, there's your atom, and it's basically a little lump. And if you go back tomorrow and take another look through the microscope, the atoms are still there in the same configuration.

Well, okay, some researchers would say, but that's because the pictures were made at a very low temperature. In a warmer environment, it wouldn't be possible to put atoms where you want them and keep them there because of thermal [heat] vibrations. You see, atoms are always quivering with energy, and the warmer they get, the more they move around. I'll admit that this could cause some problems in molecular manufacturing, but I think the difficulties can be overcome. Again, let's look at the example of the cell. Despite thermal vibrations, the molecular machinery in some cells makes fewer than one mistake in 100 billion operations when manufacturing proteins. To achieve that accuracy, cells have special error-correction systems, and we'll have to develop the same sort of mechanism for assemblers.

One thing you should understand about nanotechnology is that it's essentially an engineering challenge, and most of the critics of nanotechnology have been scientists.

Science Year: And you think that can be done?

Drexler: I don't see why not. One thing you should understand about nanotechnology is that it's essentially an engineering challenge, and most of the critics of nanotechnology have been scientists. They've raised certain legitimate questions, such as the ones we've just discussed, but they haven't presented any solid reasons why nanotechnology can't work. A lot of them just offer opinions: "I'm a scientist, and I feel this or I think that," instead of "I'm a chemist, and here's a chemical issue that's crucial to your proposal, and it's wrong." That sort of comment I would take seriously, and if anyone had such an argument, I think I would have heard it by now. But I haven't heard it, and so I claim it ain't there.

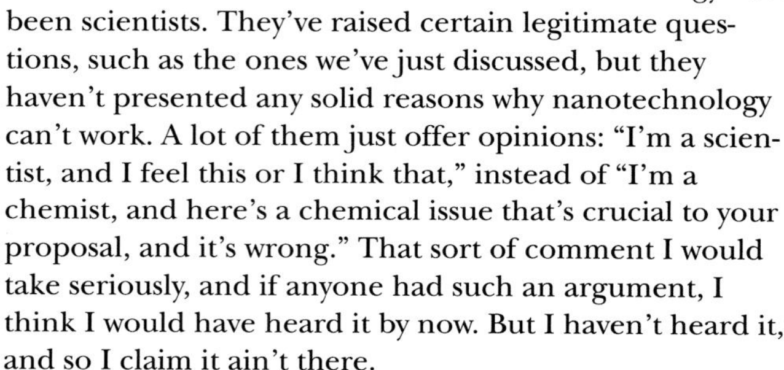

Science Year: Do you feel that molecular manufacturing is becoming a more mainstream idea?

Drexler: Overwhelmingly, yes. It's the usual story with any new idea that's different enough to be upsetting to the scientific establishment. Initially, you have lots of people saying, "This is silly, this is crazy." But then you get a new generation of young scientists and engineers coming up who are excited by the idea and saying, "Let's do it." That's the situation we have now.

Science Year: So what's the first big step?

Drexler: I think the first big step toward nanotechnology would be to create protein molecules that could be used as tools to develop better tools. Scientists are making considerable progress in designing new protein molecules from scratch on the computer. These proteins are assembled synthetically from chemical building blocks called amino acids, and the completed molecule folds itself into a particular shape.

I foresee a developmental process in which we build some crude, first-generation nanomachines from parts made of variously shaped protein molecules. We would then use those machines to build other molecular machines until we got to some level where we had a machine—still made of protein—that could do some productive things.

Although a machine made of protein molecules may sound strange, custom-designed proteins could be made into many useful shapes, such as pipes, beams, and casings.

Science Year: How would you go from making things out of proteins to making them out of more durable materials?

Drexler: We would begin by using the biological machines we've created to construct things from inorganic substances. Nature shows us in countless ways that it can be done. For instance, there are tiny single-celled plants in the ocean called diatoms that produce little glassy cases for themselves from minerals in the water. That's a good example of an intricate inorganic structure that's made by a biological system. For the purposes of nanotechnology, a similar process could be accomplished starting with our protein machines. Like the diatom, they could build structures out of nonliving materials. Then once we have nanomachines made out of inorganic molecules, they would replace the biological systems. From there, we would proceed to the development of efficient assemblers, and that would lead to molecular manufacturing.

I think the first big step toward nanotechnology would be to create protein molecules that could be used as tools to develop better tools.

Science Year: What would an assembler look like and how would it operate?

Drexler: An assembler would look something like an industrial robot of the kind used on present-day auto-assembly lines, but less than a millionth the size. Essentially it would be a rigid, jointed arm with some sort of gripping mechanism on the end, and it would be mounted on a sturdy base. An assembler would take an atom or molecule and place it in a certain position so it forms a strong chemical bond with another atom or molecule. It would conduct about a million operations a second; things happen very quickly at the nano level.

Science Year: How would assemblers be used to manufacture things?

Drexler: It depends. If you're making something really small, then having a single assembler putting one atom in place at a time is just fine. But if you want to make something big enough to hold in your hand or park in your garage, you'll need millions or billions of assemblers working in parallel. We know that molecular machines can build big things because we have redwood trees and blue whales. That's natural nanotechnology in operation on a huge scale. Enormous things like that don't get made by a single molecular machine—you have an astronomical number of them working together. To construct something artificial on that scale, such as an office building, you'd have to use assemblers with the ability to replicate [make copies of] themselves so that you could get trillions of them working on the project.

Thinking Small

A very unusual article appeared in the September 1981 *Proceedings of the National Academy of Sciences.* Entitled "Molecular Engineering: An Approach to the Development of General Capabilities for Molecular Manipulation," the article contended that it should be possible to construct materials and machines "from the ground up" by assembling individual atoms and molecules in a predetermined order. After all, argued the author—a 26-year-old graduate student at the Massachusetts Institute of Technology (MIT) named K. Eric Drexler—the cells of our bodies do that very thing every day when they carry out genetic instructions to manufacture proteins from molecular building blocks. The young engineer predicted that researchers would eventually surpass nature by learning to build custom-designed protein molecules for new tasks.

That ability, Drexler said, would offer great technological potential. As researchers gained the ability to create made-to-order proteins, he wrote, it would become possible to construct molecular-sized "protein machines." Those devices would then be used to construct equally minuscule machines out of more durable materials, such as metal, ceramic, or diamond.

Drexler gave this futuristic vision a futuristic name: nanotechnology, from nanometer, or billionths of a meter—the scale of atoms and molecules. He had no doubts that nanotechnology would eventually be as common as large-scale manufacturing is today. As our ability to manipulate matter at the smallest level matured, Drexler predicted in later writings, humanity would gain vast new powers of commercial production. Armies of tiny "assemblers" would churn out limitless quantities of goods—furniture, cars, T-bone steaks, you name it—from basic raw materials at a cost not much greater than that of the materials. An era of unlimited plenty would dawn.

While an undergraduate at MIT, Drexler, *above,* at far right, was part of a group that studied possible technologies for colonizing space.

Drexler also foresaw other dazzling applications of nanotechnology: Computers far more powerful than any now existing and microscopic robots that circulate in the body and keep us free of disease.

This view of the world's future was so utopian it was dismissed by many people as science fiction. And yet, Drexler was not the first person to suggest the possibility of making matter do our bidding. In a now-famous lecture entitled "There's Plenty of Room at the Bottom," no less an eminence than Richard Feynman, one of the greatest physicists of the 1900's, made the same prediction. Feynman commented, "The principles of physics, as far as I can see, do not speak against the possibility of maneuvering things atom by atom." When we gain that ability, he said, "We can expect to do different things. We can manufacture in different ways."

Feynman's landmark talk lent credence to the astonishing ideas advanced by Drexler, who for some years felt like a voice crying in the wilderness. To skeptics who insisted that engineering things on an atomic scale was a fantasy, he would retort, "Hey, if you don't think this is possible, go argue with Richard Feynman."

Kim Eric Drexler was born in 1955 in Oakland, California, and raised in several parts of the country. As a boy, he was more interested in outer space than the inner space of atoms and molecules. The manned orbital and moon-landing programs, which were then at high tide, fired his imagination, and he was especially captivated by the idea of space colonization as a solution to the limits of growth on Earth.

Drexler pursued his interest in space at MIT, where he enrolled as a freshman in 1973 in the interdisciplinary science program. As an undergraduate, and later as master's candidate in engineering, he wrote several papers for technical journals in which he proposed designs for electric-power-generating satellites, orbiting manufacturing facilities, and machinery to extract minerals from the lunar surface.

While at MIT, Drexler met Chris Peterson, a chemistry student from upstate New York who shared many of his interests. They fell in love and married in 1981.

Drexler chats with a participant at a 1995 nanotechnology conference in Palo Alto, California, sponsored by his organization, the Foresight Institute.

By that time, Drexler had become intrigued with the idea of nanotechnology and the boundless abundance it offered as another answer to the limits of growth. His article in the journal of the National Academy of Sciences marked a turning point in his career. He followed that piece with a number of other technical articles and two books for popular audiences, *Engines of Creation: The Coming Era of Nanotechnology* (1986) and *Unbounding the Future: The Nanotechnology Revolution* (1991), coauthored with his wife, Chris, and electronics industry entrepreneur Gayle Pergamit.

Along the way, Drexler spent five years—from 1986 to 1991—as a visiting scholar at Stanford University in Palo Alto, California. In one year, he even taught a course on "Nanotechnology and Exploratory Engineering." He also pursued further studies at MIT, and in 1991 he earned the first doctorate in molecular nanotechnology to be awarded by any institution. A year later, he published *Nanosystems: Molecular Machinery, Manufacturing, and Computation,* a highly technical volume aimed at nanotechnology researchers. That book won the Association of American Publishers award as best computer-science book of the year.

Drexler was beginning to attract attention. Although he had no shortage of critics, he also impressed many people with his ideas. In 1992, he was invited to testify before a Senate subcommittee on "New Technologies for a Sustainable World," chaired by then-Senator Al Gore. The following year, he was named the recipient of the annual Kilby Young Innovator Award, named for Dallas inventor Jack Kilby, the developer of the integrated circuit. And in 1997, *Newsweek* magazine cited him as one of "100 people to watch as America prepares to pass through the gate to the next millennium."

Drexler is frequently asked to give an estimated time of arrival for nanotechnology. At his Senate appearance, in response to that question from Gore, he replied, "Fifteen years [by the year 2007] would not be surprising for major, large-scale applications." That guesstimate may have been overly optimistic, but Drexler has never wavered from his conviction that, sooner or later, nanotechnology is inevitable.

Many other people now think so too. Nanotechnology has been rapidly gaining scientific respect in the United States and is also being hotly pursued in Europe and Japan. In America, the White House science adviser, physicist Jack Gibbons, predicted in 1994 that molecular manufacturing was a coming reality and recommended government support for research efforts in nanotechnology. And by 1998, at least one private company—Zyvex, in Richardson, Texas—had already been formed for the sole purpose of developing a molecular assembler.

Drexler has supplied much of the momentum in the drive to develop nanotechnology. He has lectured widely, steadily gaining converts to his vision, and in 1986 he established the Foresight Institute, a privately funded organization in Palo Alto aimed at furthering nanotechnology research and preparing society for "the coming revolution in molecular manufacturing." He serves as chairman of the institute; Chris Peterson is the executive director. Foresight has two sister organizations, the Institute for Molecular Manufacturing and the Center for Constitutional Issues in Technology.

Each fall, the Foresight Institute sponsors a conference on nanotechnology in Palo Alto that is attended by many of the leading lights in the field. The keynote speaker at the 1997 conference was Rice University chemist Richard Smalley, who shared the 1996 Nobel Prize in chemistry for the discovery of unusual carbon molecules called fullerenes that could have useful applications in nanotechnology.

The institute also confers two annual $5,000 Feynman prizes—named for Richard Feynman—for noteworthy research advances in nanotechnology. And it is offering a $250,000 Feynman Grand Prize to whoever first builds a molecular-sized robotic arm that could be used as part of an assembler and a tiny computing device that demonstrates the feasibility of building a nanoscale computer.

So far, the grand prize has been in no danger of being claimed. But Drexler would love for someone to win it. [DD]

Science Year: How would assemblers be given the instructions they need to carry out required operations?

Drexler: In most cases, they'd follow instructions fed to them by computers. A central computer would be used to relay information to nanoscale computers, which would in turn give instructions to various groups of assemblers. So each assembler would be following a "molecular tape"—an internal mathematical blueprint—to carry out its portion of the manufacturing process. This would be similar to the way that a ribosome in a cell follows a string of instructions from DNA to manufacture a protein.

Science Year: What do you think a nanotechnology factory of the future will be like?

Drexler: Well, first picture a traditional factory where people make things out of many parts on a large scale. But now imagine that the parts are a whole lot smaller, the area in which the parts are being moved around is also much smaller, and you have many more operations being performed at the same time. What nanotechnology will make possible is putting all the complexity of a modern manufacturing plant into a box about the size of a microwave oven that you could keep on your kitchen counter—literally.

What nanotechnology will make possible is putting all the complexity of a modern manufacturing plant into a box about the size of a microwave oven that you could keep on your kitchen counter—literally.

Science Year: A household factory?

Drexler: Yes. It would have a reservoir in which you would put various sorts of low-cost raw materials in liquid form. Inside the factory, the raw materials would be separated into individual atoms or molecules, which would be physically shunted around to where they are needed, to be fastened together by the assemblers. The tiny structures made by those assemblers would then be connected by other, larger assemblers to make progressively bigger parts until the item was completed. The various operations would be carried out in a vacuum to keep air molecules from messing things up.

To make a particular product, you'd select a desired item from a computerized catalog and specify size and style and color and whatever. Then you'd punch in your order on the machine, and an hour or so later a timer bell would chime and you'd open a door and take out the completed product—a wristwatch, for example. And the cost of the item, leaving aside such things as licensing fees and taxes, would be basically the cost of the raw materials—a couple dollars at most.

Science Year: What about the production of larger products, such as automobiles?

Drexler: For large, massive items, you obviously would need a larger production mechanism. If you wanted to make something the size of a car, you would probably want a nanotechnology factory that's maybe

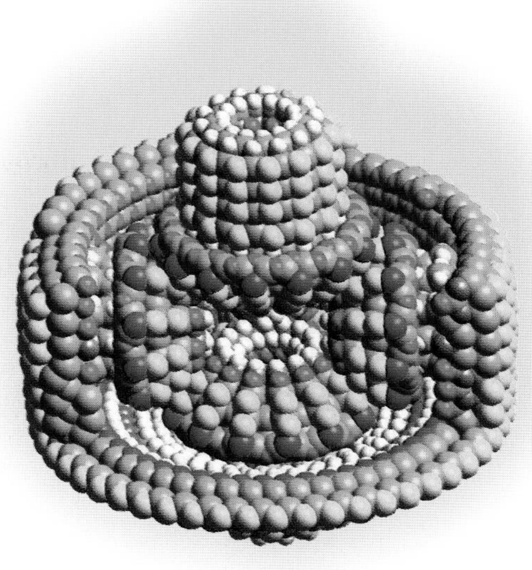

Drexler and his wife, Chris Peterson, who serves as executive director of the Foresight Institute, discuss one of his designs. The device, shown enlarged, *above,* is a differential gear, a mechanism in which different parts can move at different speeds. At present, there is no specific intended use for most of Drexler's designs. They are basically engineering exercises intended to demonstrate the possibilities of molecular-scale constructions.

the size of a garage. In general, the tools would have to be matched to the product. Big things like cars would undoubtedly continue to be made in commercial facilities.

Science Year: You mentioned nanocomputers as an integral part of molecular manufacturing. Could an entire computer be of submicroscopic size?

Drexler: A special-purpose computer for directing a particular manufacturing process could indeed be that small, and I expect to see such computers being developed in coming years. The whole thrust in microelectronics has been toward steadily shrinking components. What we haven't seen is the extension of structures very far into three dimensions, because that's technically difficult with present manufacturing methods. But it will be an easy thing to do with molecular manufacturing—you're "stacking up bricks" in a controlled way. So I see nanotechnology and computer technology developing in parallel: As we get better at making things smaller, we'll build ever-tinier computer components and eventually nanocomputers. And in turn, nanocomputers will help us construct a limitless array of products.

Nanocomputers won't be the whole story of course. We'll still have larger computers, but their components will be so small that a single desktop machine could have a memory and computing capacity equal to that of all the world's present-day computers combined. That may seem like an outlandish prediction, but consider that today we have more computing power in a home computer than existed in the entire United States in the early 1950's. And the computer age is still in its infancy. We're also going to see the development of artificial intelligence—thinking computers—and who knows where that will lead us.

Science Year: Another development you've predicted is tiny nanorobots that will cruise through our bodies and keep all our systems in good repair.

Drexler: Yes. There are many health problems that are simple in the sense that there is something you need to get rid of, such as a virus in the blood or fatty deposits on the inside of arteries. If you could just get something into the body that could recognize the threat and destroy it, you'd be okay. With cancer, for example, you have certain cells that are different from the other cells of the body. The differences are subtle, which is why cancer cells can evade the immune system and why it's so hard to kill them selectively with drugs. But what if you had a microscopic device, equipped with a powerful nanocomputer and various kinds of sensors, that could go from one cell to another looking for surface features that identify a cell as cancerous? When it found a suspicious cell, it would investigate further and make sure that it was indeed a cancer cell. Then it would take some sort of action that would kill the cell. Perhaps it would inject the cell with a bit of a toxic substance. We could have billions of these devices patrolling our bodies.

Science Year: What do you think have been the most interesting and promising developments in nanotechnology research in recent years?

Drexler: I guess I've been most impressed with two lines of development. One, which I mentioned before, is the manipulation of atoms and molecules with the scanning tunneling microscope and atomic force microscope. That ability, I think, provides the basis for a kind of robotic manipulator that could be used for assembling things on a molecular scale. A typical industrial robot in a present-day factory has two main parts. One part is an arm designed to place an object—an automobile component, or whatever—in a particular position. The other part is a hand at the end of the arm that holds the object and turns it in various ways so precise operations can be performed on it. The scanning tunneling microscope and the atomic force microscope are analogous to the robot's arm. What we don't have right now is a good "hand" on the end of the microscope probe. But that seems like a reasonable job for chemists to undertake. So I think we're fairly close there to what can be regarded as a primitive molecular assembler.

Nanotechnology will give us the ability to manufacture things, including components of nanomachines, out of diamond by laying down one carbon atom at a time in tight crystal arrangements.

The other line of development that I'm pretty excited about is the rapid advance in the custom-design and production of molecules. These include proteins and other large, highly structured molecules that assemble themselves and also smaller molecules that stick together in predictable ways to make larger structures. Researchers are now making these things with an increasing amount of control to create a wider range of building blocks that could probably be used in making nanostructures.

Science Year: Some investigators see a future in nanotechnology for the group of carbon molecules called fullerenes. What's your expectation there?

Drexler: It's hard to say. Fullerenes—or buckyballs as some types of them are often called—are a third crystalline form of carbon, the other two being graphite and diamond. Fullerenes are certainly interesting, and they may have significant uses. A variant form called buckytubes or nanotubes, which are tiny cylinders of precisely arranged carbon atoms, might be particularly useful. They're already being used as atomic probes in some microscopes. But I'm not quite sure how we're going to make use of fullerenes in nanotechnology until we find some way of making them in a controlled way, which we don't have yet. Right now, they're made by heating carbon to extremely high temperatures with lasers, and that's not quite what a chemist would regard as a precisely controlled process. But people are trying to overcome the problem, and they may well succeed. Eventually, if we find that fullerenes are truly useful, we should be able to make them with assemblers just like we would any other material.

Science Year: You mentioned diamond. Haven't you speculated about using diamond as a building material in nanotechnology?

Drexler: Yes. Diamond is a really amazing material. Its rigid crystalline structure gives it enormous strength in relation to its weight. For example, if you were to take the space shuttle and build it out of diamond-based materials instead of aluminum, you would end up with a vehicle that weighs 98 percent less than a conventional shuttle and is somewhat stronger besides. Nanotechnology will give us the ability to manufacture things, including components of nanomachines, out of diamond by laying down one carbon atom at a time in tight crystal arrangements. These would be very solid, stable structures.

Science Year: What sort of research are you yourself involved in?

Drexler: My work involves using the computer to design various kinds of mechanical components made of thousands of atoms—different sorts of gears and bearings, for example. These things can't be built yet; they exist only in the computer. Eventually, such pieces would be constructed by assemblers as parts for larger molecular machines. All these things would be made of diamondoid materials, which is to say substances with a lot of carbon atoms in a crystal arrangement but also including other kinds of atoms, such as silicon. I've also designed a diamondoid piece I call a fine-motion controller, which is intended to serve as a hand for an assembler arm.

Science Year: The big question in everyone's mind about nanotechnology is undoubtedly, when will it become a reality? Back in 1992, in testimony before a Senate subcommittee on new technologies,

What I expect to see over the next 10 years is a steady accumulation of increasingly impressive developments: new materials, new levels of complexity in the manipulation of atoms and molecules, and so on.

you predicted that nanotechnology might start to bear fruit in 15 years, or by the year 2007. Do you still think that's a feasible date?

Drexler: Well, if you're talking about products made by molecular manufacturing, then I don't think we'll be seeing that by 2007. What I expect to see over the next 10 years is a steady accumulation of increasingly impressive developments: new materials, new levels of complexity in the manipulation of atoms and molecules, and so on. Then we'll enter a sort of transition zone when we have all the various tools we need to get started, and we'll use those tools to build better tools and the first real nanomachines. The kinds of applications associated with full-blown molecular manufacturing will come some time after that—perhaps by the year 2015.

The correct scientific answer to the question of when nanotechnology will reach fruition, though, is, I don't know. But of course, that's not what people want to hear, so I try to lay out some sort of reasonable timetable. But there are a lot of uncertainties, and there are bound to be some unexpected developments. We could have amazing advances that speed up the timetable or snags that slow it down. But sooner or later, nanotechnology will be here.

Science Year: You've issued some warnings about the potential technological perils inherent in nanotechnology, such as the possibility that self-replicating assemblers could multiply uncontrollably and run amuck.

Drexler: Yes, and I do think that we need to at least consider dangers of that sort. I mean, here we have this giant steam locomotive of a technology that's bearing down on us from the future, and hardly anyone is thinking about all the possible ramifications of it. Every new technology has its pitfalls, and the way to avoid being blindsided is to see the potential problems and work to prevent them.

Now, having said that, I'll add that I'm not unduly worried about self-replicating assemblers running out of control. I think these things will be engineered so that it will be pretty hard for them to do things we don't want them to do. I am quite concerned, though, about the potential that nanotechnology will offer for deliberate misuse, such as by militaristic nations. If molecular manufacturing can be used to produce consumer products, it can also be used to make various kinds of really terrible weapons in enormous quantities. So in that respect, I think, nanotechnology could be a destabilizing force in the world.

Science Year: You've also done quite a bit of speculating about the social effects of molecular manufacturing. What impact do you think this technology will have on society?

Drexler: If we can make unlimited quantities of low-cost goods, there should be enormous wealth per capita. I would guess that a society in which everyone's materials needs and desires had been satisfied would start to center less on work and more on leisure activities. People would do things because they want to do them, not because they have to in order to support themselves. I think the best model for that sort of world is the wealthy segments of society today and in the past. People with lots of money have always had their pick of the best things available in their time. Historically, it's always been a very small percentage of the population who have been in that enviable situation. But in the future, I think, it's likely that the bulk of the world's population will live like that.

Science Year: Is there a possible downside to such a world?

Drexler: Certainly. The potential for social pathology would be enormous in a society that is no longer centered on labor and in which large numbers of people have lots of time on their hands. Boredom would be a real possibility in a world like that, and there's no telling how people would deal with it. Human beings have always been very creative when it comes to thinking up new kinds of pathological behavior. One basic observation I've made, though, is that most people seem to be pretty good at working out a decent life for themselves if they have the time and resources to do so.

Science Year: So you think that, on the balance, the social effects of nanotechnology will be positive?

Drexler: I think that, on the balance, the effects on the world of any great technological change are impossible to predict. We'll just have to wait and see how things play out. But I'm hopeful that this technology will be mostly a force for good.

Science Year: What sort of world do you foresee in 50 years?

Drexler: If things work out and we handle our choices well, the world will be made up of healthy, wealthy people making wise use of a technology that is intrinsically very clean and efficient. The Earth, I suspect, will become a park. And to a certain extent, the future will be unimaginable because people will have the freedom to do things we can't even conceive of today. The world will be a radically different place in 50 years, I think that's a certainty.

For further reading:

Drexler, K. Eric. *Engines of Creation,* Doubleday, 1986.

Drexler, K. Eric, Peterson, Chris and Pergamit, Gayle, *Unbounding the Future.* William Morrow and Co., 1991.

Additional resources:

The Foresight Institute maintains a site on the World Wide Web. The address is www.foresight.org. The full texts of *Engines of Creation* and *Unbounding the Future,* as well as information about the institute, are available at the Web site.

Museums That Make Science an Adventure

By Larry Bell

Museums filled with "hands-on" exhibits are exciting and fun to explore, but they are also places where people can learn a lot about science and technology.

"I can't believe she's getting on that thing! Is that girl crazy?"

"But Mom, she won't fall off—it's science!"

As they approach the lobby of COSI Toledo—the Center of Science and Industry in Toledo, Ohio—a family watches a young girl get onto a bicycle. A museum staff member helps her fasten a seat belt. What's all the fuss about? Anyone can ride a bike. But this is no ordinary bicycle. This bike moves across a thick steel cable 6 meters (20 feet) above the lobby floor. Why doesn't the bicycle fall off the wire? One reason is that the bike wheels are slotted so that they can't slip off the cable. But the more important reason, visitors learn, is based on the principles of gravity and counterbalance: a 113-kilogram (250-pound) weight hanging below the bicycle keeps the bike's center of gravity below the wire—and the bicycle on the cable.

COSI Toledo, which opened in 1997, is an interactive science museum—a "hands on" kind of place. It is one of the newest of some 400 interactive science museums—or science centers, as they are often called—in major cities throughout the world in 1998. In the early 1980's, there were only about 30 such interactive science museums. The growth of these institutions is aided by the Association of Science-Technology Centers in Washington, D.C., which was established in 1973 to provide professional support to U.S. science museums and science centers.

Interactive science centers have their roots in the venerable natural history museums that have long given people the opportunity to see rare and wonderful things from throughout the world. Since the early 1900's, almost every major city in the United States has boasted a natural history museum, and quite a few prominent scientists trace their interest in science to childhood visits to such a museum. Natural history museums display rocks and minerals, fossils and dinosaur bones, pre-

served animals in simulated natural habitats, and the artwork, tools, and clothing of ancient peoples.

Preceding page:
A young visitor at COSI Toledo—the Center of Science and Industry in Toledo, Ohio—learns about gravity and counterbalance as she rides a bike high above the museum lobby floor.

The purpose of a natural history museum is to store knowledge. Many of the first natural history museums were created so that scientists could share knowledge with other scientists and with their students. Even when they opened their displays to the public, the scientists organized the exhibits to suit fellow scientists' interest in the materials. The organization of displays in a natural history museum often illustrates the similarities and differences among the objects in the museum's collections. For instance, minerals that contain iron might be placed in one case and minerals that contain copper in another. Animals that lived in North America might be exhibited in one hall and animals that lived in Africa in another.

Grouping exhibits by similarities also reflects the way in which scientists classify the objects that they find in the natural world. For example, a natural history museum might set up a hall of *invertebrates* (animals that have no internal skeletons) based on displays of creatures that make up that group: insects, worms, and *mollusks* (animals with shells, such as clams and oysters).

In a natural history museum, visitors learn what items are displayed, where the items came from, and why they are interesting or important by reading labels on the exhibits written by the museum staff. The objects themselves also convey information. From a fossil dinosaur bone, for example, a visitor can gain an understanding of how big the entire dinosaur was.

Despite the many opportunities for learning that natural history museums provide, the subject matter of such museums is limited to objects found in the natural world. Modern science deals more with phenomena than it does with objects, and technology is a constantly evolving field. Out of this need to teach the public about scientific phenomena and the current state of technology, the interactive science center was born.

Early interactive science museums

The first interactive science center, the German Museum of Outstanding Achievements in Natural Science and Technology, was founded in Munich in 1903. Exhibits in the German museum allowed visitors to push buttons or turn cranks to activate physics and chemistry demonstrations in glass cases. For example, viewers could push a button that activated an electric current in a wire and see that it made the needle of a compass turn, a demonstration that the flow of electricity through a wire generates a magnetic field.

The author:
Larry Bell is the vice president for exhibits at the Museum of Science in Boston.

Staff members at the Franklin Institute Science Museum in Philadelphia, the Museum of Science and Industry in Chicago, and the Museum of Science in Boston were inspired by the new push-button exhibits at the German museum. While some of the exhibits in the American science museums continued to include objects that were of interest in and of themselves, such as locomotives, airplanes, and even a submarine, the staffs began to construct many new exhibits about scientific phenomena—the behavior of everyday things.

One exhibit introduced in the 1930's at the Museum of Science and Industry in Chicago demonstrated a principle called the conservation of momentum. The museum staff suspended 12 bowling balls from the ceiling in a single row. When a visitor pulled back and released the first bowling bowl, it struck the second and stopped moving. None of the other balls moved except the last one. The ball's momentum—a quantity equal to the ball's mass times its velocity—was transferred down the line, finally causing the 12th ball to swing upward. Other interactive exhibits that the Museum of Science and Industry developed at that time included a model train with a highway crossing signal that visitors could operate and a half-scale model of a working power shovel.

Visitors in the early 1940's learn about electricity at one of the first interactive science museums in the world, the German Museum of Outstanding Achievements in Natural Science and Technology in Munich.

Even more interactive

In 1969, a new kind of science museum, called the Exploratorium, opened in San Francisco. All of the exhibits in the Exploratorium were interactive. In addition, visitors could experiment with a number of variables in an Exploratorium demonstration, using equipment that stood out in the open rather than being enclosed in glass cases. For example, at an exhibit called the Lens Table, visitors could manipulate an image projected onto a screen by placing different lenses or boards with holes varying in size in the path of the light. At the Organ Pipe exhibit, people could learn about resonance. By adjusting the length of a tube that continually emitted sounds or changing the frequency of a speaker attached to the tube, visitors could vary the sounds.

Some of the exhibits created by the Exploratorium staff and by guest artists and scientists became so popular that other science museums

Children at San Francisco's Exploratorium in the 1960's experiment with perspective in the museum's Distorted Room exhibit. The room's trapezoidal shape causes the boys and girls to appear to shrink or grow, depending on where they are standing.

began inquiring about buying exhibits or making copies of them. The Exploratorium responded by creating a series of instruction books, which the staff called "cookbooks," explaining how to construct particular exhibits. Thus, museums began to share ideas about how to best informally educate people about scientific concepts.

One of the most copied Exploratorium exhibits, Everyone Is You and Me, introduced in 1978, explores the properties of light and mirrors. In this exhibit, two visitors sit facing each other on opposite sides of a pane of glass that has been slightly silvered to make a partial mirror. Some light passes through and some is reflected. Lights are aimed at each of the two visitors, and knobs control the brightness of the lights on each side. Suppose one is a boy and one is a girl. When the boy turns up the lights that shine on him, he sees himself and the girl also sees him—the glass acts as a mirror from his perspective and as a window from hers. If the lighting is reversed, the opposite effect occurs. But if they adjust the lights so that there is some light shining on both of them, and if they position themselves exactly opposite each other so that their faces are at the same height, they each see only one face in the glass—but it doesn't belong to either of them. The image each person sees contains features of both faces—a really weird experience. The coating on the mirror causes the glass to act partially as a window and partially as a mirror.

Do-it-yourself experiments

One of the museums that took interactive exhibits to a new level was the Science Museum of Minnesota in St. Paul. Originally a natural history museum, the Science Museum of Minnesota was well known for its traveling wildlife exhibits. In 1991, the museum opened a new exhibit area called the Experiment Gallery, which allowed visitors to explore such phenomena of physics as waves, sound, light, and electricity. While the typical interactive exhibit is designed so that people can investigate all of its possibilities in a minute or two, those at the Minnesota museum, called "experiment benches," invite visitors to conduct more elaborate, in-depth studies of scientific phenomena. At the exhibit on sound, for example, people can use a keyboard or a microphone to produce and record sounds and then analyze them with an electronic instrument called an oscilloscope.

Another museum to feature highly interactive exhibits is Portland's Oregon Museum of Science and Industry, which opened Engineer It! in 1993. At Engineer It!, visitors learn about the design and engineering process by trying out their own designs. For example, they can design and build a model sailboat out of materials provided by the museum and then test the boat in a water tank with an artificial wind source. They can also build a paper airplane and test it in a wind tunnel or construct a building out of blocks and see how well it withstands a simulated earthquake.

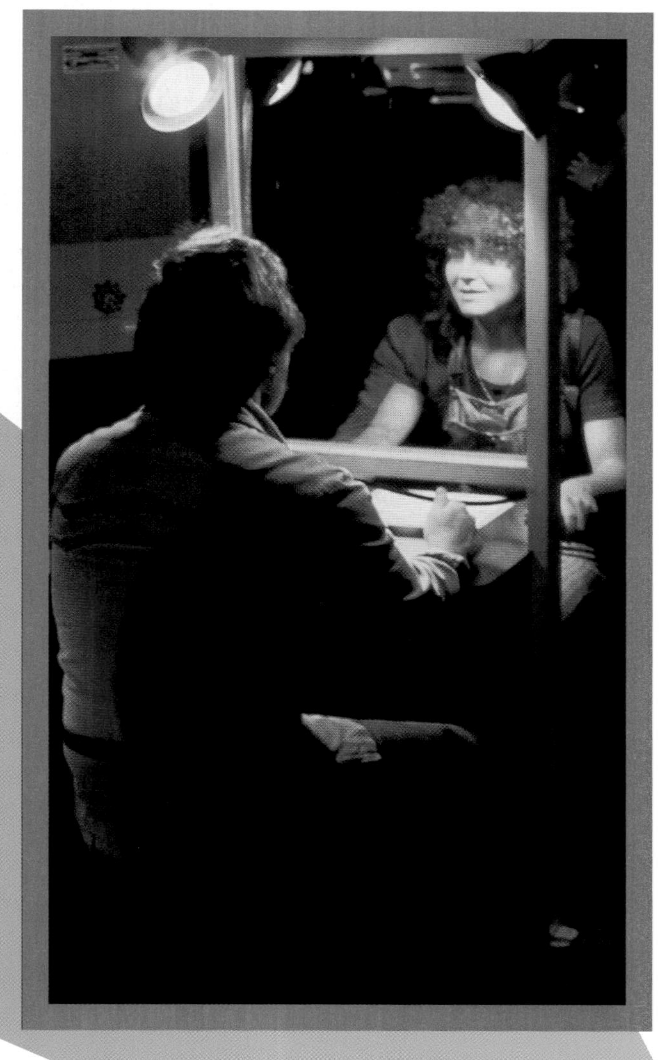

Visitors experiment with one of the most popular exhibits at the Exploratorium, Everyone Is You and Me. By adjusting the amount of light that shines on each of them, the man and woman can cause the partially silvered glass between them to act as a window or a mirror or to create a composite view that includes features from both of their faces.

The Museum of Science in Boston made interactivity the focus of a long-range exhibit plan started in 1989. Rather than planning exhibits devoted to various scientific disciplines, such as biology, chemistry, or physics, the Boston museum decided to go with demonstrations emphasizing scientific thinking skills. The result was Investigate! A See-for-Yourself Exhibit, which opened in 1996. Investigate! encourages visitors to ask questions, carry out experiments, and reach their own conclusions in various fields of science. In one activity, visitors build and test a model solar-powered car. By changing such variables as the car's wheels and tires and the amount of light shining on an indoor

At the New York Hall of Science in New York City, a museum worker called an "Explainer" uses a microscopic camera to introduce children to the world of microbes.

Visitors to the New York Hall of Science can see the varieties of microbes that live and grow on human hands, *above*. People can press their own fingers into a dish of *agar,* a jellylike substance used to culture bacteria, or examine microbes from the hands of previous visitors. Each spot is a colony of individual microbes thriving on nutrients in the agar.

race track, people can affect the performance of the car and learn which factors are most important in powering a car by sunlight.

At another Investigate! activity, visitors dig in a simulated archaeological site to find remnants of the lives of ancient peoples. By comparing the shells, bones, and stone tools they find at the site with ones at reference benches, they can draw conclusions about the daily lives of the people who left the traces of their existence. The visitors can then compare their interpretations with those tape-recorded by archaeologists or entered into a computer by previous investigators of the site.

The dig site represents another way in which science centers have been trying to heighten the learning experience: by creating exhibits that visitors can immerse themselves in. This was the main approach taken by the New York Hall of Science in New York City, which in the summer of 1997 opened one of the largest outdoor science "playgrounds" in the United States. The playground, open from spring through fall, includes exhibits demonstrating a number of physical phenomena. One popular attraction is a giant climbing web made of rope reinforced with flexible steel. No matter how many people are on it, the entire web shifts whenever someone moves. The shifting illustrates a principle of all physical structures: that continual adjustments are made to achieve *equilibrium* (a balance of forces).

Beyond exhibits

The attractions at some science museums include theatrical performances. One or two actors put on short plays in an exhibit area or at a nearby theater to shed light on important social issues, such as drug abuse, AIDS, and ethical dilemmas raised by genetic engineering. In a play entitled *Mapping the Soul* performed at Boston's Museum of Science, two characters discuss the possible consequences of mapping human genes:

> "Gus: . . . once we map and sequence every gene in our bodies, for the first time in history, we'll be able to tell—chemically, physically—what it is to be human.
> Charlotte (to audience): That's what I'm afraid of.
> Gus: What?
> Charlotte: That you are going to tell us what it is to be human."

Such plays often involve the audience, as the actors ask visitors to voice their opinions.

Another method of exploring social issues was used by the Ontario Science Centre in Toronto, Canada, which opened in 1969. The museum created an exhibition called A Question of Truth in 1996 to focus on the problem of prejudice. Visitors enter the exhibition by walking through an electronically controlled locked door, which allows some people in right away but holds others back without explaining why they cannot enter. People experience first-hand how it feels to be denied opportunities enjoyed by others. The exhibition itself contains 38 exhibits, each of which illustrates aspects of a particular knowledge system, from that of primitive cultures, such as herbal lore, to that of technically sophisticated societies, such as multimedia computing. A Question of Truth challenges visitors to honor the ideas of others.

At the Great Lakes Science Center in Cleveland, one prominent exhibit, opened in 1996, educates people about environmental hazards in and around the Great Lakes. In a section of the exhibit called the Great Lakes Situation Room, visitors at 24 computer stations face a wall of video monitors. Science Center staff members run a videotaped program on the monitors, stopping at frequent intervals to ask questions. The participants record their answers on the computers, and the results are shown on the monitors. In this way, the museum staff and visitors share their knowledge and opinions about environmental issues of critical importance to the Great Lakes.

To create the types of exhibits that will attract people to interactive science centers, staff members often develop prototypes and then test them with visitors before building a permanent exhibit. Museum officials want to know whether visitors like a particular exhibit, if it operates properly, and if people learn anything from it. At the Boston science museum, prototype exhibits are tested in an area called the Test Tube. Staff members watch people operating the exhibits and ask the people their opinions about prototypes.

At the Museum of Science in Boston, visitors at the Investigate! exhibit search for clues about the everyday lives of prehistoric people at a simulated archaeological dig site.

Museums without walls

The educational mission of science centers isn't confined to the presentation of exhibits. Most science museums also develop special programs for schools. Some of these programs are offered at the museums, while others are presented at schools or posted on the Internet. The offerings include special classes and activities, kits for conducting science projects, and teacher-training programs. Some science centers even serve as partners for public schools. The Science Museum of Minnesota, for example, is affiliated with the Museum Magnet School, where children with an interest in science learn about scientific principles and participate in the development of the museum's interactive exhibits.

During the 1990's, many science museums began to revise their programs and exhibits as part of a nationwide effort to change the way science is taught in the schools. As groups of educators at the national, state, and local levels worked to develop new guidelines and standards for teaching science, the museums adapted their presentations to support the new criteria.

Visitors at Boston's Museum of Science race model cars of their own design on a solar track. By changing such variables as the cars' wheels, tires, or the amount of light shining on the track, people can learn what factors are most important in designing an efficient solar vehicle.

In 1996 and 1997, the Exploratorium designed two exhibits to address themes that were part of the California science standards. One of these themes was "cycles," the idea that many phenomena in nature occur in patterns that are repeated over and over. The Cycles exhibit includes such displays as a chicken-egg incubator and artificial geysers made of glassware that erupt every few minutes on a set schedule.

The Orlando Science Center in Florida, which opened in 1997, was the first science center in the United States in which all of the exhibits and programs were designed for the specific purpose of supporting educational standards. But despite their serious intent, the Science Center exhibits are still interesting and fun. For instance, visitors walk through a huge mouth and over a squishy tongue to get into interactive exhibits about the human body. Take a guess how they get out!

Do people learn anything at interactive museums?

Sometimes all the fun evident in a science center makes it look as if there isn't any learning going on. And sometimes, for some people, that may be true. But many educators and organizations, including the American Association for the Advancement of Science (AAAS), contend that science centers provide unique learning opportunities. In a report titled *Science for All Americans,* published in 1989, the AAAS said

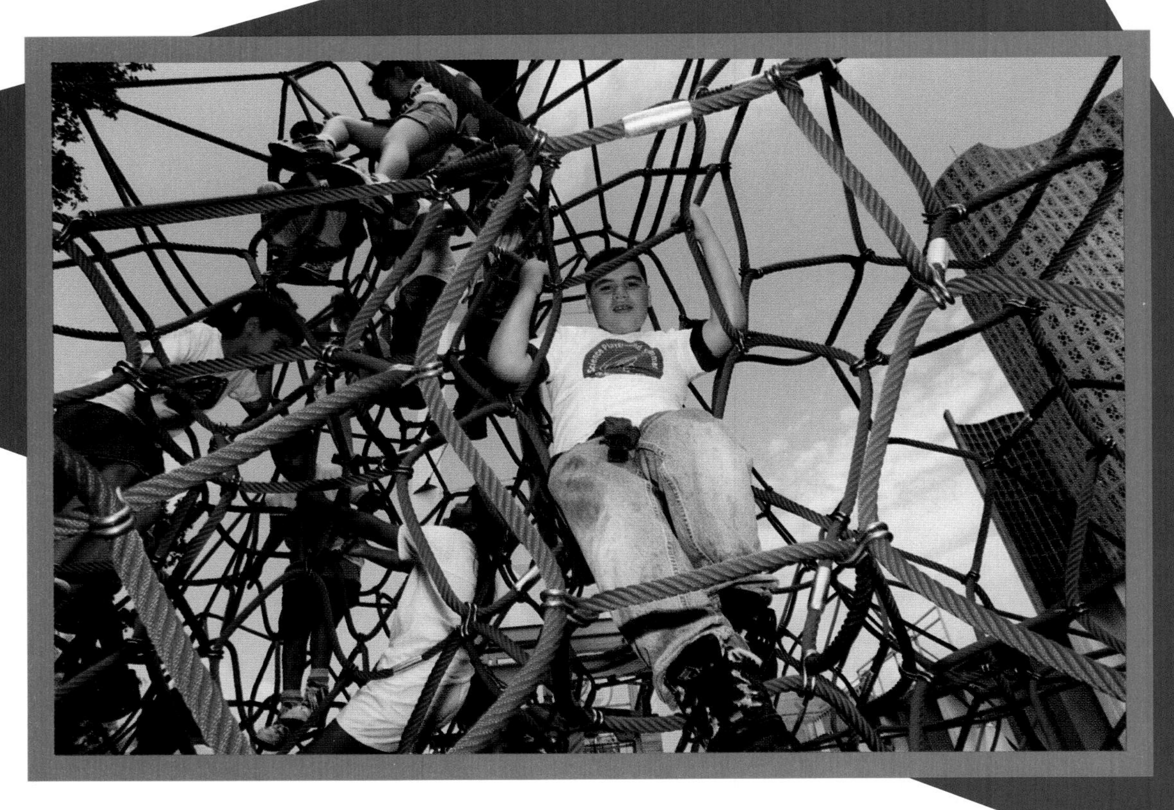

Children at the New York Hall of Science climb the 3-D Spider Web, which is part of the museum's outdoor science "playground." As they play, the children also learn about such scientific principles as *equilibrium,* the shifting that all physical structures undergo to maintain balance.

that "to understand [science, mathematics, and technology] . . . requires that students have some experience with the kinds of thought and action that are typical of those fields. . . . The present science textbooks and methods of instruction, far from helping, often actually impede progress toward scientific literacy. They emphasize the learning of answers more than the exploration of questions, memory at the expense of critical thought. . . ."

Because science centers excel at creating enthusiasm for science and technology, many educators believe these museums can serve as models for schools. And because science museums enable visitors to experience scientific concepts by seeing and touching the materials involved, these institutions can accommodate people whose learning style is visual, sensory, or exploratory, rather than *linguistic* (language-based), the approach favored by schools.

Science museums of the future

What can we expect from science museums in the future? Many of the trends begun in the 1980's and 1990's are likely to continue as museums strive to create ever-better interactive experiences for visitors. The Fort Worth (Texas) Museum of Science and History, for example, has developed a traveling exhibit called Whodunit?: The Science of

Solving Crimes, which could be a sample of things to come. The entrance to Whodunit? looks like the inside of a diner, where a murder is being reported on an overhead television set. As visitors leave the diner, they enter an alley—the scene of the crime—with various clues still in place. In the rest of the exhibit, people are challenged to figure out who committed the murder by using such scientific crime-fighting techniques as fingerprinting, genetic identification, and analysis of microscopic fibers. Visitors even observe an autopsy—performed on a mannequin—to learn how the victim died. At the end of the exhibit, they compare their own conclusion of "who done it" with that of the experts.

Science centers are also exploring how to create the kinds of special ex-

Visitors at the Franklin Institute Science Museum in Philadelphia, *above*, learn about the heart by walking through a large-scale model of one. At the California Science Center in Los Angeles, *left*, Tess, a partially transparent figure of a woman 15 meters (50 feet) in length, takes the giant-heart idea one step further. Visitors listen to an audio-visual presentation about the human body as the figure's organ systems light up to demonstrate how the various systems work together.

A microscopic slide depicting *diatoms* (one-celled plants) is included on the New York Hall of Science's Web site. Many science and technology centers have such sites, where visitors can learn about the museums' exhibits and programs or explore information prepared by the museum staffs on a wide variety of topics.

hibits that people remember 20 or 30 years later. In 1995, researchers from the University of Chicago and Columbia University in New York City conducted a survey to learn what museum professionals remembered from childhood trips to science museums. Several of the 76 adults interviewed mentioned the giant model of a human heart that visitors could walk through at the Franklin Institute Science Museum in Philadelphia and the Museum of Science and Industry in Chicago. Building on the giant-heart idea, the California Science Center in Los Angeles constructed a partially transparent, 15-meter (50-foot) woman as part of an exhibit called the World of Life that opened in February 1998. The reclining figure's organ systems are illuminated one by one as an audio-visual presentation explains how the organ systems work together.

One of the most unusual approaches to the teaching of science was being planned for Science City, scheduled to open in Kansas City, Missouri, in late 1999. Science City was envisioned as a combination science center, theme park, and theater. The museum was designed to offer a number of "you are there" experiences, such as witnessing a surgical operation or training to become an astronaut. Staff members dressed in appropriate costumes and playing rehearsed roles will create an atmosphere of realism in each situation.

Science museums are also exploring the use of modern communication technology, especially the Internet, to enhance visitor learning in museums and to reach audiences at distant sites. At newMetropolis, an interactive science center in Amsterdam, the Netherlands, computers wired to the Internet are placed among the exhibits to be used as learning tools. Many science centers also offer information and activities to schools and the public at their sites on the World Wide Web.

Science centers play an important role in a modern, technologically oriented society. By teaching people about science, and about the scientific method of discovery, they are helping to create educated citizens who can make informed choices as voters and consumers. Science museums perform that mission by making science fun, but it is fun with a very serious purpose.

Museum	Address	Phone	Web site
California Science Center	700 State Drive Los Angeles, CA	213-744-7400	www.casciencectr.org
Exploratorium	3601 Lyon Street San Francisco, CA	415-563-7337	www.exploratorium.edu
Fort Worth Museum of Science and Industry	1501 Montgomery Street Fort Worth, TX	817-255-9300	www.startext.net/homes/fwmsh
Franklin Institute Science Museum	222 North 20th Street Philadelphia, PA	215-448-1200	www.fi.edu
Great Lakes Science Center	601 Erieside Avenue Cleveland, OH	216-694-2000	www.greatscience.com
COSI (Center of Science and Industry) Columbus	280 E. Broad Street Columbus, OH	614-228-COSI	www.cosi.org
COSI (Center of Science and Industry) Toledo	1 Discovery Way Toledo, OH	419-244-COSI	www.cosi.org
Museum of Science	Science Park Boston, MA	617-723-2500	www.mos.org
Museum of Science and Industry	57th Street & Lake Shore Dr. Chicago, IL	773-684-1414	www.msichicago.org
New York Hall of Science	47-01 111th Street Flushing Meadows, NY	718-699-0005	www.nyhallsci.org
newMetropolis Science and Technology Center	Oosterdok 2 Amsterdam, Netherlands	011-31-20-531-3233	www.newmet.nl
Ontario Science Centre	770 Don Mills Road Toronto, Ontario, Canada	416-696-1000	www.osc.on.ca
Oregon Museum of Science and Industry	1945 S.E. Water Avenue Portland, OR	503-797-4000	www.omsi.edu
Orlando Science Center, Inc.	777 E. Princeton Street Orlando, FL	407-514-2000	www.osc.org
Science City	Pershing Road & Main St. Kansas City, MO	816-483-8300	www.kcmuseum.com
Science Museum of Minnesota	30 E. 10th Street St. Paul, MN	612-221-9488	www.sci.mus.mn.us

To locate science centers around the world via Internet, go to the Association of Science-Technology Centers (ASTC) Web site at www.astc.org and select the Science Center Travel Guide. For a listing, write to ASTC, 1025 Vermont Avenue NW, Suite 500, Washington, DC 20005.

Special Report | SPACE TECHNOLOGY

The International Space Station

Supporters hope that this outpost in orbit—operated by crews from 16 nations—will result in major medical and technological breakthroughs and stir the human spirit.

By James R. Asker

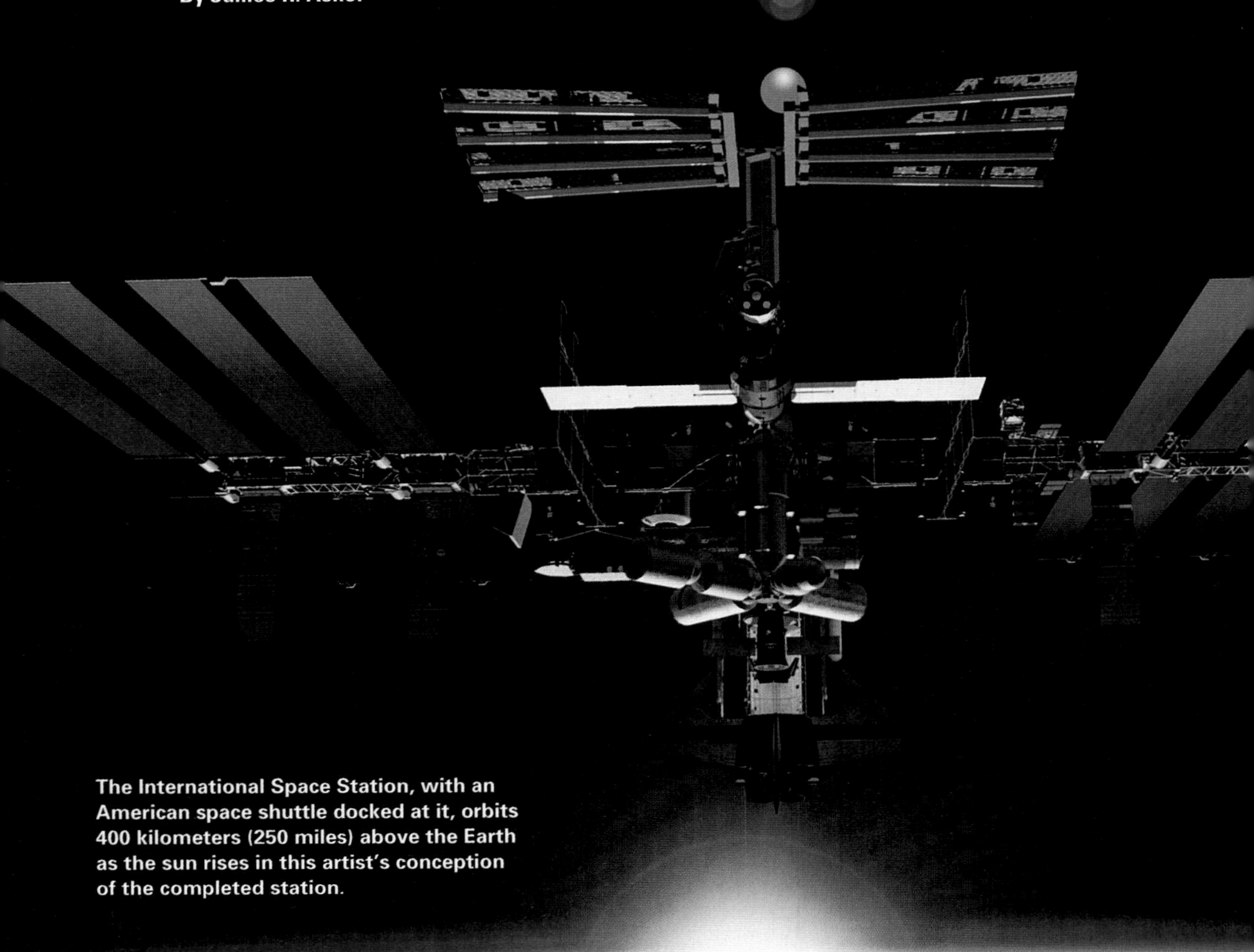

The International Space Station, with an American space shuttle docked at it, orbits 400 kilometers (250 miles) above the Earth as the sun rises in this artist's conception of the completed station.

Imagine a day in the early 2000's. In the black stillness high above the Earth, a small, wedge-shaped spaceplane carrying seven astronauts glides slowly toward a huge orbiting complex. Although it is an impressive sight, this outpost in orbit—the International Space Station—looks like an ungainly hodgepodge of cylindrical modules, metal frameworks, huge solar panels, antennas, and mechanical arms. At 109 meters (360 feet) wide and 88 meters (290 feet) long, the structure would sprawl over an area almost as big as two football fields if it were on the ground.

When the spaceplane is still several meters from the station, it stops. A long mechanical arm on the station locks onto the spaceplane and pulls it to a docking site. Finally, the astronauts enter the station. Each has a different specialty—biology, organic chemistry, combustion science, materials science, earth science, metallurgy, and space medicine—but they have at least two things in common: All of them have trained hard for this mission, and all look forward to working on the station for the next several months. The astronauts are greeted by the men and women they will replace as crew of this *microgravity* (very low gravity) laboratory. The crews are drawn from 16 nations, and the spaceships that bring them are launched from three continents. The International Space Station is not just the most expensive laboratory ever built, it is also the largest and most complex science project in history.

In 1998, this scenario was still science fiction, but it was on its way to becoming science fact. Construction of the components for the International Space Station was well underway, and their assembly in orbit was due to begin late in the year. Experts predicted that by the time the station is completed—about 2004 at the earliest—it will have cost tens of billions of dollars. Will it be worth it? Many scientists say no. Even space station enthusiasts cannot guarantee that it will produce any major scientific results. So why is the station being built? To understand the reasons, let's look at the history of human flight into space.

The early years of spaceflight

Along with exploring other planets, the establishment of a permanent outpost in Earth orbit was one of the earliest dreams of those who envisioned the era of spaceflight. But the reasons they proposed for building space stations were only indirectly related to scientific research. Two rocket theorists of the early 1900's, Konstantin Tsiolkovsky of Russia and Hermann Oberth of Germany, suggested that orbital stations might provide navigational guidance that could help the crews of oceangoing ships plot their positions more accurately and avoid icebergs. Tsiolkovsky even pointed out that space stations could solve a problem stumping rocket engineers of his time: how to build a spaceship that could carry enough fuel to reach the moon or another planet without making it so big and heavy that it couldn't get into space. Instead of taking off from Earth with all the fuel it needed, he said, a rocket could stop at a space station to refuel.

Tsiolkovsky's idea went largely unnoticed, however. Engineers kept

Terms and concepts

Atmospheric drag: The friction that results when gas molecules in space collide with a spacecraft, causing the craft to slow and fall to a lower orbit.

Centrifugal force: An outward force felt by someone on a rotating body; it could be used in space to generate artificial gravity.

Cold War: A period of often-tense relations between the United States and the Soviet Union that lasted from the end of World War II to the Soviet Union's collapse in 1991.

Microgravity: Very low gravity.

Space race: A Cold War rivalry between the United States and the Soviet Union from 1957 to 1969 in which each country tried to be the first to land a man on the moon.

Truss: A long structural framework of beams and cross members like that used in some types of bridges.

trying to devise plans for rockets that could go directly to the moon and beyond. But by the 1930's, many of them had concluded that interplanetary rockets using chemical propellants were impractical. The English scientist Phil E. Cleator suggested that nuclear energy would be the eventual key to space travel. But Willy Ley, a German American rocket engineer and science writer, again raised the idea that a space station could serve as a refueling terminal in space. In addition, Ley said that a space station could be used for what would one day be called microgravity science.

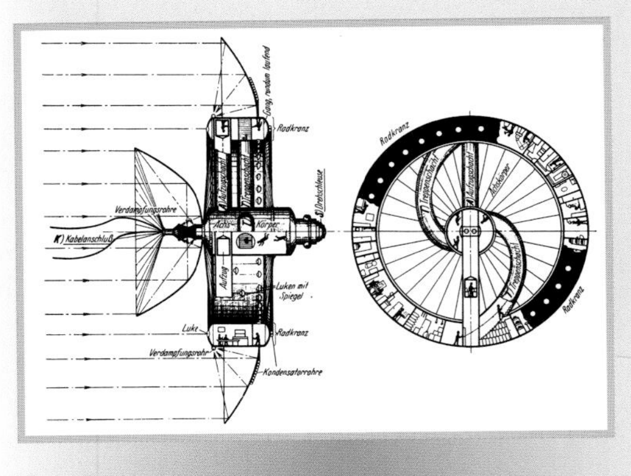

One of the first diagrams of a proposed space station was published in 1932 by Austrian engineer Hermann Potocnik, better known by his alias, Hermann Noordung. Noordung envisioned a three-part station. An observatory (not shown) would be connected to the station's powerhouse by cables, far left. The powerhouse would use mirrors, condenser pipes, and boiler pipes to supply power to the living quarters, situated around the rim of the station. Rotation of the station would create artificial gravity in the living area.

Ley's writings were included in a series of articles called "Man Will Conquer Space Soon," published in *Collier's* magazine in 1952. The series also contained the work of Wernher von Braun, a German rocket engineer who came to the United States after World War II (1939-1945). Von Braun envisioned a huge, wheellike space station that would rotate slowly, an idea also proposed in 1932 by an Austrian engineer, Hermann Potocnik. The resulting *centrifugal* (outward) force would create artificial gravity.

Immediately after World War II, the idea of a human presence in space was still little more than a fantasy, but that soon changed. By the 1950's, the United States and the Soviet Union were engaged in an intense rivalry known as the Cold War, which lasted until the Soviet Union collapsed in 1991. The two nations competed on many fronts, but no arena was more visible than the space race. The drive to achieve dominance in space was the major engine for advances in space technology.

The author:
James R. Asker is the Washington bureau chief for *Aviation Week & Space Technology* magazine.

The space race

The space race began in 1957, when the Soviets launched Sputnik (Traveler), the first artificial satellite. A shocked United States quickly launched its own satellite, Explorer 1, in 1958. That same year, Congress established the National Aeronautics and Space Administration (NASA), a civilian agency, to develop and run a U.S. space program. Although NASA officials discussed the possibility of building a space station, the idea was shelved. For the next decade, both the United States and the Soviet Union concentrated on landing men on the moon, though the Soviets placed more emphasis on operations in Earth orbit.

In 1961, Soviet cosmonaut Yuri Gagarin became the first human being to fly in space. By the time U.S. astronaut John Glenn made his celebrated 3-orbit flight around the Earth in 1962, a Soviet cosmonaut had circled the globe 17 times. But the United States took the lead in the space race in the late 1960's with the Apollo moon missions. In July 1969, astronaut Neil Armstrong of Apollo 11 became the first person to walk on the moon.

The first space stations

The Soviets' continued concentration on operations in Earth orbit paid off in 1971, when they launched the world's first space station, Salyut 1 (Salute). The Salyut, consisting of four large cylinders connected end to end, weighed about 18 metric tons (20 tons), was 14 meters (46 feet) long, and was powered by solar panels. Salyut's first crew arrived on a Soyuz (Union) spacecraft, but could not get the station's hatch open. A second crew managed to get in and spent three weeks on board. However, as the three cosmonauts returned to Earth, a valve in their Soyuz opened, allowing all the air inside the craft to escape. The cosmonauts, who were not wearing pressurized space suits and helmets because of the Soyuz's limited cabin space, suffocated.

In all, the Soviets launched seven Salyuts in the 1970's and 1980's. Cosmonauts manned the stations for varying lengths of time. They closed the last of them, Salyut 7, in 1986, but planned to retrieve it someday and refurbish it for further use. Before they could do so, however, *atmospheric drag* (friction from gas molecules in space) "decayed" Salyut's orbit, and, in 1991, the station—like the previous six Salyuts—fell to Earth, breaking apart and burning up in the atmosphere.

In the early days of the Salyut program, the United States space effort continued to focus on the moon. But after the sixth and last moon landing in 1972, NASA was faced with smaller budgets and was forced to decide where to redirect its efforts. The agency elected to concentrate on developing a

Early space stations

The first space stations, though small in size compared to the International Space Station being built, taught the astronauts of Russia, the United States, and many other countries valuable lessons about living, working, and troubleshooting in space.

The first U.S. space station, an experimental model called Skylab, *above,* was launched in 1973. Skylab was built from the empty casing of a Saturn rocket. During the launch, Skylab's meteoroid shield was torn off, taking the right solar panel with it. Without the shield, which also protected Skylab from the sun, the station began to overheat. Astronauts rigged a plastic sunshade to replace the shield and then lived and worked aboard Skylab through early 1974.

The Soviet Union launched seven Salyut space stations, the last of which, *above,* at left, was closed in 1986. By that year, the Soviet space station Mir, right, had been placed in orbit, destined to serve both Russian astronauts and those of other countries until the late 1990's. The Soviet Union's Salyut 1, launched in 1971, was the world's first space station.

Building the International Space Station

Sixteen nations will supply the dozens of parts necessary to construct the International Space Station. They include the United States, Russia, Japan, Canada, and 11 member nations of the European Space Agency (ESA). Brazil will also provide some hardware, in exchange for the opportunity to send an astronaut to the station. Construction of the station was expected to begin in space in late 1998. Plans called for the station to be completed about 2004. Some of the major elements of the station, and the countries that will supply them, are listed below.

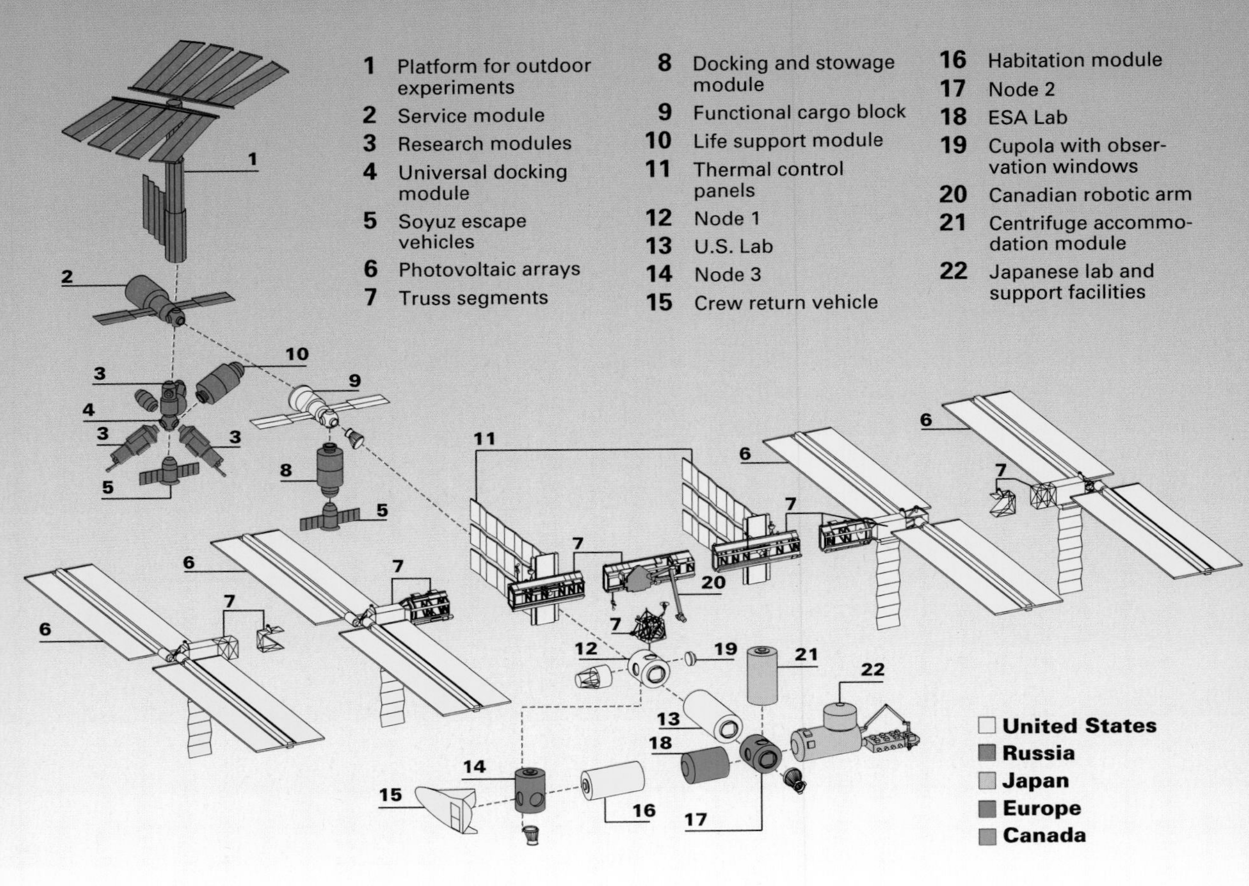

Russian aerospace workers assemble the first element of the International Space Station scheduled to be launched—the Functional Cargo Block, known by its Russian initials FGB, and later renamed Zarya, Russian for "sunrise." Plans called for Zarya—which has its own propulsion, guidance, electrical power, and thermal control systems—to be launched in late 1998.

Workers at NASA's Marshall Space Flight Center in Huntsville, Alabama, prepare to move the U.S.-built habitation module. The module, which was to serve as living quarters for astronauts aboard the International Space Station, was scheduled to be launched in 2004.

A technician tests the most important Canadian contribution to the International Space Station, the Remote Manipulator System. The robotic arm, along with a hand, will be of critical importance to the assembly of the station in space, helping to lift individual station components and maneuver them into place.

reusable spacecraft—the vehicle that would become the space shuttle. At the same time, NASA decided to put to use some of the Saturn rockets left over from the Apollo program. Engineers used the third-stage casing of a Saturn 5 rocket to build the first U.S. space station, an experimental model called Skylab. Skylab was lifted into orbit by another Saturn rocket in 1973.

By the standards of the time, Skylab was huge. It was 26 meters (85 feet) long, weighed 91 metric tons (100 tons), and had the living space of a three-bedroom house. Solar panels powered the station, just as on Salyut. Three crews of three astronauts each served stints on Skylab. The last crew spent 84 days in space from late 1973 through early 1974, a length of time that remained a world record for 4 years and a U.S. record for 21 years. By the late 1970's, atmospheric drag was slowing

Skylab down. But because money was so tight at NASA, the space agency could not afford a mission to boost the station into a higher orbit, and it fell to Earth in 1979. Much of it burned up as it reentered Earth's atmosphere, but a large piece of the station crashed in a remote area of western Australia.

In 1981, the U.S. space shuttle era began with the launching of the orbiter Columbia. Six orbiters were built—Enterprise, Columbia, Challenger, Discovery, Atlantis, and Endeavour. However, Enterprise was only a test vehicle and was never intended to fly in space.

Beginning in 1983, the space shuttle was used to carry a project called Spacelab, designed and built by the European Space Agency (ESA), an organization of 14 Western European nations. Spacelab included a manned laboratory and several *pallets* (platforms) exposed to the elements of space. On some missions, the laboratory was used. It fit into the shuttle's *payload* (cargo) bay, where it was connected to the crew cabin. Payload specialists, scientists who were not career astronauts, performed experiments in the weightless environment for up to two weeks at a time. On other missions, the shuttle's payload bay carried the pallets, to which experiments were attached.

The Soviets, in the meantime, had developed Mir (Peace or World), the first true *modular* space station—one consisting of several modules, or sections. The core module was carried into orbit in 1986 by a Soviet Proton rocket. During the next decade, five more modules were added, giving Mir a total mass and volume roughly equal to that of Skylab.

Except for two brief periods, Mir had a crew on board continuously into 1998. Three cosmonauts spent a year or more on Mir. One crew was in space when the Soviet Union dissolved in 1991 and Russia took over its space program. Astronauts from other nations also worked on Mir. Norman E. Thagard, the first U.S. astronaut to stay aboard Mir, lived on the station for 115 days in 1995. Astronaut Shannon W. Lucid spent 188 days aboard Mir in 1996, breaking both the U.S. endurance record and the international record for women.

In the mid-1990's, accidents and equipment failures on Mir began to concern the nations whose astronauts worked there. During 1997, a series of mishaps occurred. In February, fire filled the station with smoke; in June, an unmanned supply ship punctured a Mir module; and from July to the end of September, Mir's main computer failed five times. Although some nations began to wonder whether Mir had reached the end of its useful life, the Russians remained confident that Mir was safe and continued to operate the station.

The beginnings of the International Space Station

Nonetheless, Mir was due to be supplanted by the International Space Station. The construction of this mammoth orbiting complex had been a NASA objective for nearly 20 years. As early as 1981, when the space shuttle was built and flying, the agency said that a large space station was "the next logical step" in human space flight. To generate the kind of public support enjoyed by the Apollo moon landing program, NASA officials emphasized the importance of establishing a permanent presence in space. In 1984, President Ronald Reagan approved the construction of a U.S. space station, announcing later that its name would be Freedom.

NASA continued trying to generate public enthusiasm for the project, but—beginning with what would prove to be a ridiculously low price tag of $8 billion—it promised Americans too much. According to NASA, the space station would be amazingly versatile. It would be a microgravity science laboratory; a platform for Earth observations and astronomy; and a place to assemble, service, and repair spacecraft and satellites. It would also create and expand opportunities for private enterprise in space and promote cooperation among nations. Many members of Congress were skeptical of such grandiose claims. And as the projected costs of the space station skyrocketed, it appeared that they had reason to be.

In 1985, NASA invited the ESA and the governments of Japan and Canada to become its partners in building and operating the station. The involvement of other nations decreased NASA's costs. However, it also made coordination of the project more difficult.

Adding even more complexity to the project was a new NASA construction scheme. To build widespread political support for the space station, the agency divided the U.S. portion of the work into four packages and assigned each one to a different aerospace contractor, each overseen by a different NASA field center.

Station design changes—and a new international partner

Over the years, the space station went through a series of design changes. The wheel design predicted by von Braun in the 1950's proved not to be a logical choice to fill the needs of the 1980's. Von Braun's idea of having a rotating station generate its own artificial gravity required a much larger structure than NASA could hope to build. More importantly, since the most promising research in space in the 1980's was based on microgravity, a station with gravity was pointless.

The first concept NASA seriously considered, called the "power tower," consisted of a single 130-meter (425-foot) *truss,* a long structural framework of beams and cross members like that used in some types of bridges. Living quarters and work areas were to be clustered at one end of the truss. When engineers determined that the power tower would be too unstable for microgravity experiments, that design was replaced with one for a gigantic assemblage dubbed the dual-keel configuration. In the dual-keel design, two parallel trusses were joined by habitation and laboratory modules at the center, in the form of an "H." But as engineers kept dealing with changes in requirements handed down by managers, the configuration of the station kept changing. At the same time, estimated costs rose from $8 billion in 1984 to $37 billion by 1990.

Finally, in 1993, President Bill Clinton ordered NASA to totally re-

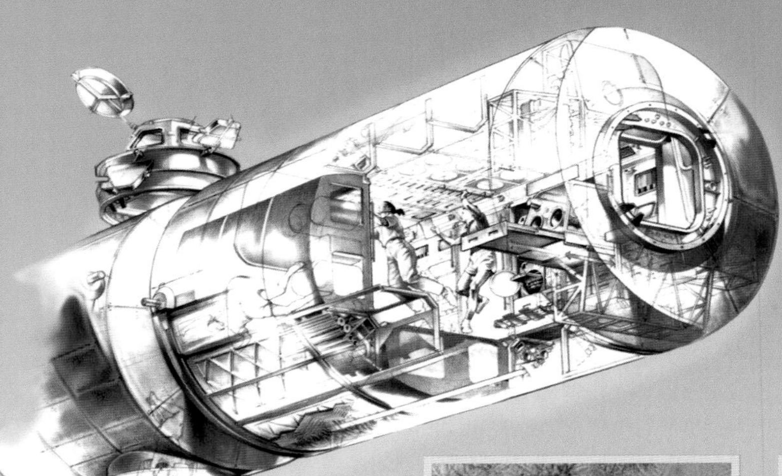

Living and working aboard the station

As many as seven astronauts will be able to live aboard the International Space Station when it is completed. Construction of the station was scheduled to be concluded by about 2004.

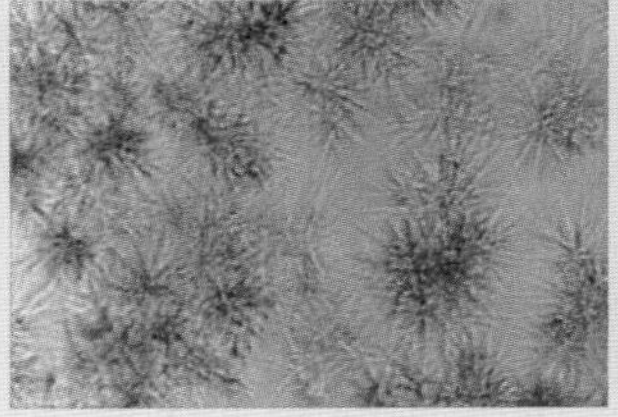

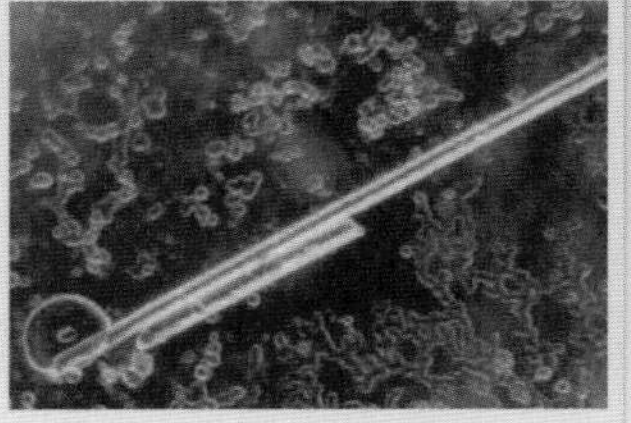

Some protein crystals important to cancer research grow much more slowly in Earth-based laboratories, *left,* than they do in microgravity laboratories in space, *right.* Such crystals were successfully grown aboard Skylab and Spacelab in the 1970's and 1980's. Researchers hoped that growing the crystals over longer periods of time on the International Space Station would lead to quicker breakthroughs in finding a cure for the disease.

The U.S. laboratory module, at top, was designed for astronauts to work on microgravity experiments, those that require the extremely low gravity of Earth orbit. Plans called for the lab to be launched in 1999. The U.S.-built habitation module, above, can accommodate four astronauts in its dining, bathing, sleeping, and exercise facilities. In mid-1998, the U.S. considered replacing the planned habitation module with an inflatable unit called TransHab that would accommodate four to six astronauts.

design the station. In June of that year, the agency showed the president three station options, called simply A, B, and C. The president chose A (renamed Alpha), a scaled-down version of a previous design. In Alpha, modules would be suspended below four parallel sets of solar arrays, linked to the arrays by a single horizontal truss. The Alpha design kept the laboratory modules near the center of the complex for greater stability for the microgravity experiments. Engineers estimated that Alpha would cost about $17 billion.

At the same time it submitted the revised designs to the White House, NASA made a dramatic proposal: Add Russia as a partner in the project. Russia had much more experience in human space flight in terms of the number of hours logged by its cosmonauts. Moreover, the hardware of its spacecraft, though typically simpler than U.S. equipment, was often more durable. If American spacecraft were the equivalent of finely tuned high-performance sports cars, the Russian vehicles were like solid, reliable trucks that always started and hauled whatever was needed.

For the Clinton Administration, asking Russia to join the station effort would serve two purposes. The money that the United States would pay Russia to build the first station module would help Russia financially as it moved from Communism to democracy. It would also keep Russian aerospace workers busy with a peaceful project instead of selling their skills to military troublemakers around the world.

There were disadvantages, too, of course. Russia's economic situation was shaky as the country struggled to replace its Communist economy with a free-market system. Some administration officials worried that the troubled nation might falter in its commitment to the space station just when it was most needed. Furthermore, adding Russia as a partner would increase some costs for the United States. The reason for that had to do with geography. The Russian Baikonur launch facility in Kazakhstan, a former Soviet republic, is much farther north of the equator than the Kennedy Space Center in Florida. To make the space station accessible to Russian launch vehicles, NASA would have to change the station's orbit from the originally planned orbit, spanning 28.4 degrees north and south latitude, to one covering 51.6 degrees north and south. Reaching the station from Florida would then be more difficult for U.S. space shuttles, requiring them to carry less cargo on each trip.

Despite the drawbacks, the Clinton Administration concluded that accepting Russia as a partner in the space station would be a positive step, and in 1993 the two nations signed an agreement to cooperate in space. Phase I of the International Space Station program—in which Russian cosmonauts and U.S. astronauts trained and flew on each other's spacecraft—began immediately. The first cosmonaut flew on a U.S. space shuttle in 1994. Meanwhile, U.S. contractors built hardware so the shuttles could dock at Mir. In June 1995, the shuttle Atlantis flew to Mir, and U.S. shuttles eventually visited the station nine times.

Final preparations

A new agreement on how the International Space Station would be built and shared was signed in January 1998 by representatives of the 15 nations taking part in the project. The participating countries included the United States, Canada, Japan, Russia, and 11 member nations of the ESA—Belgium, Denmark, France, Germany, Italy, the Netherlands, Norway, Spain, Sweden, Switzerland, and the United Kingdom. According to the agreement, the United States would be the largest financial contributor to the International Space Station project, and NASA would manage the station. Brazil signed a separate pact with the United States to provide some hardware for the station in exchange for access to U.S. facilities and permission to send a Brazilian astronaut to the station.

The components of the space station would be truly international. NASA was slated to provide connecting nodes for the station, two laboratory modules, a habitation module, the supporting truss, and part of the solar arrays. The ESA and Japan would each provide a laboratory, and Canada would contribute a robotic arm and hand. Russia would build several modules. The core module was designed as a self-supporting unit that would propel the station initially and to which other modules could be attached—a service module, two laboratory modules, and part of the solar arrays. The United States agreed to pay for the core module.

Assembly of the station, planned to orbit about 400 kilometers (250 miles) above the Earth, was scheduled to begin in late 1998. Over the projected six years of construction, about 45 launches of U.S. space shuttles and Russian Proton and Soyuz rockets were to carry some 425 metric tons (470 tons) of hardware, including more than a dozen modules, into orbit to build the station. Construction was to be carried out jointly by U.S. astronauts and Russian cosmonauts. U.S. astronauts would spend at least 1,100 hours working in space—400 hours more than all U.S. astronauts combined had spent in spacewalks or on the moon from the start of the space program to 1997.

Construction timetable

The first stage of construction—Phase 2 of the overall project—was to be the assembly of the U.S. and Russian components. The first component scheduled to be taken aloft was the core module, also called the Functional Cargo Block or Energy Block. Plans called for this module, known by the Russian initials FGB, (later renamed Zarya—Russian for "sunrise") to be carried into orbit on a Proton rocket from Baikonur. The FGB has its own propulsion, guidance, communication, electrical power, and thermal control systems. Russian ground crews were to control the FGB until the arrival of the first station crew.

Once the FGB was in place, the space shuttle Endeavour and a crew of five were to carry a unit called Node-1 (later renamed Unity) into orbit from the Kennedy Space Center. Node-1, which has six hatches, would serve as a connector for other modules. The Endeavour crew was to use the shuttle's robotic arm to connect Node-1 to the FGB and then return to Earth. The third component, the Russian-built service module, contains working and living areas for a small station crew.

Russia's economic problems, however, delayed the launch of the FGB and the service module. NASA was forced to push back the scheduled beginning of station assembly from 1997 to November 1998. The agency also began a costly effort to modify a backup vehicle—a formerly secret U.S. Navy spacecraft—in case the service module was not completed in a reasonable time.

In 1999, a U.S. shuttle flight to the station was to bring the first section of a truss that would eventually grow to 108 meters (354 feet) long. To this station "backbone," astronauts were to attach the Canadian manipulator arm, communication antennas, platforms for experiments that scientists want exposed to the vacuum of space, and vast arrays of solar cells and cables for electrical power. The station would generate an average of 110 kilowatts of power, almost four times the amount produced on Mir.

As shown in an artist's conception, the first two pieces of the International Space Station—Zarya (formerly called the Functional Cargo Block), upper left, and Unity (formerly Node 1), slightly below and to the right of Zarya—will be fitted together by astronauts aboard the U.S. space shuttle. The astronauts will use the shuttle's Canadian-built remote manipulator arm to perform the operation.

Upon completion of Phase 2, the first crew to operate the station was to arrive on a Soyuz spacecraft. The two Russians and one American were to live on the station for five months, with the American serving as the first station commander. The Soyuz was to remain docked at the station so that the crew could return to Earth in an emergency.

The final phase of station construction, Phase 3, would involve the other international partners. During this stage, five laboratory modules containing dozens of refrigerator-sized racks for experiments and laboratory equipment were to be added. Sometime in the year 2004, the last component, the U.S.-built habitation module, was scheduled to be attached. All together, the modules will give station crews a volume of 1,300 cubic meters (46,000 cubic feet) in which to work and live—about as much space as is contained in the cabins of two Boeing 747 jumbo jets. As many as seven crew members will be able to live at the station at one time.

What will life be like on the space station? Mostly, it will be long days of construction, maintenance, and scientific work. The most precious commodity on the station will be the limited time that its small crew can spend on experiments. Scientists hope that experiments on the space station will lead to scientific breakthroughs in fields such as medicine and technology.

Among the most promising classes of experiments is protein crystal growth. If scientists can grow large crystals of some of the tens of thousands of different proteins in the human body and analyze their three-

The International Space Station flies over Florida and the Bahamas in this artist's rendering of the station as it will appear after completion, in about the year 2004.

dimensional molecular structure, they may be able to develop designer drugs to enhance or inhibit the functioning of the proteins. Although researchers have been crystalizing proteins since the 1920's, growing crystals of some proteins is difficult on Earth because gravity can upset the process. Crystals were grown successfully in space aboard Skylab and Spacelab. Aboard the new space station, scientists will be able to grow crystals over longer periods and analyze them there.

Despite the demonstrated value of conducting some experiments in space, many scientists have been cool to the space station project. Critics say that when shuttle flights and other expenses are added in, the total cost of the space station could top $100 billion. James Van Allen, the U.S. physicist who discovered Earth's radiation belts, called the station's cost "far beyond any justifiable scientific purpose, or any justifiable practical purpose." According to Van Allen, anything worth doing on the station can be done on unmanned spacecraft for far less.

But there is at least one type of research that clearly cannot be done without putting people in space—gathering data about how the human body reacts to extended periods of weightlessness. Such studies are necessary if humans are to travel to other planets to explore those worlds and perhaps one day settle them. Critics such as Van Allen and U.S. physicist Robert L. Park questioned, however, whether manned exploration of the solar system is necessary at all, considering the effectiveness of robotic probes.

What are NASA's future plans for the space station? The station's design life is 15 years. But with proper maintenance and occasional replacement of some components, experts say, the station could be used indefinitely and given new capabilities. Someday, the station might even become the sort of base that visionaries like Ley and von Braun foresaw. A large spacecraft to carry astronauts to Mars could be launched from Earth in pieces and assembled at the station.

In the meantime, the station will have to prove its worth as a laboratory. In 1998, it was impossible to say whether the International Space Station will turn out to be billions of dollars wasted or a solid investment. Supporters thought that one or two major scientific breakthroughs could make it all worthwhile.

Ultimately, whether the station is judged a success or failure will depend on more than an accountant's assessment of its costs and benefits. To weigh the real value of this facility, the citizens of 16 nations will also have to ask themselves some other important questions: Did the space station inspire young people to study mathematics and science? Did it advance space exploration? Did it stir the human spirit?

For further reading:

"International Space Station Ready for Flight." *Aviation Week & Space Technology,* Dec. 8, 1997, pp. 42-91.

Harland, David M. *The Mir Space Station: A Precursor to Space Colonization.* Wiley, 1997.

Heppenheimer, T. A. *Countdown: A History of Space Flight.* Wiley, 1997.

For additional information:

National Aeronautics and Space Administration Web pages—http://station.nasa.gov and http://spacelink.msfc.nasa.gov

Space News Web page—www.outerorbit.com

Spaceviews Web page—www.spaceviews.com

Questions for thought and discussion

1. Why do some scientists believe it is important for the United States to participate in the development and building of the International Space Station?
2. Why do many other scientists oppose the space station program?
3. If you had the opportunity to work aboard the International Space Station for three months, would you do so? Why or why not?
4. What changes would you make in the design of the International Space Station if you were consulted?
5. What would you most like to see the space station used for?

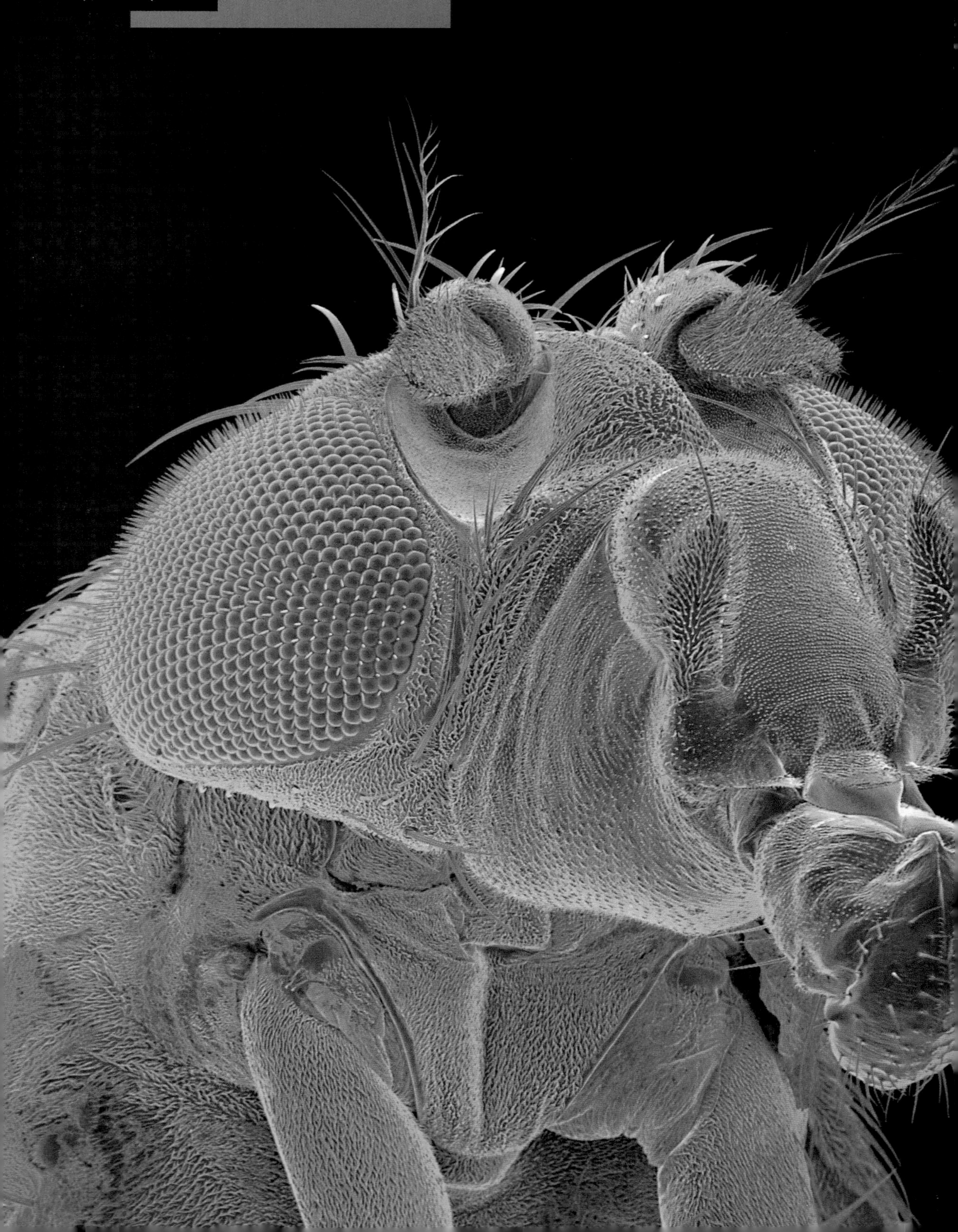

Windows on Invisible Worlds

by David Dreier

An array of high-technology microscopes is giving scientists increasingly detailed glimpses of biological systems and of matter at its smallest dimensions.

The 1990's have been called a golden age of microscopy, because a number of highly advanced instruments have enabled scientists to peer as never before into once-hidden regions of inner space. The images obtained with these marvels of technology are often more than just informational—many are works of art.

Microscopes now available to researchers include electron microscopes, scanning probe microscopes, and instruments that use magnetic fields to make images. They also include new types of *optical* (light) microscopes that provide novel ways of looking at cells and tissues.

Electron microscopes, of which there are two major kinds—transmission electron microscopes and scanning electron microscopes—use a beam of electrons to produce an image of a specimen. One noteworthy development with these devices has been the introduction of a new technique to produce naturalistic color images with the scanning electron microscope.

Scanning probe microscopes use an extremely tiny probe to scan the surface of specimens. These instruments can be used to make three-dimensional images of individual atoms and molecules.

New optical microscopes use lasers and fluorescence to produce images. Instruments based on magnetic resonance, the same phenomenon employed by advanced scanning machines in hospitals, can show organ systems of the tiny developing fetuses of small animals.

All in all, there are now very few objects, no matter how tiny, that cannot be observed in considerable detail. And different kinds of microscopes can often provide several ways of viewing the same object.

Opposite page: A scanning electron microscope photo of a fruit fly, made with a technique pioneered by California microscopist David Scharf, shows the insect in vivid hues that closely mimic its natural coloration.

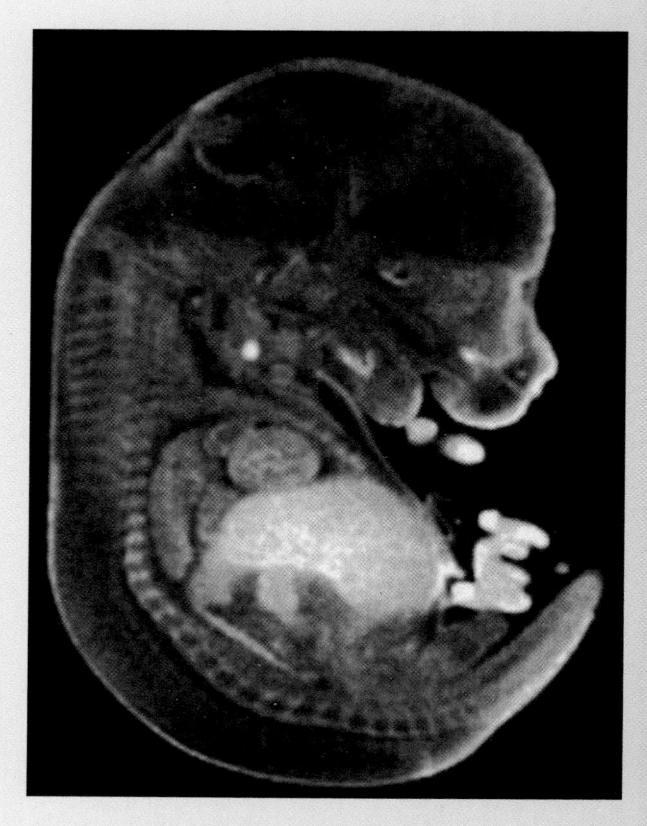

An image of a 14½-day-old mouse embryo about 1 centimeter (0.4 inch) long shows the mouse's developing organs. The picture was made with a magnetic resonance microscope, in which an organism is placed in a magnetic field, which causes the nuclei in certain of the animal's atoms to align. The specimen is then exposed to a short pulse of radio energy, which the nuclei absorb, causing them to change their configuration. When the radio pulse is turned off, the nuclei revert to their prior alignment, giving off weak radio signals of their own. The microscope detects those signals and translates them into an image.

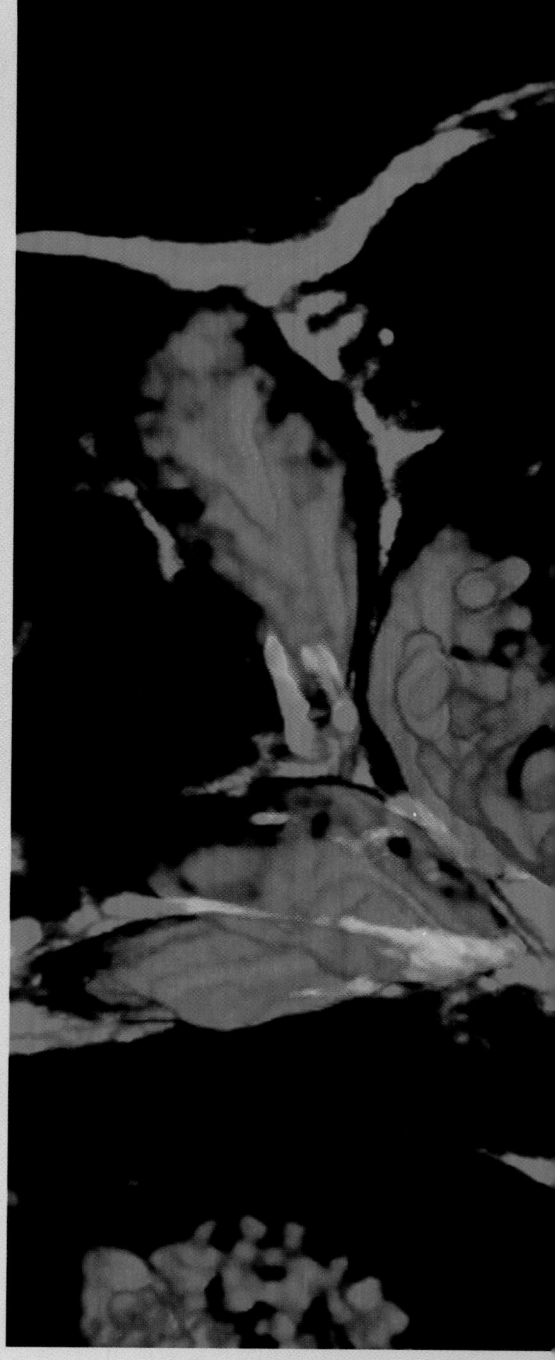

The author:
David Dreier is the managing editor of *Science Year.*

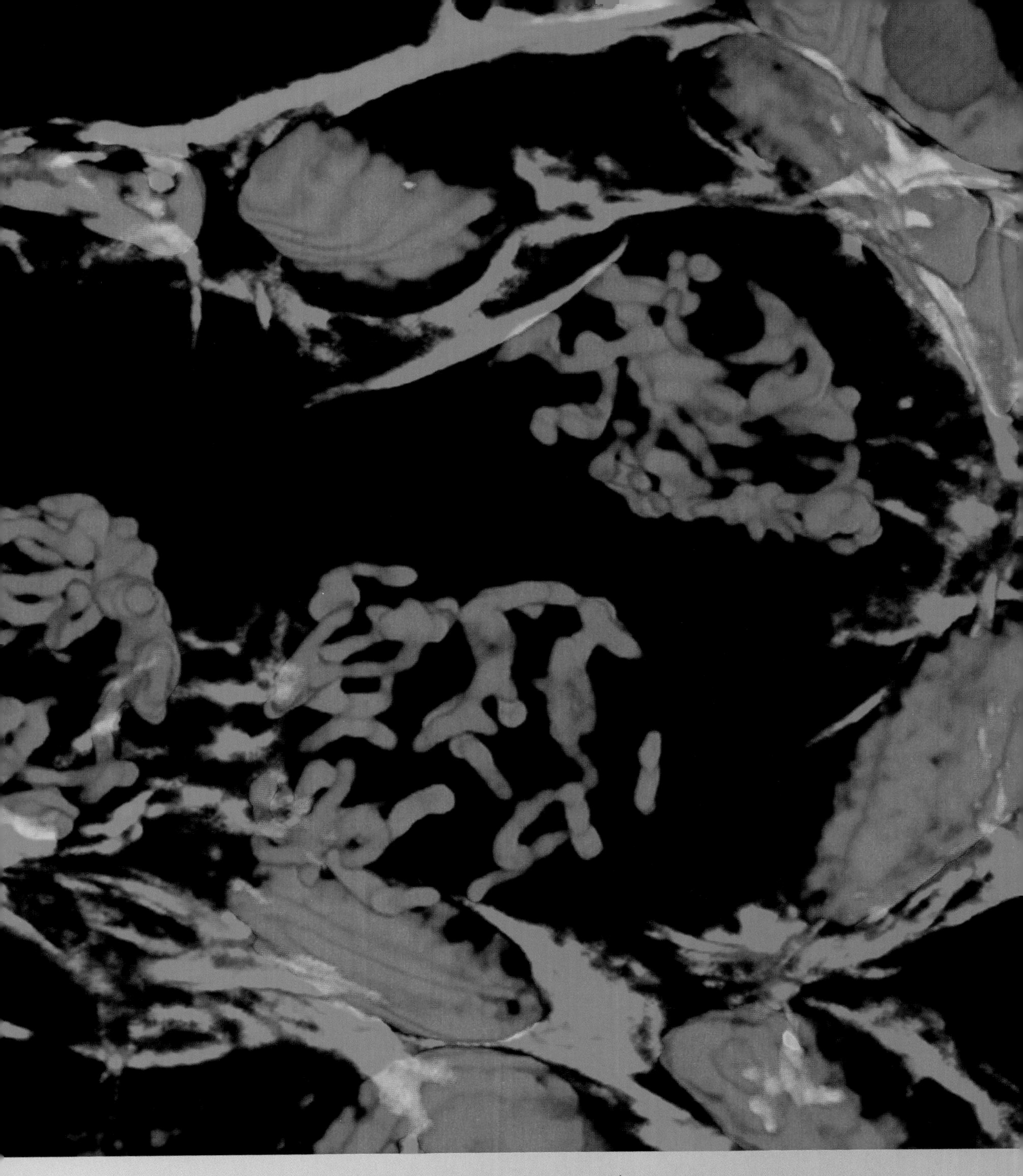

In an image made with a laser scanning confocal microscope, chromosomes (red) in frog *oogonia*—precursors of egg cells—are surrounded by protective proteins (green). To create this image, a laser was used to scan a frog ovary that had been stained with a fluorescent dye, causing the dye to glow. The microscope captured the light emitted from successive "slices" of the specimen. The slices were then combined by a computer into a single 3-D image.

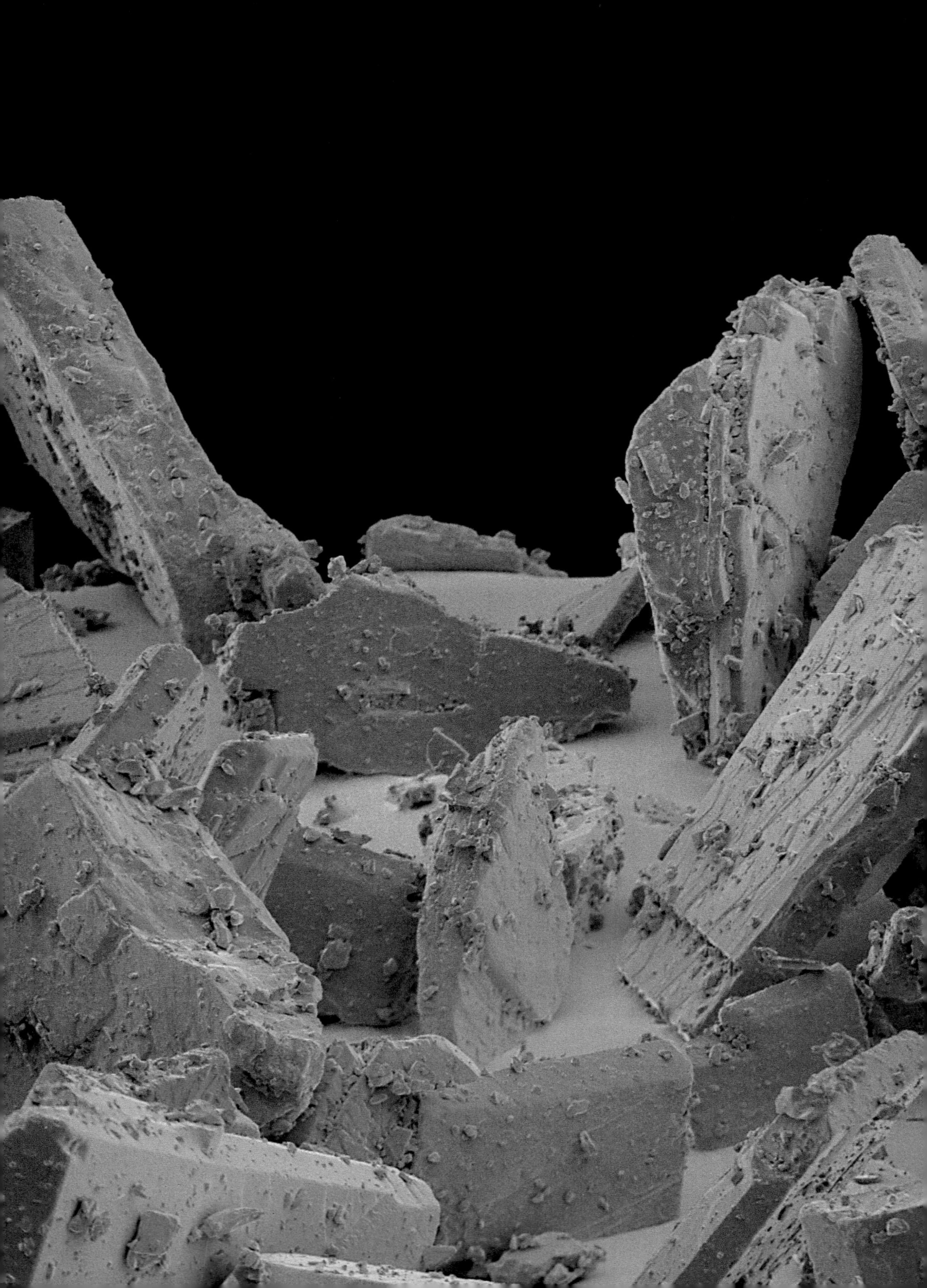

A David Scharf image of vitamin C crystals resembles a landscape on an alien planet. Scharf's patented colorization technique uses electron detectors set at varying angles to register shadowing effects from the electron beams. Each detector is assigned a color, and the interaction of the detectors produces a full-color image. The hues are false color—as they are in all tinted images made with electron microscopes—but they are close to natural color.

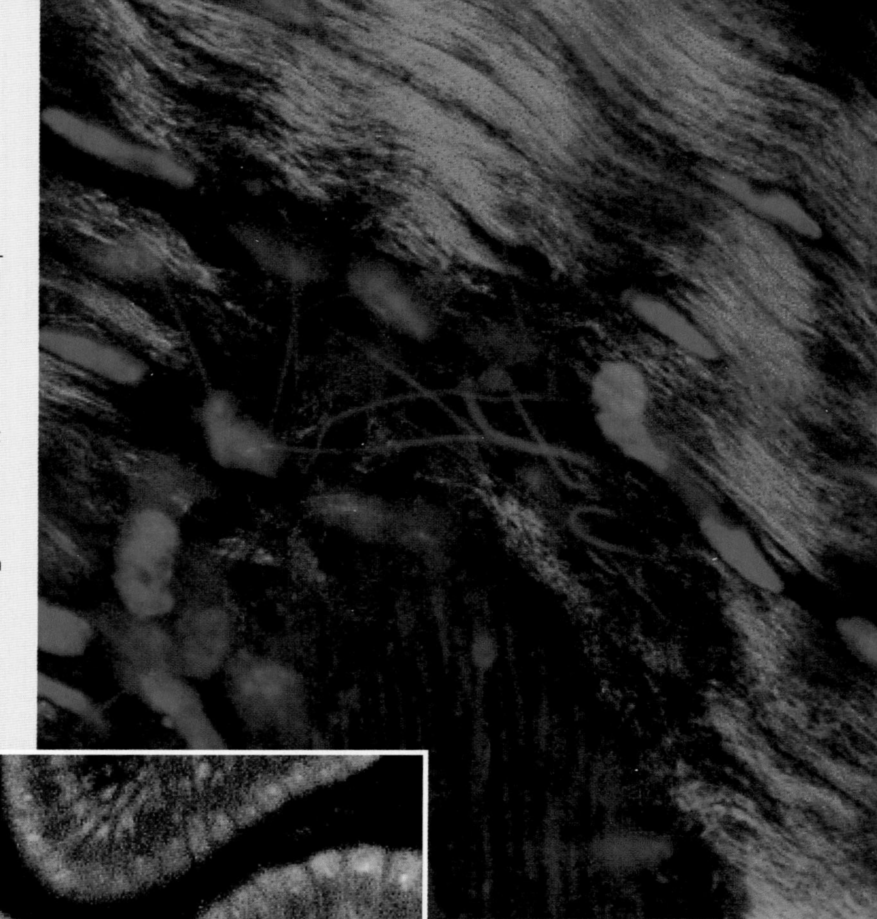

Constituents of a mouse's *femur* (leg bone)—strands of the fibrous protein collagen (green), and cells called osteocytes (red), which are responsible for maintaining the mineral content of bone—are revealed by multiphoton excitation microscopy. In this technique, a pulsed laser beam is directed into a sample under an optical microscope. The bursts of light cause certain tissues to *fluoresce* (glow). The collagen in this image fluoresced on its own, but the osteocytes had to be exposed to a special stain in order for them to fluoresce.

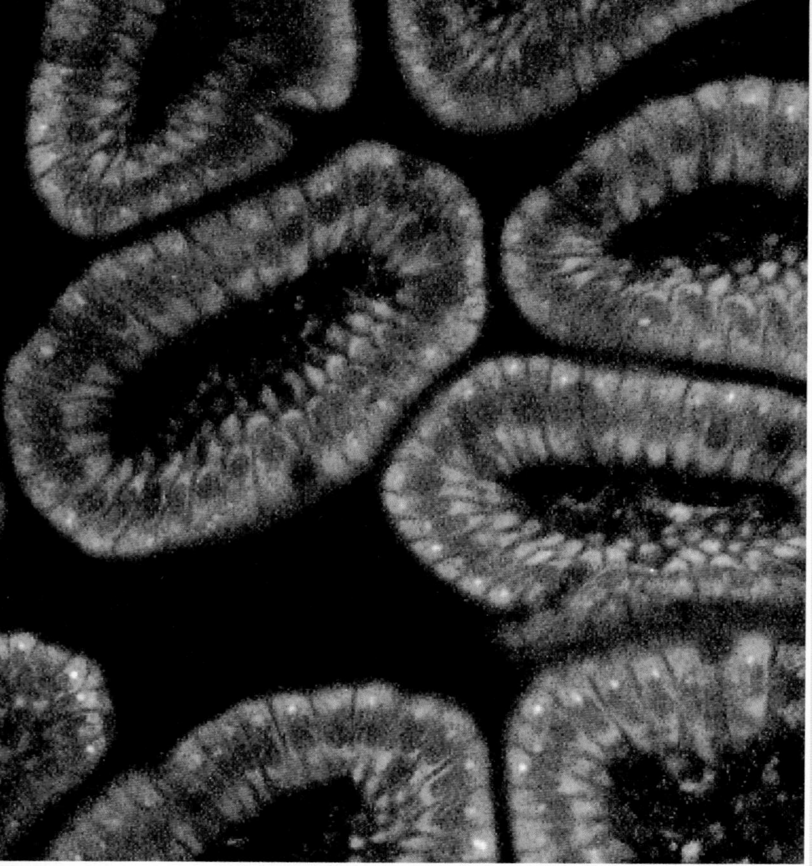

Another multiphoton excitation microscope image shows cross sections of mouse intestinal villi—threadlike structures on the inside wall of the small intestine through which nutrients are absorbed. The villi fluoresced naturally under the pulsed laser beam.

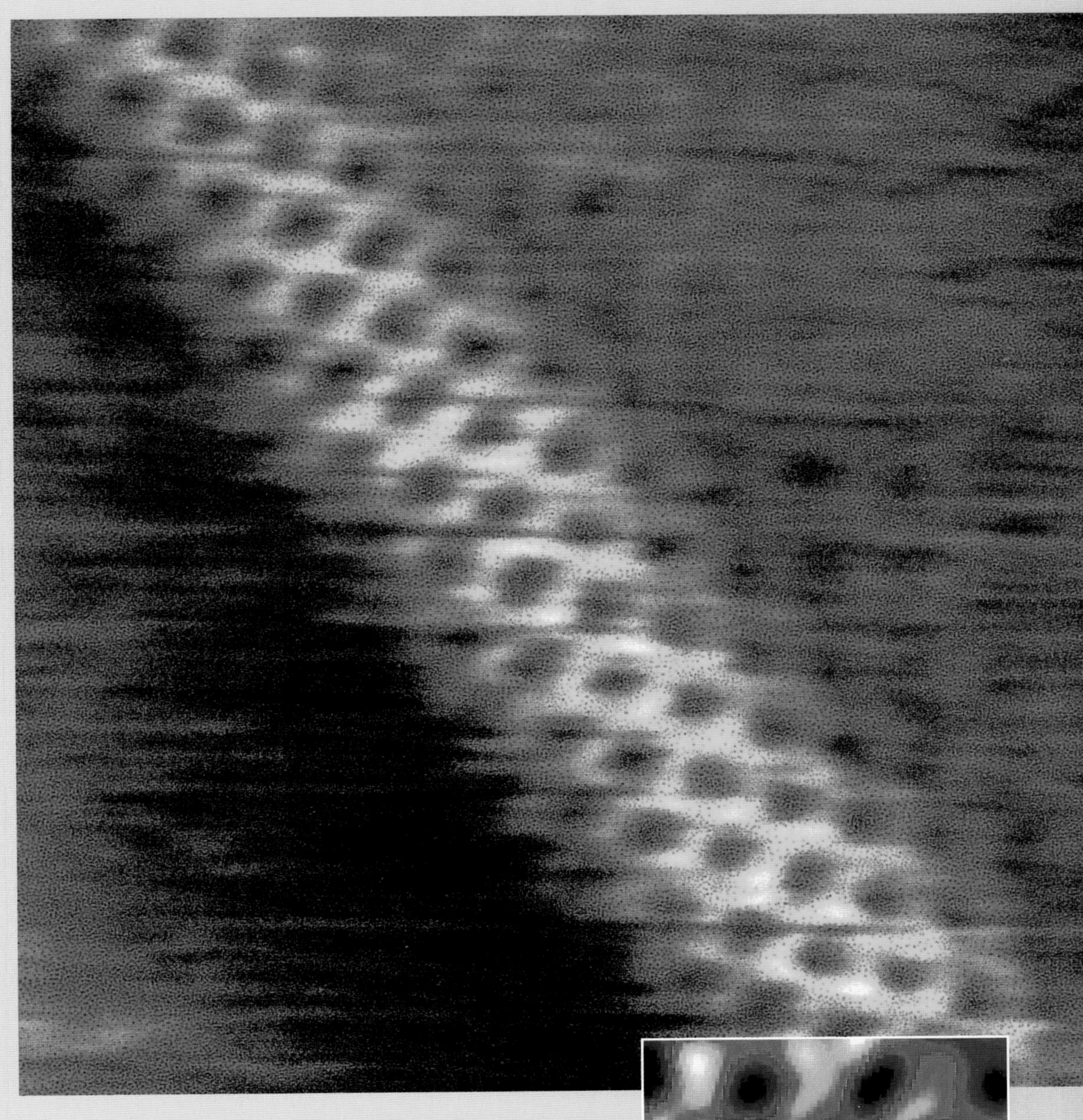

An image of a carbon *nanotube*—a one-atom-thick sheet of graphite rolled into a hollow cylinder—shows its honeycomb lattice structure, *above*. The image was made with a scanning tunneling microscope, a type of scanning probe microscope in which variations in an electric current flowing between a probe tip and a specimen are converted by a computer into a picture. *Right:* Organic—carbon-based—molecules (red) are bundled around tiny crystals of silver (blue). The molecules and crystals have been arranged into a superlattice, a structure that might be useful for optical communication. The image was made with a transmission electron microscope, which passes a beam of electrons through a thin specimen.

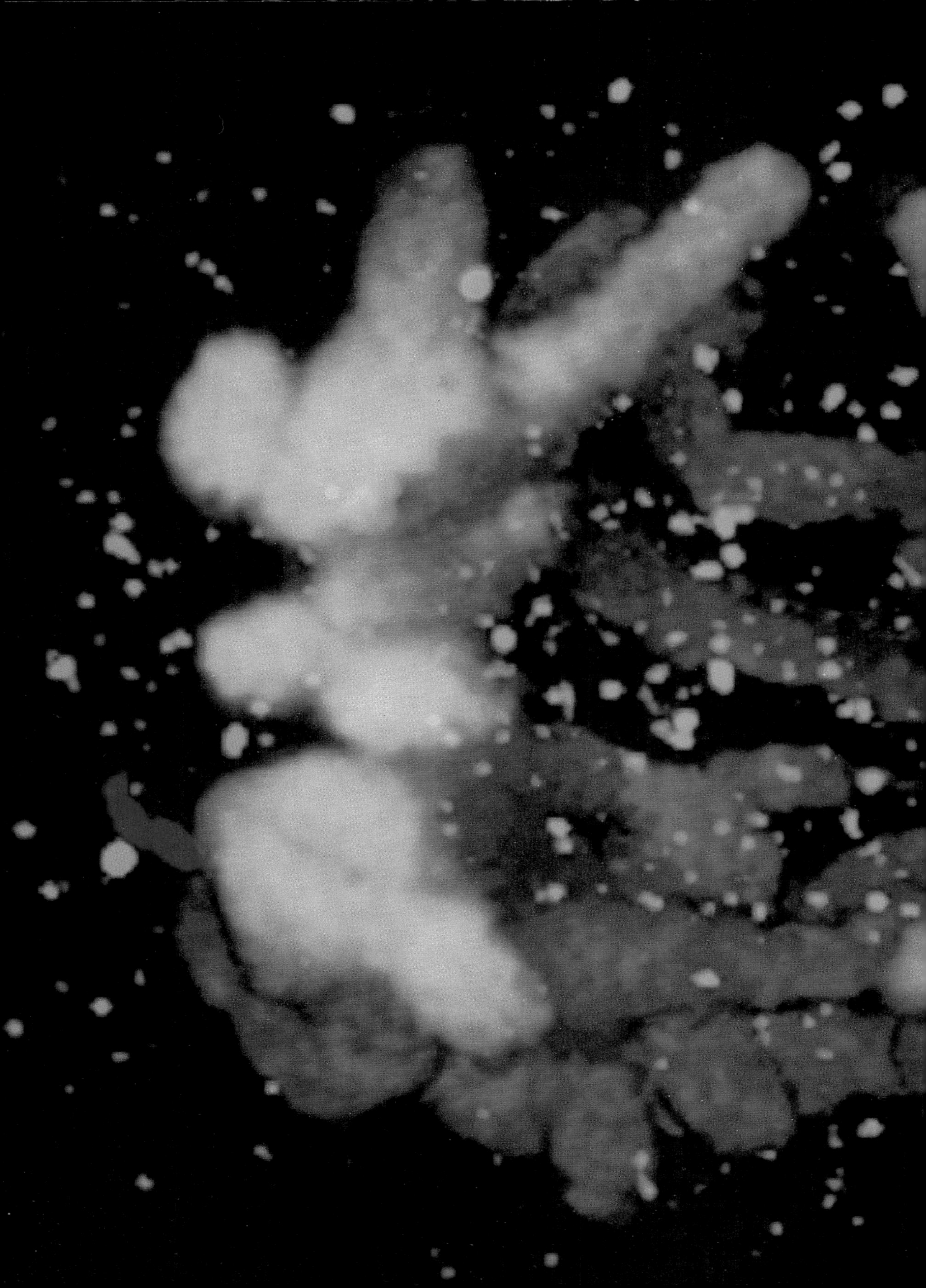

In a stunning three-dimensional close-up, two sets of newly formed chromosomes in a dividing human cell, *left,* are captured with an optical-microscope technique called epifluorescence/deconvolution microscopy. The chromosomes are surrounded by proteins that help translate genetic information into the production of other proteins. Another image made with the same process, *below,* shows a protein called pericentrin (yellow), a major component of the centrosome, a cellular structure that plays an important role in the splitting of chromosomes during cell division. Pericentrin provides a location for the formation of long strands called microtubules (red). During cell division, the centrosome splits apart, forming two new centrosomes, which move to opposite ends of the cell. Microtubules connecting the two centrosomes guide the movements of the replicated chromosomes into the two new, dividing cells. These pictures were made by "tagging" the samples with special stains that fluoresce when exposed to light. A mathematical process called deconvolution was used to enhance details in the images.

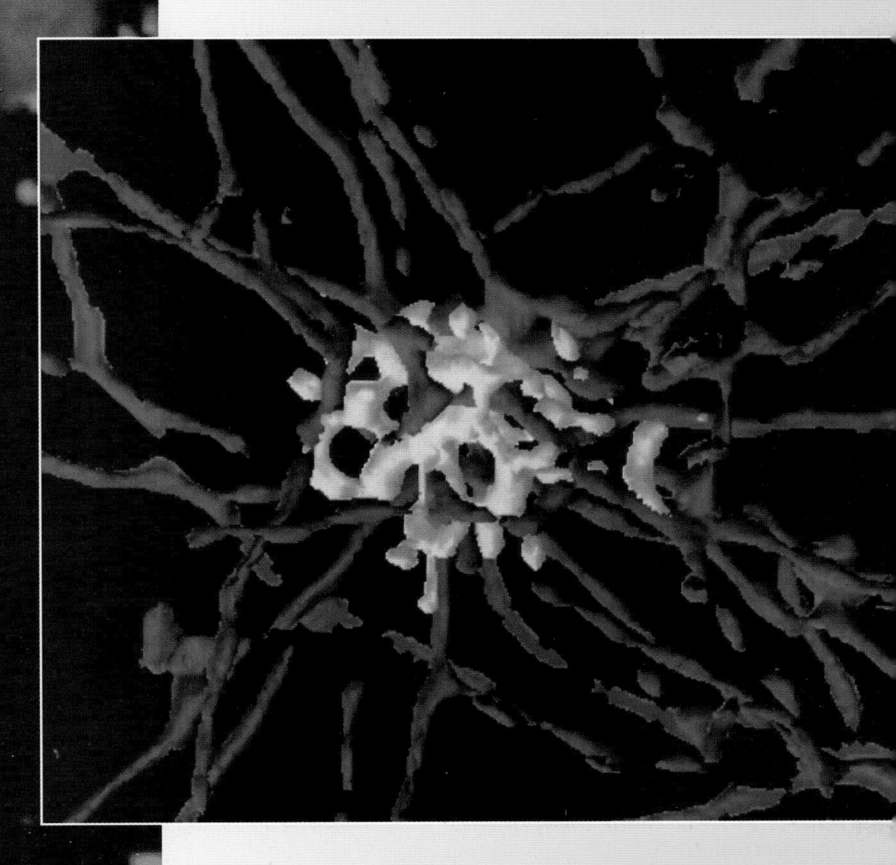

An image of the edge of a razor blade, *right,* made with an atomic force microscope, shows the bumpy layer of the soft *polymer*—a chemical compound made up of two or more simpler molecules—that coats the stainless-steel surface to make the blade shave smoothly. The atomic force microscope is a type of scanning probe microscope in which the probe tip scans the surface of a specimen by gently touching it and sensing its contours. Electronic devices monitor the probe's movements and relay the information to a computer, which translates the data into a picture.

Parts of *bacteriophages* (viruses that infect bacteria) are aligned in a repeating array. The bacteriophage segments are connectors, structures through which the viruses inject their genetic material into bacteria. The connectors are bunched together in an alternating manner, so that the top end of one connector is next to the bottom of an adjoining connector. On each connector, the pore through which the genetic material passes is clearly defined. This atomic force microscope image was made as part of a study of viral structures.

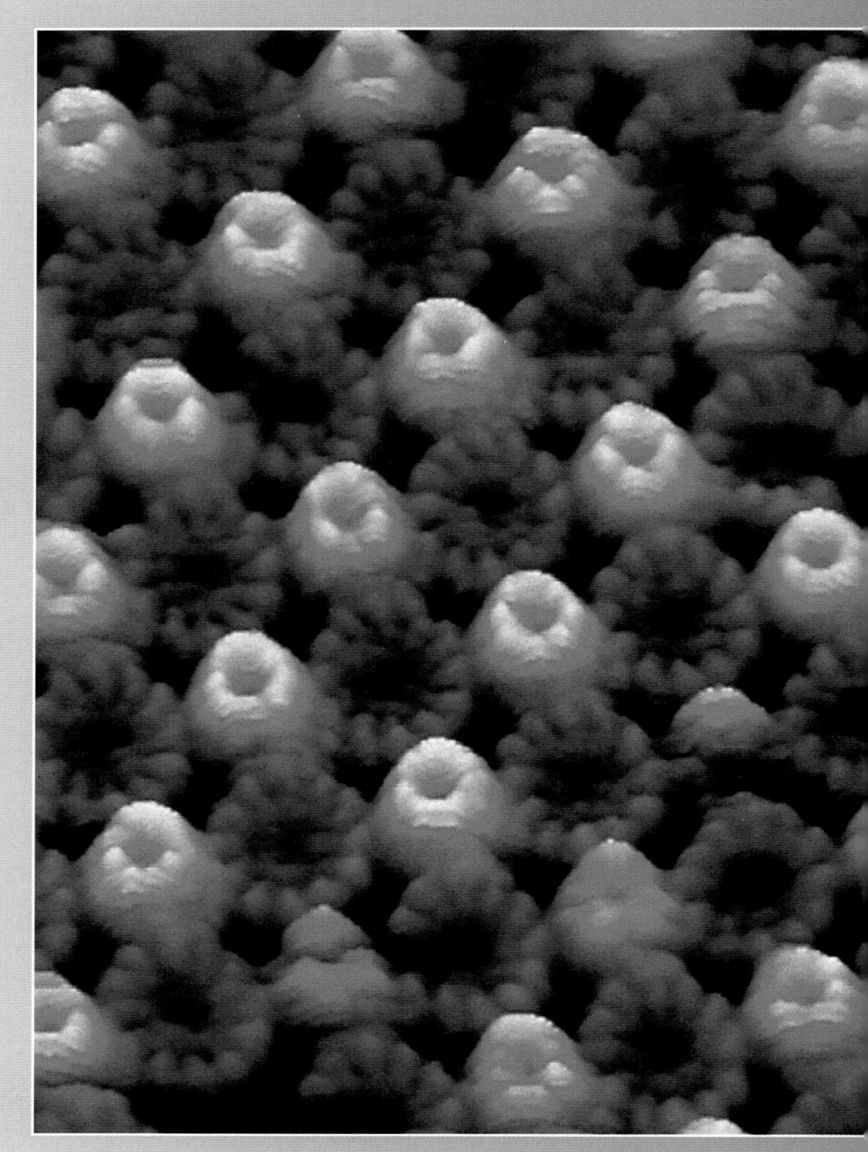

SCIENCE STUDIES

Science Versus Crime

Criminal investigators are increasingly relying on science and technology to solve crimes and bring suspects to justice.

BY JOHN I. THORNTON

Introduction

One hundred years ago, forensic science—the science of crime-solving—scarcely existed. Unless a witness saw a crime occur or a person confessed to the deed, police had little chance of bringing the lawbreaker to justice. That is no longer the case. Today, police investigators wield powerful scientific tools as they unravel crimes and track down the culprits.

Such tools include high-intensity lights and special chemicals that reveal the presence of invisible fingerprints and computers that can search police databanks to identify the person who made the prints. Evidence that is too tiny to be seen—such as hairs, fibers, and specks of soil—can be examined with powerful microscopes. This sort of analysis may show, for example, that fragments of glass on a suspect's shirt came from a window shattered during a burglary. Or, on the other hand, it may clear a suspected murderer's name by showing that his hands are free of the tell-tale particles produced when a gun is fired. Even crimes that destroy a great deal of evidence, such as an arson fire or an explosion, can be solved through sophisticated laboratory techniques. One method can break down the remnants of burned fuel into its separate components, revealing which fuel was used in an arson.

The greatest advance in scientific crime-solving thus far has been the development of tests that analyze the genetic material of blood, semen, saliva, and other biological evidence. Such tests can show that only one person in the world could have committed a particular crime.

These and many other techniques, together with advanced technology used to locate suspects, are part of the modern crime-fighting tool kit. Forensic science has many aspects but just one goal: to make our world a safer place by defeating crime.

The author:
John I. Thornton is a professor emeritus of forensic science at the University of California at Berkeley.

Nikon

The Evolution of Forensic Science

Since about 1900, *forensic science*—the science of crime-solving—has made great strides. As a result, police investigators today wield powerful tools as they bring crime to light and criminals to justice.

The word *forensic* comes from the Latin word *forensis,* meaning *of the forum.* In ancient Rome, the Forum was where the government debated policy and held trials. Thus, the Forum was Rome's courthouse. Over the years, forensic science has come to mean using scientific methods to resolve legal conflicts, mostly criminal cases.

The beginnings of forensic science

Although forensic science is largely a modern development, using scientific methods to solve crimes did not simply spring into existence overnight. The roots of scientific crime solving can be traced back for hundreds of years.

Some crime-history experts place the origins of forensic science in early writings on *forensic medicine*—medical knowledge used to solve crime. This, they believe, was the first direct application of science to criminal investigations. For example, a Chinese work titled *The Washing Away of Sins,* published in about 1250, described ways to distinguish between accidental death and murder.

In the Western world, a breakthrough in forensic science occurred in 1374, when French doctors received Pope Gregory XI's permission to perform autopsies. An autopsy is the medical examination of a dead body to determine the cause of death. Until then, because of restrictions by the church, investigations of suspicious deaths often did not include an autopsy.

In 1621 Paolo Zacchia, the physician of Popes Innocent X and Alexander VII, collected and published information on forensic medicine. But such knowledge remained fragmented until 1806, when training in this field was first offered by Andrew Duncan, a professor of anatomy at the University of Edinburgh, in Scotland. By the mid-1800's, the teaching of forensic medicine was well established in Europe and the United States.

Advances in detecting poisons in dead bodies also helped establish forensic science. In the early 1800's, murder by poison was much more common than it is now. Murderers of that day often favored poisons because physicians could detect few lethal compounds in a body after death. Doctors might suspect that a person had been poisoned if the individual had suffered particular symptoms, such as convulsions, but they had no way to verify their suspicions.

That unfortunate situation changed abruptly in 1813, thanks to a French physician, Mathieu Orfila, who published an extensive account of his research into analytical methods for detecting poisons. Orfila pointed out that poisons were chemicals and thus could be detected with chemical tests. For example, he explained, arsenic poisoning could be detected by heating specimens of a dead person's organs or stomach contents and holding a cold mirror over the vapors. If arsenic was present, a black layer of the compound would form on the glass. Orfila did not invent this test, but he alerted many doctors to its existence.

Orfila was involved in solving a number of poisoning cases. In one sensational case, in 1840, he helped convict a newlywed named Marie Lafarge of killing her husband after he found arsenic throughout the dead man's body. As a result of his work, the science he pioneered, forensic toxi-

A police investigator dusts a car door with powder to bring invisible fingerprints to light. This technique has been used since the early 1900's to link criminals to the scenes of their crimes. Fingerprint analysis is a very reliable way to identify criminals, because every person's prints are unique.

Investigators at a crime laboratory examine casts of tire tracks, which can show that a particular car was used during a crime, and of human teeth, which can aid in identifying human remains. Major cities in the United States began establishing scientific labs for the examination of evidence in the 1920's, and such labs were common by the 1950's.

cology, became widely recognized by law courts.

A branch of forensic science that was slower to win acceptance was document examination. Today, document examiners are often called upon to verify or discredit the authenticity of a letter or other piece of writing. In a fraud case, for instance, they might show that a signature on a check is forged. These specialists may also compare two supposedly unrelated documents to see if the documents might actually have been written with the same ink or on sheets of paper from the same batch of paper.

Document analysis began in the mid-1600's, after several books on examining documents and comparing handwriting had been published in France. For a long time, document examination—and handwriting analysis, in particular—remained very subjective, relying greatly on the examiner's personal judgment. But by the mid-1800's, document examination had become a fairly sophisticated profession.

The field suffered a setback as a result of the 1894 treason trial of Alfred Dreyfus, a French army officer. Dreyfus was convicted of spying for Germany after an expert testified that Dreyfus's handwriting matched that on a letter referring to the sale of French military secrets. In 1906, France's highest court reviewed the case and overturned the conviction, ruling that the letter was not written by Dreyfus.

Document analysis began to redeem its reputation in 1910 with the publication of the book *Questioned Documents* by an American handwriting expert, Albert Osborn. Osborn tried to make handwriting comparison as objective as possible by showing how a person's handwriting normally varies, such as in its slant. He did much to place document examination on a firm scientific footing, thereby making it acceptable to courts.

Another vital part of forensic science, dating from the late 1800's, is the identification of crime suspects and victims. For example, police may need to verify that a suspect has given his or her real name. Or they may have to identify a dead body.

One of the pioneers of this branch of forensic science was Alphonse Bertillon, a French crime expert. In 1881, Bertillon suggested to the Paris police that they take 11 precise measurements of every criminal arrested, including the width of the skull, the length of the forearms, and the maximum span of the outstretched arms. If a criminal was later rearrested and gave a false name, Bertillon said, investigators would be able to establish the criminal's true identity by comparing his or her measurements with those on file. This system, called Bertillonage, was adopted in France and a number of other countries, and it was used as a means of identifying people until about 1920.

A flaw of Bertillonage was that more than one person could—and sometimes did—have the same 11 measurements. Such coincidences actually led to the conviction of several innocent people, including an Englishman named Adolf Beck, who in 1895 was convicted of a jewelry theft and sentenced to seven years' hard labor. Beck was eventually cleared of the crime and compensated for false imprisonment.

Despite the shortcomings of Bertillonage, the police still make limited use of it today. When human bones are found, investigators may measure them to find the dead person's height. That measurement can be compared to the height of missing people and so may help identify the remains.

Fingerprint and firearms identification

But for the prosecution of criminals, Bertillon's system of identification was eventually eclipsed by one based on fingerprints. In contrast to body measurements, fingerprints are unique and provide a foolproof method for identifying suspects.

One of the first people to suggest using fingerprints to identify criminals was a British civil servant in India named William Herschel, who got

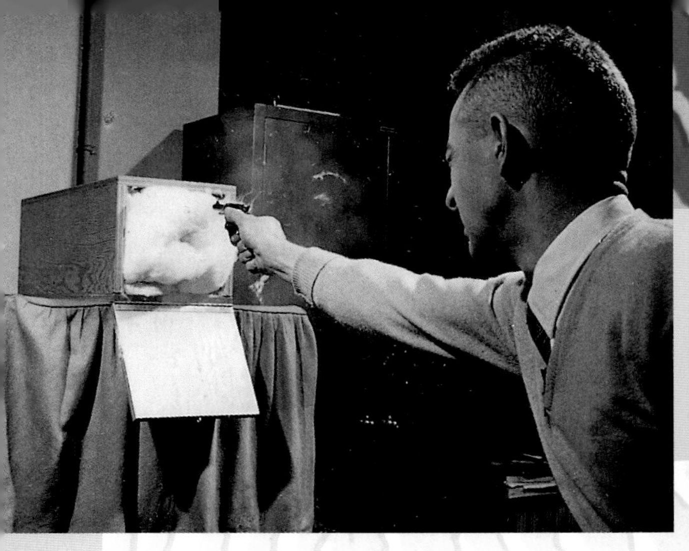

Did this gun fire a fatal bullet?
In the 1920's, firearms identification—proving that a particular gun was used in a crime—became a part of forensic science. To make a firearms test, an investigator fires a suspect's gun into cotton or water, *above.* Microscopic examination reveals whether tiny marks made by the gun on the test bullet match the marks on a bullet from the crime scene. If the marks match, *right,* the bullets were fired by the same gun.

the idea of using fingerprints to prevent payroll and merchandising fraud. Herschel discovered that some Indian pensioners and suppliers were collecting payments twice. To counter that abuse, he asked payment recipients to dip a finger in ink and make a fingerprint on the receipt.

Pleased with the success of his innovation, Herschel suggested that prison authorities in India use fingerprints to identify criminals. Although his idea was rejected as impractical, Herschel persisted. He compiled evidence showing that fingerprints stay the same throughout life and that no two fingerprints are the same. He then presented the information to Sir Francis Galton, a renowned British anthropologist and statistician. Galton developed Herschel's ideas into a workable system of fingerprint identification and published an authoritative work on the subject. By the beginning of the 1900's, fingerprint evidence was being allowed in British courts.

It was not until 1911, however, that fingerprint evidence was introduced into a court of law in the United States. In the following years, fingerprints became increasingly accepted by U.S. courts and law enforcement agencies. In 1930, the Federal Bureau of Investigation (FBI) opened a central fingerprint file to serve the entire nation.

By that time, crime investigators had still another valuable tool at their disposal: firearms identification. The essence of firearms identification is proving that a particular gun was used to commit a crime being investigated. This is done by matching bullets recovered from the crime scene with test bullets fired in a lab with the suspect weapon. Studies have shown that every gun leaves a unique pattern of tiny marks on each bullet it fires.

The St. Valentine's Day Massacre in Chicago in 1929 is often cited as the event that made firearms identification an important part of forensic science. This murder of seven members of the Bugs Moran gang, allegedly committed by gunmen working for the infamous mobster Al Capone, stirred a wave of public indignation and demands for action against organized crime.

As one result of the outcry, a Scientific Crime Detection Laboratory was established that same year at Northwestern University in Evanston, Illinois. The Evanston facility was one of the first comprehensive crime laboratories in the United States. Although the laboratory was unable to link Capone to the Moran gang slayings, it showed that several Thompson submachine guns found by the police were used in the crime.

The director of the lab, Calvin Goddard, contributed greatly to placing firearms identification on a sound scientific basis. Under Goddard's directorship, the Scientific Crime Detection Laboratory served as a model for similar facilities elsewhere. By the early 1950's, most large cities in the United States had crime laboratories.

Thanks to the work of forensics pioneers, police investigators today can rely on a powerful ally—objective scientific evidence—in their efforts to bring criminals to justice. It is an ally that has sent many a guilty person to prison.

Crime-Fighting Scientists

Whenever a person commits a crime, whether it's a swindler forging a check or a drunk driver fleeing a car crash, a small army of investigators is standing by, ready to use their knowledge to bring that criminal to justice. A great many of those investigators are scientists. Others are investigators who, though not actually scientists, use scientific methods. All are highly trained individuals able to find and interpret the tiniest clues that can point to the guilt of a crime suspect. And all are prepared to testify in court about their findings and to be cross-examined by defense lawyers.

Among the scientists who form the crime-fighting team are *forensic pathologists,* doctors who specialize in uncovering what caused a person's death. When someone is murdered or dies under suspicious circumstances, a forensic pathologist performs a detailed examination of the corpse, called an autopsy. The purpose of the autopsy is to determine the cause of death and how any injuries that are detected were inflicted. The autopsy involves an external inspection of the body followed by an examination of the internal organs. If the person was shot, the pathologist traces the path of each bullet through the body. If the person was stabbed with a knife or beaten with a blunt object, the precise nature of each injury is carefully noted.

Depending on what the forensic pathologist finds, he or she may be able to determine that a sudden death was due to natural causes, murder, or suicide. For example, if the pathologist finds no evidence of injuries but discovers that the arteries leading from a dead person's heart are blocked, the probable conclusion would be that the person died of a heart attack. On the other hand, if the pathologist finds a bullet wound in the person's chest, and no traces of gunpowder near the wound, the evidence points to a shooting from some distance away—in other words, a murder. But, if the gunshot wound is in the person's temple and is surrounded by gunpowder residues, the pathologist would suspect suicide.

When an autopsy fails to reveal the cause of death, the forensic pathologist collects samples of the dead person's blood, urine, and stomach contents for examination to determine whether drugs or poisons were in the person's system. This type of analysis is performed by a *forensic toxicologist,* a chemist trained in the effects of drugs and poisons on the human body. A forensic toxicologist can often learn whether a person who died suddenly succumbed to poison or a drug overdose or if a driver who was killed in a car crash was drunk at the time.

Other scientists who help solve crimes include *forensic odontologists,* dentists who specialize in identifying human remains. If a human skeleton is found, a forensic odontologist tries to identify the dead person by comparing the teeth in the skull with the dental records of missing people.

The crime-fighting team

A variety of scientists and other specialists help police investigators solve crimes. Some of these experts work only in crime-solving. Others work primarily in other areas but assist police investigations when their particular expertise is needed.

Forensic pathologist
Performs an autopsy to determine cause of death and method in which injuries were inflicted.

Forensic odontologist
Identifies human remains by comparing teeth with dental records; identifies a suspect by comparing bite on victim with bite made by suspect.

Forensic toxicologist
Analyzes a dead person's blood, stomach contents, and urine to see if drugs or poisons are present.

Forensic psychiatrist
Interviews and tests a suspect to gauge whether he or she is mentally ill.

Forensic engineer
Reconstructs accidents, such as automobile collisions, to determine who was at fault.

Document examiner
Investigates whether a suspect was the author of a particular piece of writing by comparing it with the suspect's handwriting.

Latent-fingerprint examiner
Brings invisible fingerprints at a crime scene to light; compares these with a suspect's fingerprints.

Firearms examiner
Investigates whether bullets found at a crime scene were fired by a suspect's gun.

Criminalist
Analyzes physical evidence found at a crime scene, such as hairs, fibers, bloodstains, and soil.

Botanist
Identifies plant material on a dead body to help pinpoint where a murder took place.

Entomologist
Examines insects infesting a decaying body to learn when the person died.

Meteorologist
Analyzes weather conditions and the phase of the moon when a crime occurred to determine how well a witness would have been able to observe it.

Examining microscopic evidence
A forensic scientist uses a powerful scanning electron microscope (SEM) to examine evidence. The SEM is used to make out details of gunshot residues, hairs, pollen, fibers, and other very small kinds of evidence.

He or she will look especially for signs of fillings and other dental restorations. If the skeleton's teeth match the dental records of a missing individual, the skeleton is almost certainly the remains of that person. The size and placement of the fillings in a person's teeth, as recorded in dental X rays, are very nearly unique and so are unlikely to be duplicated in anyone else's teeth.

Forensic odontologists are sometimes able to identify not only crime victims but criminals as well. Many people have teeth that are irregular in size, spacing, and configuration. These irregularities can help send a suspect to jail. For example, in a significant number of sexual assault cases, the attacker bites the victim, leaving plainly visible bite marks on the skin. If police suspect a particular person committed the attack, a forensic odontologist will compare photographs of the bite marks with test bite marks made by the suspect. If the marks are the same, this evidence can help convict the suspected attacker.

Another member of the crime-solving team who often helps in identifying skeletal remains is the *forensic anthropologist.* Anthropology is the study of human characteristics, cultures, and customs. Forensic anthropologists, who are trained in physical anthropology—the study of human physical traits—identify skeletal remains by carefully examining the bones to determine the dead person's sex, height, and weight at the time of death. They may also compare the bones with X-ray pictures taken of a missing person while alive to see if the bones match. In addition, they may be able to tell the police whether broken bones were broken at the time of death or afterward.

One specialist who works primarily with criminal suspects is the *forensic psychiatrist,* a doctor trained to recognize mental illnesses that can cause pathological behavior. Forensic psychiatrists help courts decide whether a suspected criminal is legally insane. Criminals found by a court to be insane are considered not responsible for their actions and are likely to be committed to a mental hospital rather than sent to prison.

A forensic psychiatrist may assess a suspect's mental health by first interviewing the person, encouraging him or her to freely express thoughts and emotions. The suspect's responses to questions help the psychiatrist understand how the person thinks and acts. The psychiatrist may also administer diagnostic tests such as the Rorschach test, in which people are asked to tell what 10 standardized inkblots look like to them. Some kinds of responses are considered indicative of a disturbed mind. Another tool often used by the forensic psychiatrist is the electroencephalogram brain-wave test, which detects brain damage.

In some legal investigations, the police solicit the help of *forensic engineers,* engineers with special expertise that may help decide a case in court. For example, an engineer with an in-depth knowledge of vehicles may be able to determine which of two drivers caused a car crash. To do this, the engineer may examine the wrecked cars for damage that shows how the impact occurred. The engineer will also measure skid marks to deduce how fast each car was going. The information obtained from the vehicles and the crash site enables the specialist to reconstruct the accident, revealing the sequence of events that led to the crash.

Whenever a sample of handwriting is central to a criminal investigation, the police turn to *document examiners.* A document examiner is trained to compare a piece of handwriting that has played a part in a crime—such as the words and numbers written on a forged check—with a suspect's handwriting. This expert may also judge, for example, whether a suicide note is genuine or was written by someone other than the deceased—an indication that the "suicide" was actually a murder.

In comparing two samples of handwriting, the examiner pays close attention to certain characteristics of the writing. These include the slant of the writing, the size of the letters in relation to one another, and the way the letters are formed. Document examiners also have an extensive knowledge of paper types, inks, and other distinguishing characteristics of documents.

An important member of any crime-solving team is the *latent-fingerprint examiner.* This expert finds and photographs fingerprints left at a crime scene. Then he or she compares the fingerprints with those already on file with law-enforcement agencies, to see if they can be traced to a particular person.

At a crime scene, some fingerprints are visible, such as those left by fingers coated with blood, grease, or paint. These visible fingerprints are simply photographed. Other fingerprints are *latent* (hidden). A latent-fingerprint examiner *develops* (makes visible) such fingerprints by brushing them with a special powder or chemicals or by exposing them to laser light to make them glow. Once a latent fingerprint has been revealed, it too is photographed.

Fingerprints are unsurpassed at linking criminals to their crimes. If a latent fingerprint matches a file copy of a suspect's inked fingerprint, the examiner will testify in court that the fingerprint at the crime scene was made by the suspect and could not have been made by any other person.

The only rival to fingerprints in showing that someone was present at a crime scene is DNA "fingerprints." *DNA* (deoxyribonucleic acid) is the molecule genes are made of. It can be extracted from any bit of blood, skin, or hair that a criminal leaves behind. DNA fingerprints are detailed patterns of DNA fragments that are created when DNA is broken apart in the laboratory by special chemicals. A suspect can be strongly tied to a crime scene if the DNA fingerprints made from a sample of the person's blood match the pattern made by DNA from a tissue sample found at the crime scene. Because of the importance of DNA fingerprinting in modern criminal investigations, geneticists have become an indispensable part of the crime-solving team.

The investigation of crimes involving guns, which include about two-thirds of all murders committed in the United States, is likely to require the help of a *firearms examiner.* The firearms examiner compares bullets and cartridge cases found at a crime scene with test bullets and cartridge cases from a suspect's gun, fired in a police laboratory. This analysis is made with a special microscope called a comparison microscope.

Criminalists and other experts

Whatever evidence falls outside the expertise of the various specialists working on a criminal case is assigned to a general investigator known as a *criminalist.* Criminalists help solve crimes by analyzing, identifying, and interpreting many kinds of physical evidence. They are trained to build a case from hairs and fibers, bullets and cartridge cases, bloodstains, soil, paint, glass, paper, drugs—an endless number of objects and substances that might link a suspect with a crime scene. For example, a criminalist may show that fibers found on a murder victim came from the suspect's clothes.

Other kinds of scientists are occasionally also called on to help bring criminals to justice. Although they don't devote all their time to solving crime, these scientists help to do so in special circumstances.

If a murder victim's body was moved after the killing, for instance, a botanist may help show where the murder took place by identifying unusual plant material on the body. An *entomologist* (scientist who studies insects) may help establish when a murder victim died by studying the insects infesting the person's body. Different insects invade a body at various stages of its decomposition. Flies are apt to lay eggs on a body only minutes after death, but beetles infest a body only after it has been dead for some time.

A *meteorologist* (scientist who studies weather and climate) may explain what was visible on the night of a crime by describing the weather conditions, phase of the moon, and time the moon rose and set that night. This information may help the police gauge the reliability of a person who claims to have witnessed the crime.

Together, these various scientists use their expertise to help police investigators uncover crimes, identify suspects, and obtain convictions. Without their assistance, the numbers of unsolved crimes and unconvicted criminals would be far greater than they are.

Procedures at the Crime Scene

A burglar breaks into an art museum and escapes with a valuable canvas. There are no witnesses to the crime, the burglar alarm doesn't go off, and the video surveillance cameras capture no glimpse of the intruder. The canvas is quickly passed to a private art collector, who will be discreet about this latest acquisition. What, if anything, can connect the thief to the crime?

Quite a lot, actually. Whenever a crime is committed, the culprit almost always leaves behind—or takes away—various kinds of evidence that can prove his or her guilt. These incriminating details, called *physical evidence,* may be as small as a pollen grain or as large as a truck. Investigators use scientific techniques at a crime scene in search of clues to who committed the crime and how.

In the case of a museum burglary, for example, the police usually find marks made by pry tools wherever the burglar entered the building. Likewise, if the burglar cut wires leading to the security alarm, investigators look for shearing marks on the wires. They then use silicone rubber to make casts of all the marks. Later, if the police suspect a particular person of committing the burglary, they may seize tools belonging to that person. Every tool leaves tiny, unique marks on any surface it is used on. Scientists in a laboratory can use a special microscope, called a comparison microscope, to determine whether the suspect's tools made the marks seen at the museum.

Police investigators at the museum will also carefully examine any broken window glass scattered at the scene. The location and pattern of the glass fragments will reveal whether the window was broken from inside or outside the building. If the glass was shattered from the inside, detectives might suspect that the crime was committed by a person with a key to the building, who broke the window to make it appear that the burglar entered that way.

Finally, the investigators at the museum will

Forensic scientists gather *larvae* (immature insects) from the clothing of a decaying body found at a crime scene. The larvae will be grown in a laboratory, so that investigators can identify which insect species they belong to. This information can help determine how long the person has been dead, because different kinds of insects infest a body at different stages after death.

carefully gather and store the glass fragments. Later on, if the police have identified a suspect, scientists in a laboratory will examine the suspect's clothing for microscopic bits of glass. Often, they will find fragments whose characteristics match those found at the crime scene.

In any crime, the physical evidence that will be interpreted by scientists falls into one of two categories: evidence left behind by the criminal, such as the pry marks at the burgled art museum, and evidence carried away, such as glass fragments from the broken window. Either way, the evidence associates the criminal with the crime scene. For this reason, it is often called *associative evidence.* Many crimes are solved by associative evidence alone. It can reveal not only who committed a crime but also how.

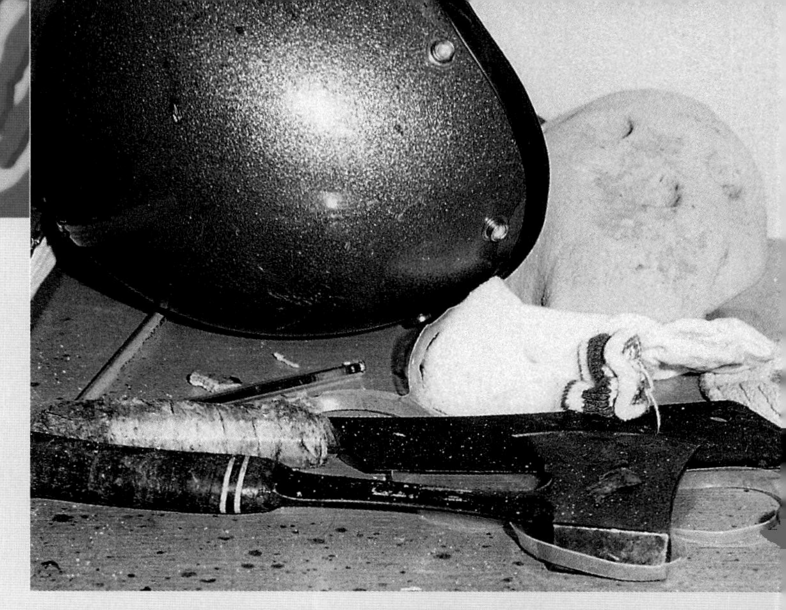

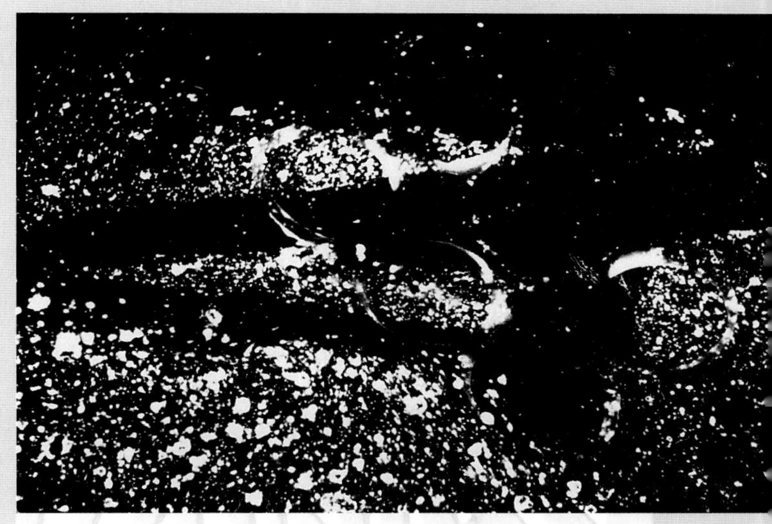

Blood reveals the truth
Chemicals that reveal the presence of blood can solve crimes. One such crime was a fatal shooting in which the suspect claimed he acted in self-defense. The victim, he claimed, had attacked him with a hatchet and cleaver. The hatchet and cleaver, *top,* were found lying on a dresser near the body. Police sprayed the dresser with the chemical luminol, which reacts with blood to give off light. The chemical showed that the dresser was splattered with the victim's blood—except for the part underneath the hatchet and cleaver, *above.* This finding showed that the weapons were on the dresser during the shooting, not in the victim's hands, thus contradicting the suspect's story.

Slow, careful methods

The first step in any investigation of a crime scene is to protect the area against contamination or alteration. The police cordon off the scene, and only authorized persons are allowed inside that boundary. Investigators wear rubber gloves, paper booties, and hair nets to avoid sprinkling the scene with new soil particles or fibers.

Then the investigators slowly and carefully document the crime scene. They first photograph and videotape it from several angles to show the location of visible evidence. Next, they take precise measurements that allow an artist to make a drawing that records where each piece of evidence was found.

Only after the crime scene has been documented do investigators remove any physical evidence. They generally remove larger pieces of evidence with forceps to avoid getting their fingerprints on them and place each piece in a clean, unused plastic container or envelope. As each item of evidence is gathered, investigators note its location, along with the date and time and the name of the person who collects it.

To collect *trace evidence*—tiny particles, such as fragments of paint, hairs, and fibers or bits of soil or glass—investigators use devices called vacuum traps. These are plastic containers that fit on the end of a vacuum hose and are designed to prevent contamination of evidence.

One of the most important procedures at a crime scene is finding and photographing the culprit's fingerprints. Some fingerprints, made when the criminal's hands were coated with a substance like blood or grease, are easily visible and can be photographed directly. Most prints, however, are *latent* (hidden) and must be made visible. The common method for finding latent prints is *dusting*—lightly brushing a surface with powder or chemicals. The material sticks to the prints and makes them visible so they can be photographed.

Many investigators now search for latent fingerprints with lasers or ultraviolet light. Both kinds of light cause perspiration residues in some prints to *fluoresce* (glow) so the prints can be photographed.

Tool marks left at a crime scene are preserved in a cast made with silicone rubber. Many other kinds of marks—such tire tracks in soft dirt—are captured in molds made with plaster of Paris. Investigators lift shoeprints left in dust or other fine powder with an electrostatic lifter. This is a battery-powered device to which is attached a large piece of black film made of vinyl or polyester coated on the back with a thin layer of metal. When the device is switched on, the electric current creates a static electric charge that causes the dust—and the shoeprint it contains—to stick to the film. The shoeprint shows up against the black surface and is then photographed.

On rare occasions, a criminal will try to eliminate physical evidence at a crime scene, for example by mopping up bloodstains. In many such attempts at deception, the person succeeds only in creating additional evidence. Cleaning a bloody floor with detergent and water actually spreads traces of blood over a much wider area, making it easier to detect with special chemicals, such as luminol. Traces of blood sprayed with luminol—even blood that has been diluted in water to a millionth its normal density—emit light.

Different crimes, different evidence

Certain kinds of crimes tend to generate particular types of evidence. In a violent crime such as murder, the assailant is likely to leave a great amount of physical evidence, including many fingerprints. Even if a killer wears leather gloves, there will be many distinctive prints—from the gloves themselves—left at the scene. That is because leather gloves are made from animal skin, which, like the skin on human fingertips, contains tiny natural ridges. These ridges leave latent prints that can be collected in much the same way that human fingerprints are.

In addition, a murder victim may have fought the attack, drawing blood from the killer that can be collected from the victim's cloth-

Gathering specks of evidence
A forensic scientist uses a vacuum trap to collect trace evidence—tiny bits of dust, glass, or other materials—from a murder suspect's pants. The presence of a particle, such as a speck of coal dust, *below right,* may show that a suspect was at a crime scene.

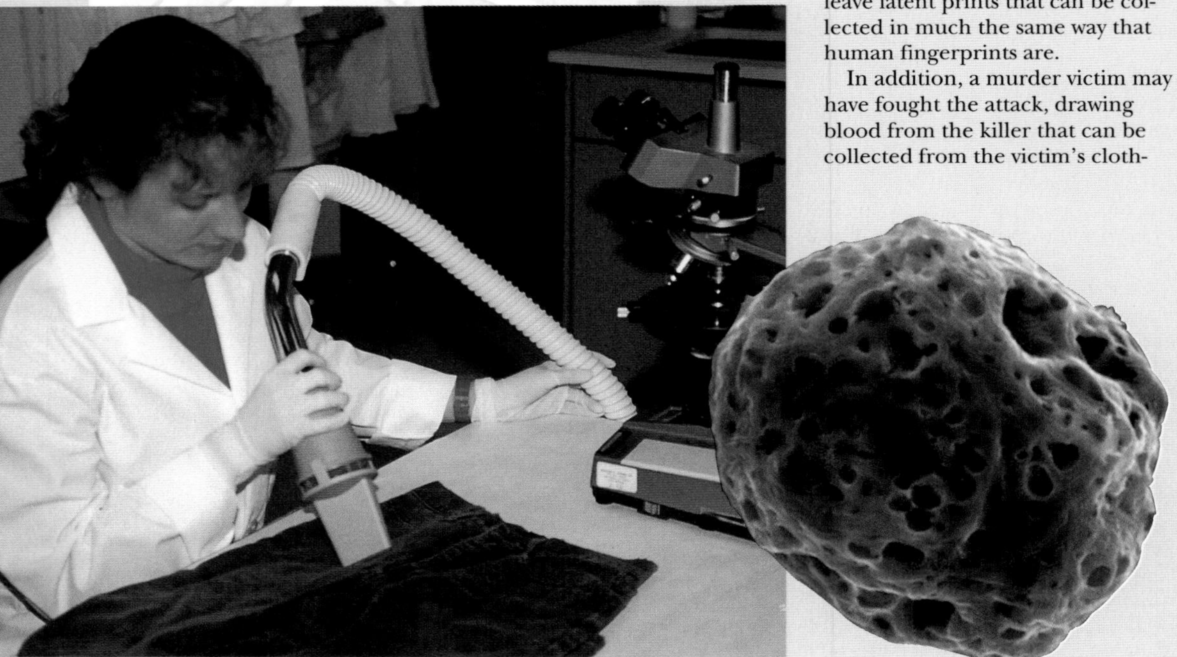

A high-intensity light reveals a latent, or invisible, fingerprint by causing perspiration in the print to fluoresce, or give off light. The fingerprint can then be photographed. Searching for prints with lights, including lasers and ultraviolet lamps, has replaced dusting prints with powders or chemicals at many crime scenes.

ing or fingernails. An analysis of the *DNA* (genetic material) taken from cells in the blood may show that the blood came from the prime suspect in the murder.

In murders committed with firearms, a bullet found at the scene may later be linked to a gun owned by the suspect. Investigators will test-fire the gun in the laboratory and compare the test bullet under a microscope with the bullet recovered from the scene. All bullets fired by the same gun contain the same unique pattern of marks.

The main piece of evidence in a murder is the victim's body. Investigators handle a corpse carefully, covering the hands and feet with plastic bags to protect any evidence—such as blood or flesh under the victim's fingernails—that can be recovered later for examination. Before the body is removed, a pathologist takes its temperature. This reading can help determine when the person was killed, because a body cools at a regular rate (which varies according to the surroundings).

Some crime scenes are more difficult to examine than others. Collecting evidence at the site of a bombing, for example, is a daunting task. The explosive material used to make a bomb is largely consumed by the blast. Materials that are not destroyed by the explosion may be chemically altered and scattered widely.

Investigators at a bombing site collect shovel loads of debris that is later sifted through a screen to reveal small fragments of the bomb. Investigators also try to collect residues of the explosive material by wiping surfaces with filter paper that has been moistened with a solvent. The filter paper is analyzed in a lab for traces of the explosive.

Arson creates one of the greatest challenges for the forensic scientist. After a fire is put out, much or all of the scene is charred and soaked with water. Finding evidence is still possible, however. Even when a fire has consumed most of a building, there are likely to be residues of petroleum products—the usual materials used in arson fires—which can be found with a portable instrument called an aromatic hydrocarbon detector. This instrument draws in a sample of air, then registers the presence of combustible substances.

Experts in 1998 predicted further advances in crime scene investigation, such as the ability to transmit an image of a fingerprint found at a crime scene over phone lines to a computer. The computer would compare the print with those filed in its memory and, if it found a match, identify the criminal immediately. This and other technology was expected to play an ever-increasing role in evidence gathering.

Analyzing the Evidence

Forensic scientists use sophisticated laboratory tests to reveal how a crime was committed, and by whom. Traces of gasoline, identified with a technique called gas chromatography, may show that a fire was arson. A drop of semen left by a rapist, subjected to genetic analysis, could reveal whether a suspect is guilty or innocent. A white powder, analyzed by an instrument called a mass spectrometer, may prove to be cocaine. And with the aid of computers, many kinds of evidence analysis are now being expanded and simplified.

DNA fingerprinting

Probably the most important innovation ever introduced to forensic science is a technique called *DNA "fingerprinting."* This method enables investigators to analyze the *DNA* (deoxyribonucleic acid)—the molecule genes are made of—in biological evidence such as blood, hair, or semen recovered from a crime scene. DNA fingerprinting provides an extremely precise way of identifying criminals, because everyone's genetic makeup is unique. (The one exception to that rule is identical twins, who share all the same genes.) The most accurate form of DNA fingerprinting can show that a particular bit of skin tissue or drop of blood could have been left at a crime scene by only one person in the world.

There are two primary methods of DNA fingerprinting: restriction fragment length polymorphism (RFLP) testing and polymerase chain reaction (PCR) testing. Each has advantages and disadvantages. RFLP (pronounced *RIF lip*) is more accurate, but requires a larger sample—several strands of hair or a dime-sized drop of blood—and typically takes several weeks. PCR, though less precise, can analyze very small bits of blood or semen, or even a single hair, overnight. For those reasons, PCR is the more commonly used technique.

In RFLP testing, laboratory technicians use chemicals called restriction enzymes to cut the DNA extracted from a sample of evidence, such as a blood stain, into numerous pieces. Then, in a procedure called electrophoresis, a solution containing the DNA fragments is placed on a gel and exposed to an electric current. The electric field causes the variously sized fragments, which themselves carry an electric charge, to move through the gel. The smaller, lighter pieces move farther than the longer, heavier pieces. In this way, the many fragments get sorted by length.

The result is a pattern of dozens of parallel bands (made visible on photographic film after additional steps), somewhat like a supermarket bar code, that reflects the composition of the person's DNA. Geneticists have estimated that there are more than 10 billion possible patterns resulting from the individual assortment of *polymorphisms*—variable lengths of DNA between restriction enzyme cutting sites—in each person's genes. Many experts say it is therefore all but impossible that the DNA pattern of one person would match that of another. So if the pattern made by the

Marks of guilt
Microscopic evidence tied conspirator Terry Nichols to the Oklahoma City bombing of April 1995. Forensic scientists showed that a bit on a drill owned by Nichols was used to force open a padlock at a quarry from which explosive devices used in the bomb were stolen. Drill marks on the lock are invisible to the naked eye, near *right,* but are apparent in a microscope image, *second right.* When a forensic scientist drilled into metal with Nichols's drill bit, the bit left tiny marks on the surface, *third right.* These marks matched those on the padlock, *far right,* showing that Nichols's drill bit was used on the lock.

DNA in a blood sample taken from a suspect is identical to the pattern of the DNA from the crime scene, there can be little doubt about the suspect's guilt.

The other type of genetic fingerprinting, the PCR test, zeroes in on specific genes that have many different forms among the general population. PCR is a genetic "copying machine" that can quickly make multiple copies of a gene in a DNA sample from a crime scene.

A DNA molecule is shaped like a twisted ladder. The two sides and rungs of the ladder are made of interconnected strands of molecules called nucleotides. PCR is performed with a machine called a thermal cycler, which heats a solution containing a sample of DNA, causing the double-stranded molecule to separate into individual strands. After a few minutes, the machine lowers the temperature of the solution. With the aid of an *enzyme* (biochemical catalyst) called DNA polymerase, free nucleotides in the solution then join to form new strands of DNA, and those come together to re-form the double-sided molecule.

This procedure repeats many times, with each cycle doubling the amount of DNA in the solution—which is why the process is called a chain reaction. PCR can create millions of copies of the original DNA sample overnight.

Next, scientists determine which version of a particular gene is in the enlarged quantity of the DNA from the crime scene. They do this by placing the DNA on a membrane that contains copies of all known forms of the gene. Each copy of that gene in the multiplied DNA then attaches to its matching version on the membrane, thus identifying itself.

The same test is performed on genetic material in a blood sample taken from a suspect. If the person's DNA contains a different version of the gene than was found in the DNA from the crime scene, investigators can rule out that particular suspect. PCR is a valuable tool for narrowing down a list of suspects, but only RFLP testing can establish with absolute certainty that a biological specimen recovered at a crime scene was left by a particular suspect.

When RFLP fingerprinting cannot be carried out, a conviction may depend on whether conventional fingerprints were found at the crime scene. Like one's genetic makeup, a person's fingerprints are unique. Most fingerprints left at a scene are *latent* (invisible). They are made visible by being brushed with a special powder of chemicals or being exposed to light from a laser or an ultraviolet lamp.

The computer—an invaluable aid

An important technological advance in forensic science, second only to DNA analysis, has been the use of computers to identify fingerprints. Images of latent fingerprints found at a crime scene are scanned and translated into *digital* information (a series of 0's and 1's), which is fed into a computer. Using a program called the Automatic Latent Print System (ALPS), the computer compares the digitized fingerprints with millions of others stored in its memory and suggests a list of

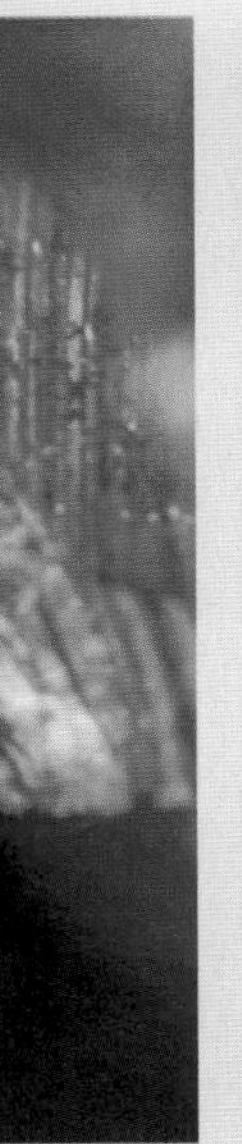

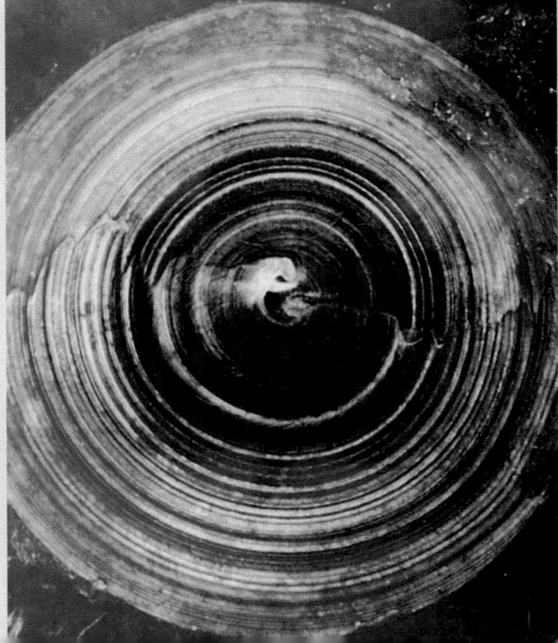

people who may have committed the crime. A human fingerprint examiner then calls up those suspects' fingerprints one by one on a computer monitor and compares them with images of the ones found at the crime scene to look for an exact match.

One of the first criminals caught with the help of ALPS was the so-called Night Stalker, who committed a series of murders in Los Angeles in the early 1980's. Police investigators were stymied in their search for the Night Stalker until they found a single fingerprint on an orange Toyota that they thought had been used by the killer. The print was scanned into a computer using ALPS. Three minutes later, the computer reported that the print had most likely been made by Richard Ramirez, a 25-year-old drifter with a long record of drug offenses and auto thefts.

Two days later, Ramirez was arrested and charged, and he was eventually convicted of 13 murders. The Los Angeles Police Department estimated that it would have taken 67 years for one person to manually compare Ramirez's fingerprint against the 1.7 million inked fingerprint cards in the department's files.

Computers can also be used to reenact certain types of crimes. To learn how a fatal shooting probably occurred, for example, investigators supply a computer with data culled from the crime scene: the path that bullets tore through the person's body and surrounding walls and doors; the victim's movements, as revealed by the trail of blood; and more. The computer then creates an animated restaging of the murder, depicting the likely sequence of events. It might conclude, for instance, that the person was shot in the chest, turned and staggered toward a door, and then was shot in the back. This interpretation of the crime may be presented in court to help the jury understand how the murder was most likely carried out.

Computers have also eased the work of many investigators, such as firearms examiners, whose job is to determine whether a particular gun was used in a crime. These specialists can often link bullets or cartridge cases recovered from a crime scene to a suspect's gun, because every firearm gouges tiny, unique marks into the ammunition it fires.

When investigators think that a suspect's gun was fired during a crime, the firearms examiner uses the gun to fire a test bullet into a container filled with water or cotton. The examiner then uses a special microscope to compare the marks on the test bullet and cartridge with the marks on the bullets and cartridge cases found at the crime scene.

Computers have made firearms examination a much more far-reaching law enforcement tool. With the aid of the computer, investigators are often able to find the likely perpetrator of a violent crime before any suspects have been identified by other means.

A fingerprint examiner uses a computer to see whether a fingerprint found at a crime scene matches any filed in a database. Computers can make such comparisons extremely rapidly. First, the computer searches the database to produce a list of possible matches. Then the fingerprint examiner takes a close look at those prints on the computer monitor to see if any of them do, indeed, match. If the examiner finds a match, the police know the criminal's identity.

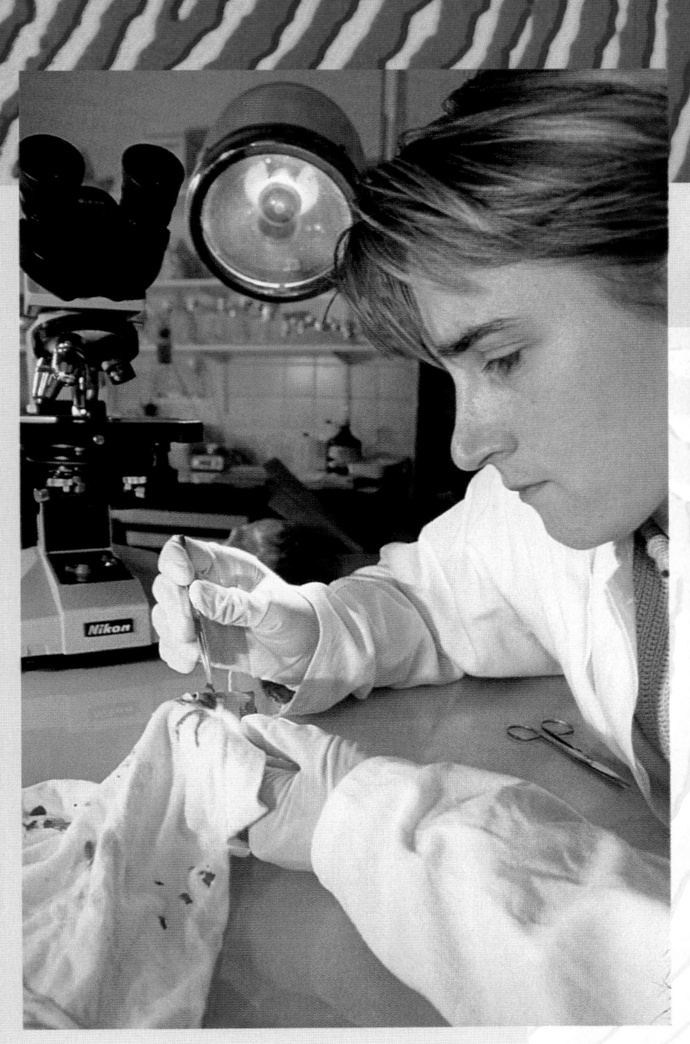

Incriminated by one's genes

A scientist removes a fragment of blood-stained material from a murder victim's clothing, *left,* so that *DNA* (deoxyribonucleic acid—the molecule genes are made of) can be extracted from it. The DNA will be examined with a technique called DNA fingerprinting to determine if some of the blood came from the suspected killer. One form of DNA fingerprinting produces a unique pattern of dark bands, *above.* Scientists can compare this pattern with that made by DNA in a blood sample taken from a suspect. If the two patterns match, it is a virtual certainty that the suspect committed the crime.

With a digital scanner connected to a microscope, investigators create a computerized image of the tiny marks on a bullet and cartridge case from a crime scene and enter the information into a computer. The computer then compares that image with thousands or millions of others stored in its memory banks and suggests possible matches. The firearms examiner then reviews the images of the proposed matches on a computer screen and decides whether any of the bullets and cartridge cases look identical to the ones from the crime scene.

Computers can speed up other methods of analyzing evidence as well. One way to identify an unknown *organic* (carbon-based) substance, for example, is with a technique called infrared spectrophotometry. A spectrophotometer is an instrument that detects and records the absorption of infrared light by a material. Infrared light, often called "heat rays," is light beyond the red end of the visible spectrum. Every organic substance absorbs infrared light at a specific set of wavelengths, and the absorption pattern of any material is unique.

Thus, a substance's infrared-absorption pattern, or spectrum, provides a highly reliable way to analyze organic evidence obtained at crime scenes. A computer can identify a plastic, paint, or other substance in a few seconds by comparing its spectrum against a digitized library of thousands of spectra in its memory.

Other crime-detection tools

Another important tool in chemical analysis—particularly for identifying illegal drugs—is a device called a mass spectrometer. This instrument can show that a white powder found at a crime

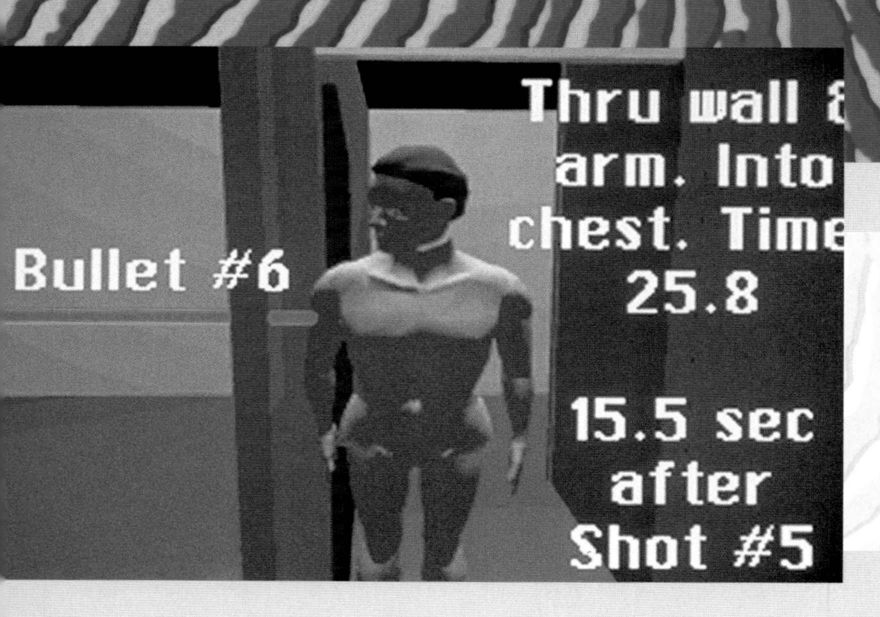

A computer-generated animation based on information collected at the scene of a shooting shows how the crime most likely occurred. Computers can help police reenact the events of a crime by analyzing a variety of data, such as the path of bullets through a victim's body and through surrounding walls, floors, or doors.

scene is cocaine or that a substance isolated from a person's blood is heroin. A mass spectrometer breaks down an organic substance into *ions* (electrically charged atoms or molecules) according to their *mass* (quantity of matter). A detector records the distribution of ions and translates it into a pattern called a mass spectrum, consisting of a series of peaks of differing height.

An organic substance can be recognized from its mass spectrum, which is different from the spectra created by all other materials. Mass spectroscopy is therefore a very precise identification method. The procedure can also be used to identify substances such as gasoline and other petroleum products used to start arson fires.

Arson evidence, as well as the residues of explosives recovered from bombing sites, can also be analyzed with a process known as chromatography. This procedure uses *adsorption* (attraction to a surface) to separate the various components of a substance. The substance being analyzed passes through a material that adsorbs its different constituents at varying rates, causing them to separate from one another.

Chromatography can be used to analyze both liquids and gases. In the investigation of a suspected arson fire, for example, scientists use gas chromatography to learn whether a particular fuel—such as gasoline, charcoal lighter, or kerosene—was used to start the fire. The result of this analysis is a *chromatogram,* a printout of the mixture's ingredients that looks like a series of peaks and valleys. The chromatogram—which detects components of *hydrocarbon compounds* (compounds made up of hydrogen and carbon) in the gas sample being tested—is compared with those of known flammable liquids to identify the fuel used in the arson. Sometimes, investigators use mass spectrometry along with gas chromatography to increase their chances of identifying a fuel.

Another sophisticated instrument used in many crime laboratories is the scanning electron microscope (SEM), which uses a beam of electrons to produce images of extremely small objects. One common use of the SEM is the examination of gunshot residues obtained from people's skin or clothing.

Gunshot residues can reveal if someone has recently fired a gun. When a person has apparently committed suicide with a firearm, for example, police investigators rely on gunshot residues to determine whether that individual did indeed fire the shot.

When a gun fires, the bullet is followed by unburned gunpowder and smoke. The smoke consists of tiny particles, typically about a micron (1 millionth of a meter) or less in diameter. These particles, which are blown back onto the shooter's hand or face, take on a spherical shape from flying through the air while very hot. The SEM makes the particles' spherical form visible.

Investigators also use the SEM to examine other forms of microscopic evidence. The smallest details of hairs, fibers, pollen, sand, soil, and other tiny particles are made clearly visible with this valuable tool.

Modern investigative techniques such as these do more than simply identify the person who is likely to have committed a particular crime. They also make an essential contribution to the ultimate conviction of guilty individuals. In our system of justice, in which a person must be found guilty "beyond a reasonable doubt," the physical evidence in a case, properly analyzed and interpreted, aids the jury in making a judgment about the guilt or innocence of the defendant.

Tracking Down Suspects

In 1725, a human head was found lying near London's River Thames. Authorities at that time had only one hope of catching the murderer: They displayed the head on a pole in a churchyard and waited for someone to recognize it. Eventually, it was identified as the head of a man named John Hayes. Hayes, it was discovered, had been killed and dismembered by two men, Thomas Wood and Thomas Billings, at the urging of Hayes's wife, Catherine. All three were convicted of the crime. Wood died in prison, Billings was hanged, and Catherine Hayes was burned alive.

Today, of course, finding and capturing criminals is a much more sophisticated business than it was in the England of the 1700's. High-technology equipment, including computers, surveillance cameras, night-vision equipment, and even satellite systems, is now routinely used by police investigators to bring criminals to justice.

Computers help the police track down criminals in several ways. The international linking of computers through the Internet now enables investigators around the world to easily gather and share information on criminals, including receiving tips from private citizens. One prominent feature of the Internet, the World Wide Web (the Web), has proved especially useful, enabling law enforcement agencies to post pictures and detailed descriptions of wanted criminals. Anyone with a computer and an Internet connection can call up a law enforcement agency's Web site, view photos of fugitives, and then send any information they have to investigators by e-mail.

In 1996, this approach to law enforcement helped the Federal Bureau of Investigation (FBI) capture Leslie Isben Rogge, a bank robber on its 10-most-wanted list. The FBI Web site provided photos and details about Rogge, which prompted one Web surfer to tell the FBI that Rogge was living in Guatemala—a tip that led to his arrest.

Computers can also help in the apprehension of criminals by creating lifelike pictures of suspects. First, a witness to a crime browses through a catalog of facial features, compiled according to race, gender, size, and shape. The witness selects the features—such as the nose, hairline, and mouth—that best match those of the culprit. For each choice, an investigator keys an assigned number into the computer.

The computer, running a special software program, uses a databank of facial features to produce a rough sketch within seconds. Then the investigator and witness refine the picture by adjusting features, such as lengthening the nose or raising the hairline. After a total of about 30 minutes the picture—almost as detailed as a photograph—is ready. A separate computer program allows the police to compare the picture with ones stored in databases, including mug shots of suspects charged in the past for other crimes.

Surveillance cameras are another tool that can help nab criminals. Surveillance cameras sometimes record crimes on videotape, giving the police an image of the criminal. Such cameras are routinely used in retail stores, parking garages, and other public places to record people's com-

Members of a French police unit train with thermal-imaging goggles, which allow them to see suspects hiding in dark places at night. Thermal imaging devices use sensors to detect invisible infrared light, or "heat rays," emitted by a person's body. The infrared light is converted into electronic signals, which are used to create a video image. Night-vision equipment has become a useful police tool.

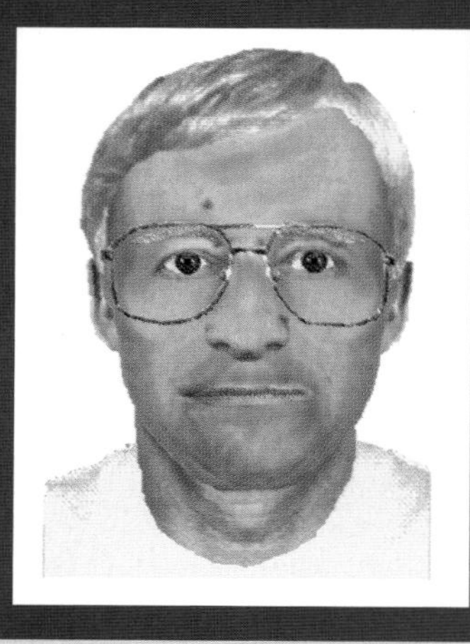

Making a sketch of a suspect by computer
Computer programs help police investigators create lifelike composite pictures of suspects from descriptions provided by witnesses to crimes. The witness looks through a catalog of facial features and chooses ones that best fit the suspect. Code numbers for those choices are keyed into the computer, which then makes an image. A computer-generated face of a man (a producer of this type of computer software), *above right,* is similar to a photo of the man, *left.* In contrast, a traditional composite sketch made with transparent overlays, *center,* is less realistic.

ings and goings, along with the time and date.

After the Oklahoma City federal building was destroyed by a bomb in April 1995, killing 168 people, investigators replayed videotape recorded by a surveillance camera at a nearby apartment complex. The tape showed a rental truck parked in front of the building just before the explosion.

Investigators traced the truck to a rental agency in Junction City, Kansas, through a vehicle identification number on the remains of one of its axles. The agency's owner identified Timothy McVeigh as the man who had rented the truck. A surveillance camera at a McDonald's restaurant in Junction City showed McVeigh leaving the restaurant shortly before he arrived at the rental agency. McVeigh was later convicted of 11 charges in connection with the bombing, including 8 counts of first-degree murder, and sentenced to death.

Satellite surveillance and thermal imaging

Surveillance with satellites is expected to play a growing role in law enforcement. One of the first applications of satellite technology in fighting crime is a tracking system that enables the police to find stolen cars and the thieves driving them. The heart of the system is a device about the size of a paperback book that is installed in a car. It includes a small computerized receiver that uses signals broadcast by satellites to calculate its position. The satellites are part of the Global Positioning System (GPS), a world-wide navigation system consisting of 24 satellites orbiting 17,500 kilometers (10,900 miles) above the Earth.

Signals from three of the satellites enable the receiver to determine its exact location, which it conveys to a monitoring center by sending a radio signal over a cellular phone in the car. The stolen vehicle's location is displayed on a screen at the monitoring center. Staff members at the center can then tell the police where to find the car.

In 1998, more than 25 U.S. police departments and law enforcement agencies were using decoy cars equipped with tracking devices to catch car thieves. Officers would simply park a decoy car in a high-theft area and wait for it to be stolen. The tracking device would then lead them to the stolen car, with the thief still at the wheel.

The system can be used in other ways as well. In December 1997, a combined city, county, and federal task force in San Antonio arrested a group of drug dealers whose cars had been secretly outfitted with tracking devices, enabling officers to follow their every move.

Some police departments now use *thermal* (heat) imaging night vision equipment to hunt down suspects hiding in the dark. Some of these devices are handheld and resemble a camcorder. Others are larger units mounted on patrol cars, boats, or helicopters.

Thermal imaging devices work by detecting invisible waves of *infrared radiation* ("heat rays") emitted by a suspect's body. These waves are translated by tiny detectors into electronic signals, which are used to create an image of the suspect on a video screen. A person crouching under bushes or hiding next to a building is easily visible with this equipment, even on very dark nights.

Criminal profiling

Often, the police not only don't know where a suspect is, they don't even know who they should be seeking. In such cases, they may put technology aside and enlist the help of a *criminal profiler,* a forensic scientist trained in psychology who can describe the type of person most likely to have committed the crime being investigated. The profiler studies the criminal's personality and behavioral traits, as revealed in his or her methods, and develops a probable description of the person.

Criminal profiling has received the greatest attention for its role in the investigation of serial murders. These crimes often stir much public fear because of their cruelty and apparently random nature. Profiling, however, is based on the premise that no one acts without motivation.

Serial killers generally plan their crimes, select their victims carefully, and attempt to control the event. Knowing this gives the profiler insight into the killer's psychological makeup. Profilers typically review the physical evidence at the crime scene, noting the position of the victim's body and the kind of wounds inflicted, along with the type of weapon used. The profiler also considers the background and habits of the victim. By analyzing all this information, the profiler may be able to suggest reasons why the crime was committed in a particular fashion.

Profilers typically do not provide the murderer's identity. But, by deducing the killer's probable motivation, a profiler may supply the police with an array of likely character and behavior attributes to guide their investigation.

Criminal profiling has had some notable successes. In 1981, for example, work by an FBI profiler led to the arrest of Wayne B. Williams, a young black man who had killed 28 people—also young blacks—in Atlanta, Georgia. The Atlanta police had thought the murderer must be a white racist. But the FBI profiler, after analyzing information from crime scenes, insisted that the police should look for a young black man who was a police buff and drove a van, a description that fit Williams perfectly.

Criminal profiling is somewhat controversial, however. Some critics maintain that while virtually all crimes are indeed driven by a motive, with severely disturbed criminals the motive may not be understood by anyone else. Also, if the profiler's description of the likely suspect is so general that it could describe many people, it may be of little value to harried police investigators.

Profiling is a tool that may or may not lead to an arrest, but investigators must use whatever tools they have at hand to solve a crime. Catching criminals is often a hit-or-miss affair, though with the aid of science and technology, the "long arm of the law" is getting longer all the time—and its grasp is increasingly sure.

Reading and Study Guide

Questions for thought and discussion:

1. Why did murderers once favor poison, and how did pioneering efforts in forensic science begin to thwart many of these criminals?
2. Discuss why fingerprinting eclipsed previous methods of identifying criminals.
3. What does an autopsy typically consist of, and how can it help in a death investigation?
4. How can teeth identify a dead body? Why is this a reliable form of identification?
5. Cite techniques investigators use to gather evidence from a crime scene.
6. How do investigators avoid contaminating a crime scene?
7. Discuss examples of evidence that can link a criminal with a particular crime.
8. Explain why the development of DNA fingerprinting has been important to forensic science.
9. In what ways do computers and the Internet help investigators in their work?
10. What, in your opinion, has been the most interesting development in forensic science?

For further reading:

DiMaio, Dominick J. and Vincent J. M. *Forensic Pathology.* CRC Press, 1993.

Fisher, Barry A. J. *Techniques of Crime Scene Investigation, 5th Edition.* CRC Press, 1993.

For additional information:

The Forensic Science Web Pages:
users.aol.com/murrk/index.html

Federal Bureau of Investigation Web page:
www.fbi.gov

Zeno's Forensic Page:
users.bart.nl/~geradts/forensic.html

SCIENCE NEWS UPDATE

Contributors report on the year's most significant developments in their respective fields. The articles in this section are arranged alphabetically.

Page 191

Page 197

Page 219

Page 283

The results of the first economic analysis of *site-specific farming,* in which farmers use satellites and computers to help them apply fertilizers more efficiently, were announced in September 1997 by agricultural economist Jess Lowenberg-DeBoer of Purdue University in West Lafayette, Indiana. Lowenberg-DeBoer reported that he and graduate student Anthony Aghib found that although site-specific farming (also known as precision farming) does not increase profits in normal growing years, it does reduce the risk of growing a poor crop when conditions are bad.

In traditional farming, a farmer generally applies an even spread of fertilizer over an entire field. However, due to varying soil conditions, different areas of a field need different amounts of fertilizer to maximize the total yield of crops. Site-specific farming allows a farmer to deliver just the right amount of fertilizer to each area.

With this method, a farmer first creates a "fertilizer demand" map showing the soil conditions of every section of the farm. The map, which is based on previous years' measurements of soil fertility, is *digitized* (converted into a series of 0's and 1's) and stored in a computer in the cab of the farmer's chemical spreader. The chemical spreader also has a radio receiver tuned to pick up signals from Earth-orbiting satellites in the Global Positioning System. The satellite signals are used by the computer to determine the precise, second-by-second location of the spreader in a crop field. As the spreader moves through the field, the computer correlates its position with the stored map to determine the exact amount of fertilizer to apply at every point.

Lowenberg-DeBoer and Aghib compared site-specific farming with traditional farming on six farms in Indiana, Ohio, and Michigan from 1993 to 1995. They examined the use of phosphorus- and potassium-based fertilizers on corn, soybean, and wheat crops in the two types of farming. The economists found that the financial return per acre from site-specific farming was normally about the same as, or even less than, that from traditional farming.

However, the economists reported, under adverse conditions such as drought or floods, farmers using site-specific techniques suffered much smaller drops in revenue than farmers relying on traditional fertilizing methods. This finding indicated that site-specific farming's targeted use of fertilizer helps stressed crops survive.

The economists added that the technology used in site-specific farming might also be useful for other agricultural purposes, including the monitoring of crop diseases.

Natural killer for corn borers. A well-known, naturally occurring fungus might help solve the problem of pesky corn borers, according to a November 1997 report by scientists with the United States Department of Agriculture (USDA). These moth caterpillars feed inside cornstalks, weakening the plants and reducing crop yields by as much as 30 percent.

The scientists, at the USDA's Corn Insects and Crop Genetic Laboratory in Ames, Iowa, found that the fungus, *Beauveria bassiana,* can live in the corn plant in a relationship that does not hurt the plant but is lethal to corn borers. The fungus kills the caterpillars by spreading throughout their bodies when ingested.

In laboratory tests, the researchers seeded corn plants with the fungus to increase the fungus's natural occurrence. They found that up to 97 percent of the corn borers in the seeded plants were killed. In field tests, corn borer mortality was somewhat lower than in the laboratory tests. Nevertheless, *Beauveria*-enhanced plants had up to 53 percent less damage from caterpillars than plants that were not protected by the fungus.

Although the USDA researchers had previously known that *Beauveria* is lethal to corn borers, they were surprised to learn that the fungus, in large concentrations, is not also deadly to corn plants. The investigators planned additional studies to try to better understand how the fungus interacts with corn.

Artificial gene helps potatoes. The development of a genetic weapon to protect potato plants against soft rot, a bacterial disease of potatoes and other crops, was reported in January 1998 by scientists at the USDA's Western Regional Research Center in Albany, California. The soft-rot bacterium can infect potatoes in the field or in storage, caus-

Agricultural scientists at the National Soil Erosion Research Laboratory at Purdue University in West Lafayette, Indiana, monitor a flume, or water trough, *below,* to obtain data on how soil is transported by flowing water. *Bottom:* Researchers at the lab observe runoff patterns from simulated rain that has been dyed green to make it more visible. These studies led to the development of a computer model that other agricultural scientists began using in late 1997 to better understand and prevent soil erosion on farms.

ing them to become soft, slimy, and foul-smelling.

The USDA researchers, led by plant physiologist William Belknap, developed a synthetic gene that carries instructions for producing a protein that makes a plant resistant to soft rot. They then combined this gene with a special "promoter" gene, extracted from potatoes, that turns the synthetic gene on when it is inside a potato plant.

Finally, the scientists inserted the combined gene, called a transgene, into a bacterium that was allowed to infect potatoes. The type of bacterium used by the scientists has the natural ability to combine its genes with those of plants. (However, the particular bacterial strain that was used had been stripped of its disease-causing genes.) Once inside the potatoes, the bacterium inserted the transgene into the potatoes' genetic material. The potatoes then began to produce the protective protein.

The scientists reported that the genetically altered potatoes successfully resisted soft rot in the laboratory and greenhouse. And preliminary results from a field trial in the fall of 1997 in Idaho also indicated that the technique is effective. However, the researchers said that results from the 1998 growing season were needed before the method could be judged a success.

"Diet pill" for crop pests? An insect "diet pill" that might lead to better control of agricultural pests was reported by biologist Dov Borovsky of the University of Florida's Institute of Food and Agricultural Sciences in December 1997.

Borovsky designed the "pill," consisting of a mosquito digestive-control hormone inside cells of a certain type of algae, to destroy malaria-carrying mosquitoes. Borovsky showed that when mosquito *larvae* (immature insects) eat the algae containing the hormone, their digestion is blocked and they starve to death. Although Borovsky began his research using a natural hormone obtained from mosquitoes, he was later able to synthesize the hormone.

In 1998, Borovsky was conducting research on other insect digestive hormones that, when placed inside crops, would kill pests that feed on them. He said that such a product could be both safer and more effective than traditional chemical pesticides. [Steve Cain]

In July 1997, a team of German and American scientists made the spectacular announcement that they had recovered genetic material from an arm bone of the original "Neanderthal man" fossils, which were discovered in Germany in 1856. The team of scientists, led by geneticist Svante Paabo of the University of Munich, concluded from their analysis that Neanderthals were not ancestors of modern humans.

The work of Paabo's team supported a theory held by many anthropologists that Neanderthals were a side branch of human evolution that contributed little or nothing to the genetic makeup of modern populations. Although most scientists agreed that the results of the study were valid, some researchers cautioned against drawing conclusions about an entire population based on genetic evidence from one specimen. Nonetheless, the success of retrieving fossil genetic material was expected to spur efforts to continue such research.

A better look at australopithecines. The discovery of the most complete skull ever found of the extinct species *Australopithecus boisei* was reported in October 1997 by a team of researchers from Japan, Ethiopia, and the United States. The skull, which was found in Konso, Ethiopia, was one of nine *A. boisei* fossils that the scientists—led by paleontologist Gen Suwa of the University of Tokyo—had found in 1.4-million-year-old *strata* (rock layers).

A. boisei is one of the early *hominids* (members of the human lineage) known as australopithecines. Like other hominids, australopithecines were *bipedal*—that is, they walked on two legs.

But australopithecines differ from the genus *Homo* (which includes modern humans) in a number of important ways. They had relatively small braincases, large projecting faces with apelike noses, and large back teeth, usually with thick enamel. Moreover, their limb bones indicate that they had an apelike ability to climb. They were also relatively small, with males weighing about 40 to 45 kilograms (90 to 100 pounds), and females weighing about 30 to 35 kilograms (65 to 80 pounds).

The criteria by which australopithecines are classified as one species or another are still controversial. Most authorities agree, however, that there were at least six different kinds of australopithecines, all of which lived only in Africa. *A. boisei* (also known as *Paranthropus boisei*) lived from roughly 2.3 million to 1.2 million years ago. *A. boisei* and its close relative, *A. robustus* (or *Paranthropus robustus*) are often called "robust" australopithecines because of their large teeth and jaws. These two species had enormous molars and premolars, bigger than those of the largest living male gorillas. The powerful chewing apparatus of the robust australopithecines suggests that these hominids ate a diet consisting mostly of tough, fibrous plants.

The relationship between the robust australopithecines and other hominids is not fully understood, but most anthropologists agree that *A. boisei* and *A. robustus* represent a highly specialized branch of the human evolutionary tree, one that died out without descendants. These two hominid species apparently shared a common ancestor with other australopithecines but were not direct ancestors of modern humans.

The *A. boisei* specimens, discovered in south-central Ethiopia, were important for several reasons. First, the fossils extended the known geographic range of the species in the Great Rift Valley of eastern Africa farther to the northeast. Second, the find was the first to include both a *cranium* (the upper portion of the skull) and *mandible* (lower jaw) from a single individual. This enabled the scientists to study how the two pieces fit together and functioned.

Finally, Suwa's team noted that the fossils demonstrate the range of anatomical variability within *A. boisei* populations. The cranium has many characteristics that place it firmly within the species *A. boisei,* but the cheekbone looks more like that of *A. robustus* and the crest of the skull more like that of yet another australopithecine, *A. aethiopicus.* The *palate* (roof of the mouth) actually resembles that of early *Homo.* The researchers cautioned that scientists are sometimes too quick to name a new species, often basing their arguments on just one specimen. They argued that scientists must use restraint when considering whether to name a new species, particularly if the fossil sample is small.

Seafaring *Homo erectus*? Stone tools at Mata Menge, a site on the eastern Indonesian island of Flores, may be as

South African researchers make a cast of footprints left by an anatomically modern human 117,000 years ago. In August 1997, the scientists reported the discovery of the prints at Langebaan Lagoon on the west coast of South Africa. Many anthropologists agreed that the discovery supported the theory that modern humans originated in Africa.

much as 800,000 years old, according to a March 1998 report by a team of Australian researchers. The scientists—led by paleoanthropologist Mike Morwood of the University of New England in Armidale, Australia—argued that 14 tools discovered in 1994 provided evidence that *Homo erectus* populations lived on the island and had traveled there by boat. Those conclusions challenged general assumptions about what this species was like.

H. erectus was an early ancestor of *H. sapiens,* the species to which modern humans belong. There is abundant evidence that *H. erectus* populations lived on other Indonesian islands west of Flores. Those populations, according to most anthropologists, reached the western islands by crossing land bridges when ocean levels were low. The presence of tools at Mata Menge, however, could not be explained so easily. Even at the ocean's lowest levels, Flores was separated from other land by at least 19 kilometers (12 miles) of water.

Morwood and his colleagues argued that some *H. erectus* populations were seafarers. But traveling by boat would have required a certain degree of cooperation, planning, and language skills—traits not usually associated with *H. erectus.* The researchers' conclusions, therefore, promised to be controversial. Although some scientists found the research valid, others argued that the artifacts were not tools but simply shattered pieces of rock and, therefore, that the dates of the specimens are irrelevant.

Ancient footprints. The discovery of 117,000-year-old footprints made by an anatomically modern human—the oldest such prints known—was reported in August 1997 by paleoanthropologist Lee Berger of the University of the Witwatersrand in Johannesburg, South Africa. The pair of footprints—found in 1995 by geologist David Roberts of the Council for Geoscience in Bellville, South Africa—are preserved in rock at Langebaan Lagoon, about 100 kilometers (60 miles) north of Cape Town.

The rock, which the scientists dated using several techniques, had originally been soft sand. The researchers determined that the person who left the

The discovery of the most complete skull ever found of *Australopithecus boisei,* an early prehuman species, was announced in October 1997 by an international team of researchers. The specimen, found in Ethiopia, allowed anthropologists to examine for the first time how an *A. boisei* skull and lower jaw fit together and functioned. The specimen was also important because it had features that resembled those of other australopithecines that have been classified as separate species. The skull indicated that some physical features that scientists had thought belonged to different species may simply represent variations within a single species of *Australopithecus.*

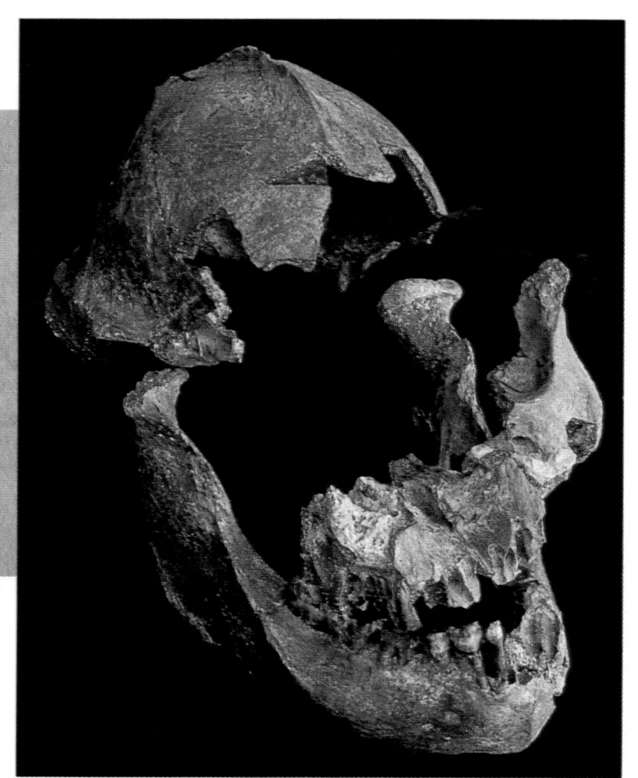

prints in the sand was probably a woman or a small man, judging by the size of the prints and the apparent length of the individual's stride.

Although the origin of modern humans is a controversial topic among anthropologists, many believe that Africa was humanity's cradle. Another site in South Africa, the Klasies River Mouth, has yielded anatomically modern human fossils from about the same time as the Langebaan Lagoon footprints. Roberts and Berger argued that the footprints confirmed the presence of *H. sapiens* in Africa at that time.

Controversial remains. In April 1998, the U.S. Army Corps of Engineers buried a site on the Columbia River where a 9,300-year-old skeleton was discovered in 1996. The action was the latest episode in an ongoing debate among scientists, Native Americans, a Norse pagan sect, and the U.S. government over the remains of "Kennewick Man," the oldest skeleton ever found in the Pacific Northwest and one of the few examples of skeletons of this age in the Americas.

Scientists are interested in studying Kennewick Man because it could contribute to their understanding of the first inhabitants of the Americas. The researchers who initially examined the skeleton reported that the facial features appeared to be Caucasoid. To arrive at more concrete answers about the skeleton, scientists would like to conduct genetic tests on the remains and investigate the Columbia River site.

The right to conduct research, however, remained tied up in court in mid-1998. The Native American Graves Protection and Repatriation Act of 1990 requires that human remains and cultural objects be given to culturally affiliated tribes. Kennewick Man originally was to be turned over to the Confederated Tribes of the Umatilla Indian Reservation. The Asatru Folk Assembly, the pagan sect, also claimed rights to the remains, arguing that the man had descended from Norse settlers. As of June 1998, legal rights to the remains had not been settled. [Kathryn Cruz-Uribe]

See also ARCHAEOLOGY. In the Special Reports section, see RETHINKING THE HUMAN FAMILY TREE.

The discovery of a previously unknown temple and other ruins in the ancient city of Angkor in northern Cambodia was announced in February 1998 by archaeologist Elizabeth Moore of the University of London. Moore's important find was made possible by a radar system developed by the Jet Propulsion Laboratory in Pasadena, California, a branch of the National Aeronautics and Space Administration (NASA).

Angkor, first discovered more than 200 years ago by French explorers, was the capital of the ancient Khmer empire, which ruled from the 800's to the 1400's. The city, with an estimated population of 1 million people by the 1100's, covered about 160 square kilometers (60 square miles) and included about 1,000 Hindu and Buddhist temples. Angkor Wat, the biggest temple, which was built in the 1100's, is one of the largest religious structures in the world. Khmer engineers also designed a complex network of canals and reservoirs to divert water during monsoon rains and store it for use during the annual dry season.

Although parts of Angkor have been studied thoroughly, dense jungle vegetation still covers much of the ancient city. NASA researchers overcame this obstacle by mapping the area from an airplane with an Airborne Synthetic Aperture Radar (AIRSAR). The AIRSAR system uses a technique called radar interferometry, which combines two radar images to create a three-dimensional map of the ground.

The maps produced by NASA provided new details about known features of the city. For example, the maps revealed that the water-management features at Angkor were much more extensive than had been suspected. The maps also helped archaeologists find several undocumented sites, including a mound outside the Angkor Wat area.

Exploring this mound in December 1997, Moore found that it included two temples that had been written about in the early 1900's and the remains of at least two and possibly four previously unknown temples. The largest of the newly found temples is about the size of a football field and, according to

An aerial radar image, released in February 1998 by the National Aeronautics and Space Administration, shows the ancient Cambodian city of Angkor, much of which is covered by dense jungle. The city's main temple, Angkor Wat, appears as a small square near the center of the image. The large dark square is a moat. As well as showing these well-known features, the radar revealed a mound (circled) which led archaeologists to discover previously unknown temples. The image also provided more details about Angkor's complex system of canals, identified as blue lines.

Roman treasures under the sea

Clay containers, *below*—which ancient Romans used to transport olive oil, wine, and other food products—were discovered in the wreckage of ships on the floor of the Mediterranean Sea. The discoveries were reported in July 1997 by oceanographer Robert D. Ballard.

Members of Ballard's crew inspect pieces of marble columns, *above,* retrieved from one of the shipwrecks. The researchers theorized that the columns were created as part of a prefabricated temple that was to be assembled after the vessel carrying the stonework reached its destination. The ships were discovered off the northwest coast of Sicily, *below.* A total of eight ships, including three from more recent times, were found at depths of about 760 meters (2,500 feet).

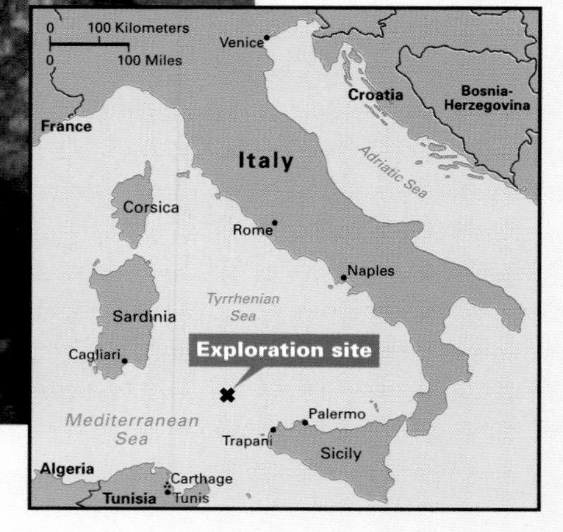

Moore's estimates, about 300 years older than Angkor Wat. Other sites identified on the radar maps proved to be even older. These findings and further explorations were expected to alter the chronology of the area's ancient history and provide more evidence about the complex Khmer society.

Sacred caves of the Maya. Studies of vast caves in Mexico and Central America suggest that they were a vital part of religious life and a source of political power for ancient Maya societies. Those interpretations were proposed in November 1997 at the annual meeting of the American Anthropological Association by researchers, led by archaeologist James E. Brady of George Washington University in Washington, D.C.

Maya civilization began about 800 B.C. and lasted until the Spanish conquests in the mid-1500's. Archaeologists have known for decades that caves, located in southern Belize, northern Guatemala, and the Yucatan Peninsula of Mexico, were used by the ancient Maya. Most scientists, however, believed that the caves had served primarily as temporary dwellings. Brady's work, the first long-term study of the caves, found evidence that caves played a much more significant role in the people's religious beliefs. Indeed, those conclusions are supported by the fact that many modern Maya groups continue to make religious pilgrimages to the mouths of the caves.

Brady began studying several caves in the late 1980's. The abundance of rock crystals that he found there provided a clue to the spiritual significance of the caves, as the modern Maya still use crystals in healing ceremonies or in rituals designed to read the future. Further evidence of religious rituals was discovered by archaeologist Keith Prufer of Southern Illinois University in Carbondale. He found small chambers in caves that contained benches, which are identified in Maya *hieroglyphics* (picture words) and historical records as objects that belonged to *shamans,* religious leaders who communicated with the spirit world.

Hieroglyphics on the walls of Naj Tunich, a cave in Guatemala, recorded pilgrimages to the site from people who lived about 65 kilometers (40 miles) away. The pottery found in Naj Tunich was considered by the researchers to be ritual offerings.

Brady also joined forces with the Petexbatun Regional Archaeological Project, a research group directed by archaeologist Arthur Demarest of Vanderbilt University in Nashville, Tennessee. They studied a system of 22 interlinked caves and passages that cover more than 11 kilometers (7 miles) below Dos Pilas, a Maya city in Guatemala.

After mapping the caves and tunnels, Brady and Demarest concluded that these caves played a role in the design of major architectural projects during the Classic Period (A.D. 250 to 900). The huge El Duende pyramid, which dominates Dos Pilas, was built over an enormous cave that had been carved out by an underground river. A large royal structure, the Bat Palace, was constructed over the opening of another cave that was linked by a passageway to the pyramid's cave. Brady and his colleagues interpreted the placement of the buildings as a "sacred geographic link" between the political leaders and the sources of their power in sacred underground water and the spirit world within the Earth.

The world's first city revisited. Research at the 9,000-year-old city of Catalhoyuk in central Turkey, led by archaeologist Ian Hodder of the University of London, yielded new insights into the religious beliefs and daily lives of the inhabitants of what may have been the world's first city. The results of the field work begun in 1993 and the new interpretations were reported in March 1998 by Orrin C. Shane III, the archaeology curator at the Science Museum of Minnesota in St. Paul, and Turkish archaeologist Mine Kucuk—both members of the Catalhoyuk research team.

Catalhoyuk (*CHA tahl hu yook*), located on both sides of an ancient riverbed, dates back to the Neolithic age, the period in which farming and settled life began to dominate human cultures. Portions of the city were first excavated between 1961 and 1965 by British archaeologist James Mellaart. He estimated that the city housed several thousand inhabitants, who raised domestic sheep and goats, planted crops, hunted wild cattle, and gathered marsh plants. The people made excellent pottery, flint sickles and knives, and mirrors made of polished *obsidian* (black volcanic glass).

Mellaart excavated 150 buildings,

which accounted for an estimated 4 percent of Catalhoyuk East, the part of the city on the eastern side of the riverbed. The sun-dried brick buildings were tightly clustered and had to be entered and exited by ladders through openings in their flat roofs. Some of the buildings were richly decorated with heads and horns of wild bulls, oxen, and rams—either the remains of the animals, plaster-covered remains, or plaster works of art. Murals on the walls depicted scenes of both life and death—such as women giving birth and vultures pecking at headless corpses. These rooms also included female figurines, which Mellaart believed were related to the fertility cult of the "Mother Goddess." He concluded that these buildings were a part of a special "priestly quarter" in the city.

Mellaart's work made significant contributions to the understanding of early settled life, but the work was canceled after a dispute with the Turkish government. Field work did not resume until Hodder began excavating the site in collaboration with the British Institute for Archaeology in Ankara, Turkey.

This work included reexcavating a room known as Mellaart's Building 1. After clearing the room, Hodder and his colleagues excavated beneath two raised platforms and the floor of the building. There, they found 67 human skeletons, representing people of all ages, from infancy to more than 60 years old. The graves beneath the dwelling were apparently opened repeatedly to receive the dead. The researchers believe that this burial custom may have been a tradition of ancestor worship.

Excavation in 1997 in Building 3 uncovered a room decorated with red and black paintings and bulls' heads, another example of Mellaart's "shrines." New excavations in other parts of Catalhoyuk, however, revealed several other "shrine" buildings scattered across the site. Additionally, microscopic evaluations of materials on the rooms' floors revealed that people cooked in the same room, or at least in the same building, that they worshiped in. Material from a space between two buildings, which Mellaart had concluded was a courtyard in the priestly quarter, was also subjected to microscopic analysis. The debris on the floor was found to contain animal dung and straw, indicating that the courtyard was most likely a stable.

Based on these findings, Hodder's team concluded that the buildings did indeed hold religious significance for the people. However, they disagreed with Mellaart's conclusion of a priestly class. Instead of a central authority controlling the religious practices of the people, religious ceremonies were apparently conducted in all of the homes.

The conclusions challenged some basic theories about how early cities were organized and ruled. Archaeologists had assumed that a large number of people could live together only when they were ruled by a central authority. The evidence at Catalhoyuk, however, suggests that this early city was organized more simply with no clear power structure.

The scientists working at Catalhoyuk planned to continue the project for 20 years in order to conduct further excavations, preserve the site's artifacts, and make the site accessible to visitors. The ambitious effort was expected to include additional researchers from the University of California at Berkeley, the Natural History Museum of London, the University of Thessaloniki in Greece, and numerous Turkish institutions.

Earthen mounds. In September 1997, a research team led by archaeologist Joe W. Saunders at Northeast Louisiana University in Monroe reported that a group of 11 earthen mounds at a site near the Ouachita River in northeastern Louisiana were constructed by Native Americans about 5,400 years ago. The mounds, at a site called Watson Brake, are among the oldest structures created by people in the Americas.

Mounds also exist at other locations in North America, but researchers have not yet determined their purpose. Some have suggested that they served as religious sites, and others believe they may only have been platforms for dwellings.

The Watson Brake mounds, first discovered in 1981 by an amateur archaeologist, Reba Jones, are arranged in an oval with a maximum diameter of about 400 meters (1,300 feet). The largest mound is 7.5 meters (25 feet) high, while the heights of the others range from 1 meter (3 feet) to 4.5 meters (15 feet). Ridges, averaging 1 meter in height, link all of the mounds together.

Saunders began studying the site in 1993, first carrying out soil-coring to

Unearthing a Roman mansion

Italian archaeologists excavate the Villa dei Papiri, *below,* a mansion in the ancient Roman city of Herculaneum that was buried by the eruption of Mount Vesuvius in A.D. 79. The first glimpses of the large complex appeared in reports published in June 1997.

Limited excavations in the 1700's revealed the approximate design of the Villa dei Papiri, *below.* Those digs also uncovered valuable papyrus rolls, which gave the mansion its name. The Italian team hoped to uncover more of these ancient manuscripts and perhaps a complete library. The modern excavations have revealed most of an area called the atrium and the rooms that surround it. A worker removes debris from one of the recently excavated mosaic floors, *bottom.*

sample the depth and composition of the mounds. The initial studies were essential for proving that the mounds were human constructions and not natural features of the landscape.

During this coring, he hit a *midden,* (trash accumulation) of fish and mammal bones. *Radiocarbon dating* of the materials indicated that construction of the mounds began about 5,300 to 5,400 years ago. (Radiocarbon dating involves measuring the amount of naturally occurring radioactive carbon 14 remaining in substances that were once alive to determine how long ago they died and stopped absorbing the element.)

The dates proved to be a major surprise, because nearby mounds at Poverty Point, Louisiana, which were believed to be the oldest ones in the region, were built about 3,500 years ago. Additional core samples and test excavations at Watson Brake—conducted by specialists in the fields of geology, soil sciences, botany, and advanced dating technology—revealed new insights into the early Native Americans who constructed the cluster of mounds.

First, the team used other dating methods to confirm the results of the radiocarbon tests. Dating specialist James K. Feathers of the University of Washington in Seattle tested soil sediments with a technique known as *thermoluminescence dating.* When objects that were heated in the past, such as the soil around a campfire, are reheated, they emit light. The longer the interval since the object was first heated, the stronger the light. Feathers's tests indicated an age of about 5,540 years for the site. An experimental dating method called the *oxidizable carbon ratio* (OCR) test was also employed. OCR measures the degradation, through natural biochemical processes, of carbon-based molecules in the soil. This test yielded an only slightly later date for the beginning of Watson Brake—5,000 to 5,100 years ago.

The research team also determined, judging by the objects recovered from middens, that the builders of Watson Brake were hunters and gatherers. Various species of fish and mussels appeared to be the predominant food source, but the middens also included the remains of numerous land animals. The scientists determined that the fish were most likely caught from spring to early fall, based on analysis of *otoliths,* pieces of cartilage in the ear that grow seasonally and reveal the age of fish. The researchers also recovered the charred remains of wild plants: goosefoot, knotweed, and marsh elder. The evidence from the fish and the seasonal availability of the plants suggested that the mound builders occupied the site seasonally.

Studies of the changes in the kind and amount of plants and animals in the middens at different levels of the mounds suggested that the people's diets changed over time. A decline in the number of fish and mussels in the upper middens led the scientists to conclude that the environment had changed over time. Those changes may even have led to the abandonment of the site.

The research at Watson Brake raised a significant question about the mystery of earthen mounds. Such massive and complex construction would most likely depend on organized labor under a strong leader and on a food supply capable of supporting numerous workers. This kind of social organization, however, is normally associated with settled agricultural peoples, not hunters and gatherers. The scientists hoped that further research at Watson Brake would help them address the issue.

Ancient shipwrecks. The discovery of eight sunken ships, including five ancient Roman vessels, was reported in July 1997 by a team of researchers led by oceanographer Robert D. Ballard, president of the Institute for Exploration in Mystic, Connecticut. The ships were discovered 760 meters (2,500 feet) below the surface of the Mediterranean Sea off the northwest coast of Sicily.

The researchers used a remote-controlled underwater vehicle, called *Jason,* to photograph the remains of the ships and collect artifacts. *Jason* retrieved 115 items from the Roman ships, which were dated from about 100 B.C. to A.D. 400. The artifacts included bronze vessels, anchors, and *amphoras*—clay containers used to transport olive oil, wine, and various preserved foods. The researchers also collected sections of marble columns that were apparently part of a prefabricated temple. The team identified the three other wrecks as vessels from the late 1700's to the early 1800's. [Thomas R. Hester]

Scientists reported in March 1998 the high probability that there is water on the moon. The researchers, led by astronomer Alan B. Binder of the Lunar Research Institute in Gilroy, California, reached that conclusion after analyzing data beamed to Earth by Lunar Prospector. The moon-orbiting, robotic spacecraft was launched in January 1998 by the U.S. National Aeronautics and Space Administration (NASA).

Some scientists had reported probable water on the moon even earlier. In 1996, radar studies of the lunar surface by the U.S. Department of Defense's Clementine spacecraft seemed to show water at the moon's south pole. However, a study using Earth-based radars contradicted that finding. Lunar Prospector was sent, in part, to resolve the matter.

Binder and his team used Lunar Prospector's instruments to measure the speed of *neutrons* (uncharged particles in an atom's nucleus) dislodged from the lunar surface by *cosmic rays* (high-energy particles from space). Neutrons slow down when they strike hydrogen atoms, and since each water molecule contains two hydrogen atoms, a neutron slowdown could indicate the presence of water. As Lunar Prospector passed over the moon's poles, its instruments recorded a slowdown in neutrons.

The researchers estimated that as much as 300 million metric tons (330 million tons) of water in the form of ice crystals mixed with dirt may lie at the lunar poles, away from the sun's heat. The team suggested that the water may have come from comets, which have bombarded the moon over billions of years.

Some scientists thought the study results offered hopes for future lunar colonies, which presumably could satisfy their water needs from the lunar soil. Others cautioned that other elements on the lunar surface may produce an effect similar to that caused by hydrogen. Nevertheless, Lunar Prospector's findings sparked international interest in additional studies of the moon.

Reassessing a Martian meteorite. Two research teams published reports in January 1998 that cast doubt on the claim that a meteorite from Mars contains evidence of past life on the red

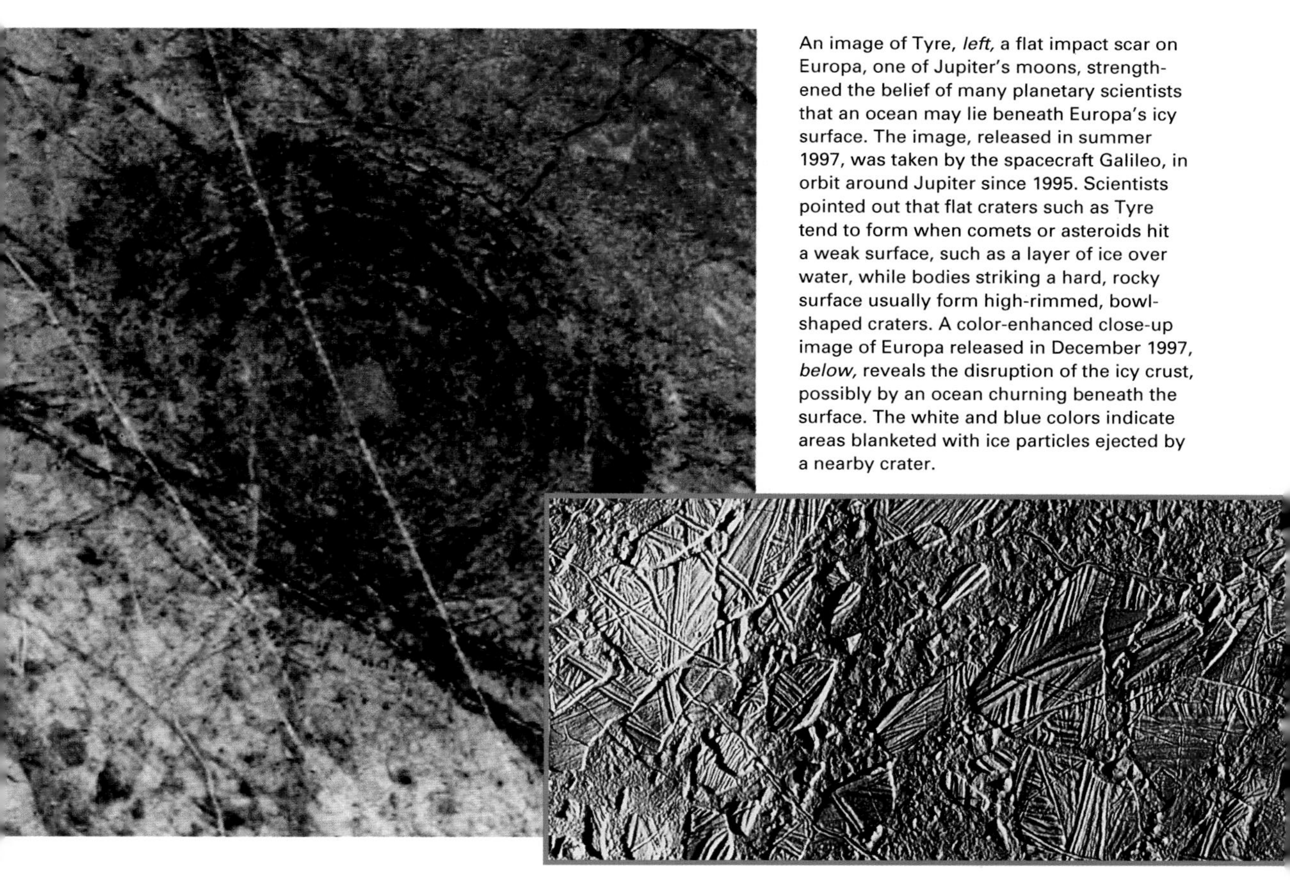

An image of Tyre, *left,* a flat impact scar on Europa, one of Jupiter's moons, strengthened the belief of many planetary scientists that an ocean may lie beneath Europa's icy surface. The image, released in summer 1997, was taken by the spacecraft Galileo, in orbit around Jupiter since 1995. Scientists pointed out that flat craters such as Tyre tend to form when comets or asteroids hit a weak surface, such as a layer of ice over water, while bodies striking a hard, rocky surface usually form high-rimmed, bowl-shaped craters. A color-enhanced close-up image of Europa released in December 1997, *below,* reveals the disruption of the icy crust, possibly by an ocean churning beneath the surface. The white and blue colors indicate areas blanketed with ice particles ejected by a nearby crater.

planet. The meteorite—ALH84001—was found in the Allan Hills of Antarctica in 1984. It is believed to have originated on Mars because it contains gases resembling those in the Martian atmosphere. Researchers think that the meteorite was blasted into space millions of years ago when an asteroid struck and shattered the rocky surface of Mars.

In 1996, geologist David S. McKay of NASA's Johnson Space Center in Houston and his colleagues had reported that chemical and physical clues in the meteorite could be signs of biological activity. For example, ALH84001 contains *organic* (carbon-containing) molecules. Because all known life is based on carbon, this evidence suggested to McKay's team that the meteorite, formed about 4.5 billion years ago, could have harbored primitive life forms.

Not so, responded research teams led by geochemists A. J. Timothy Jull of the University of Arizona in Tucson and Jeffrey L. Bada of the University of California at San Diego. Both teams analyzed the organic materials in ALH84001 and found them to be mostly contaminants from the Antarctic soil in which the meteorite lay for about 10,000 years.

Jull and his colleagues measured the *isotopes* (forms) of carbon in the meteorite. They found that much of the organic material in the rock contained a relatively high portion of carbon 14—an isotope made in Earth's atmosphere by cosmic rays that gets incorporated into organic molecules. Carbon 14 *decays* (breaks down) into nitrogen at a regular rate. The very fact that carbon 14 was found in the meteorite pointed to a terrestrial origin for the organic material in the rock. Moreover, the amount of carbon 14 relative to other carbon isotopes suggested an age of less than 8,000 years for most of the meteorite's organic content. Thus, the material must have been picked up on Earth.

Bada's group examined a particular kind of organic molecule in the meteorite, amino acids. Amino acids are the building blocks of proteins, upon which all life on Earth depends. The team found that the structures and quantities of the amino acids in ALH84001 were very similar to those extracted from the

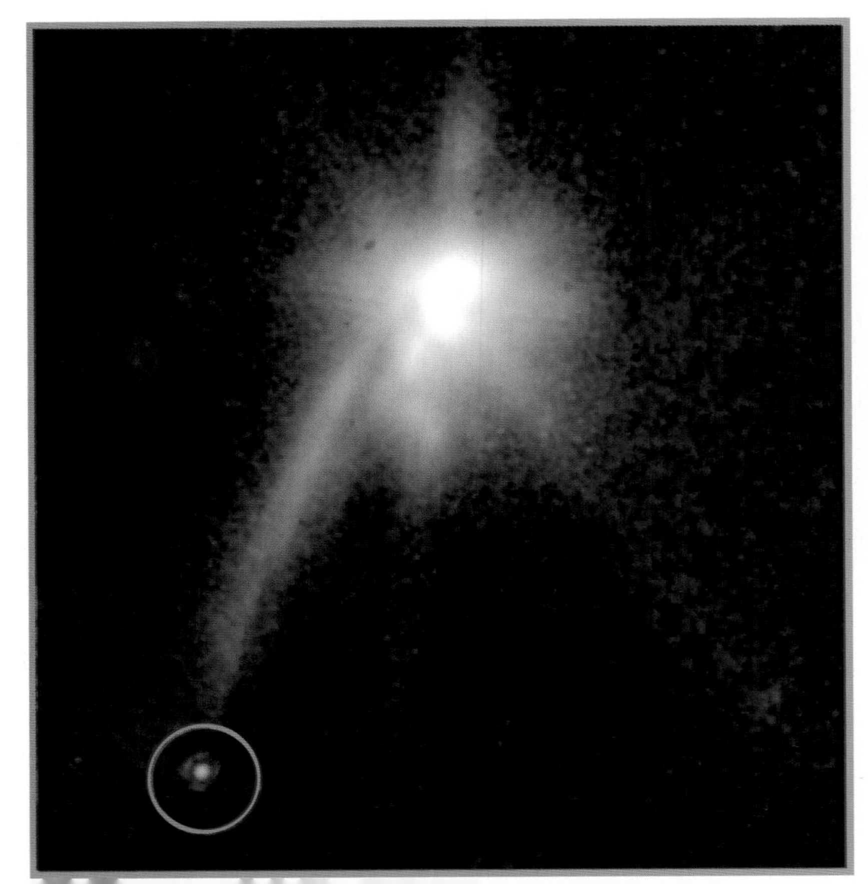

A Hubble Space Telescope image of an object (circled) near a double-star system not far from the center of the Milky Way Galaxy may be a large planet. If so, the object—reported in May 1998 by astronomer Susan Terebey and her colleagues at the Extrasolar Research Corporation in Pasadena, California—became the first planet outside the solar system ever observed directly by astronomers. Terebey speculated that the body is a large, gaseous planet with more than twice the mass of Jupiter that was ejected from an orbit around one of the stars by gravitational effects generated by the two-star system.

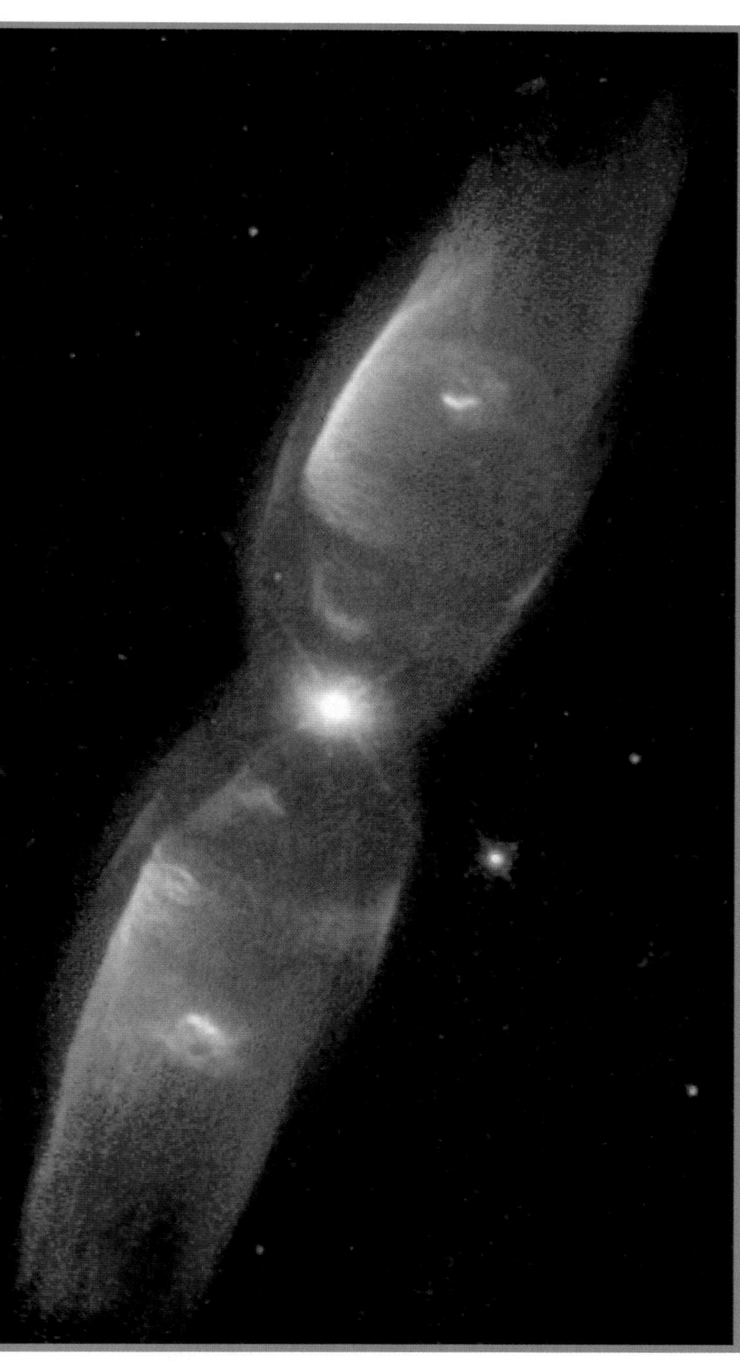

A star called Minkowski 2-9, located 2,100 *light-years* from Earth, glows in its death throes at the center of a Hubble Space Telescope (HST) image released in December 1997. (A light-year is the distance light travels in one year, about 9.5 trillion kilometers [5.9 trillion miles].) The HST caught the star in a transition phase. From a bright, cool red giant, Minkowski 2-9 changed to a butterfly-winged structure known as a planetary nebula, shooting its outer layers of gas into space. The star will eventually become a small, fading star called a white dwarf.

glacial ice in the region of Antarctica surrounding the meteorite. This finding suggested that the meteorite picked up the amino acids on Earth.

The work of neither group, however, completely ruled out the possibility that some of ALH84001's organic molecules came from Mars. In fact, the results of Jull and his team suggested that a small fraction of the organic molecules in the rock were not of Earthly origin.

New moons at Uranus. The discovery of the 16th and 17th moons known to be orbiting Uranus was reported in September 1997 by astronomers using the Hale Telescope at the Mount Palomar Observatory in California. The two moons, named Caliban and Sycorax, were the dimmest such objects ever seen with a ground-based telescope.

The astronomers, led by Philip Nicholson of Cornell University in Ithaca, New York, reported that one moon is a mere 80 kilometers (50 miles) in diameter and the other is about twice that size. Unlike Uranus's 15 other moons—which have nearly circular orbits—the new moons have highly *elliptical* (oval) orbits that are tilted relative to Uranus's equator. Because of their unusual orbits, the moons are called "irregular." The new moons are also farther from Uranus than the planet's other known moons—about 6 million to 8 million kilometers (4 million to 5 million miles) from the planet. Uranus's other moons orbit at distances of 50,000 to 584,000 kilometers (30,000 to 362,000 miles).

The astronomers speculated that Uranus's newly discovered moons may be space debris captured by the planet's gravity. Uranus's previously known moons are thought to have developed as the planet formed. Nicholson and his team noted that Jupiter and Saturn also have both irregular and regular moons.

Organics on Jupiter's moons. Researchers in the United States reported in October 1997 that they had detected organic molecules on the surfaces of Jupiter's two largest moons, Ganymede and Callisto. The scientists' discovery was based on data transmitted to Earth by NASA's Galileo spacecraft, in orbit around Jupiter since December 1995.

A team led by planetary scientist Tom B. McCord of the University of Hawaii in Honolulu analyzed data from a Galileo instrument called the Near Infrared

Mapping Spectrometer (NIMS). The researchers used NIMS to detect sunlight reflecting from the surfaces of Ganymede and Callisto and to break up the light into its various wavelengths. They then examined the light to see which wavelengths were absorbed by molecules on the surfaces of the moons. NIMS focuses on wavelengths in the *infrared*—light beyond the red end of the visible spectrum—because those wavelengths are absorbed by molecules that interest planetary scientists.

When they compared the absorption pattern that NIMS detected on Ganymede and Callisto with absorption patterns established by laboratory experiments, the researchers identified carbon dioxide and hydrogen-bearing carbon compounds on the moons. McCord and his colleagues theorized that the organic molecules may have been brought to the moons by comets that have crashed into them from time to time over billions of years. The researchers suggested that the presence of organic molecules on the moons hints at the way Earth may have been supplied with such molecules early in its history.

Kuiper Belt discovery. Astronomers reported in June 1997 that they had found the most distant known object in the Kuiper Belt. The Kuiper Belt (pronounced *KOY per*) is a region just beyond the orbit of Neptune that contains some of the comets that occasionally enter the inner region of the solar system.

The new object—1996 TL66—was found by a team led by astronomer Jane Luu of the Harvard-Smithsonian Center for Astrophysics in Cambridge, Massachusetts. The object is about 500 kilometers (300 miles) wide—less than one-fourth the diameter of Pluto.

Astronomers were most intrigued by the fact that the newly found object's orbit around the sun was more elliptical than that of other planets and Kuiper Belt objects and strongly tilted relative to the sun's equator. When it was discovered, 1996 TL66 was at its closest point to the sun, about 40 *astronomical units* (AU—the distance from the Earth to the sun). However, further observations revealed that its average distance from the sun is 84 AU. Pluto, which many astronomers consider the innermost member of the Kuiper Belt, never gets more than 50 AU from the sun.

Why is 1996 TL66's orbit so elliptical? Luu and colleagues suggested that the new find may be the first direct evidence of a new class of undiscovered objects orbiting the sun between the Kuiper Belt and the Oort Cloud, a huge source of comets even farther beyond the orbits of the planets. The scientists proposed that these objects may be remnants from the early days of planet formation. This icy debris, according to the researchers, could have seeded Earth with large amounts of water and organic molecules to form the starting point for our oceans and life.

Mysterious bands in space. Swiss astronomers announced in April 1998 that they may have identified one of the molecules that form spectral bands in *interstellar* (between the stars) space. Scientists had been puzzled by the bands since 1921, when astronomer Mary Lea Heger of the Lick Observatory at Mt. Hamilton, California, discovered them.

Heger noted features in the light spectra of several stars that were not created by the stars themselves but instead must have resulted as the starlight passed through gas or dust clouds. By the mid-1990's, astronomers had discovered nearly 200 of the unidentified diffuse interstellar bands, as the features discovered by Heger came to be called, but they could not identify the particles that formed them.

When light from a source such as a star is analyzed with an instrument called a spectrometer, each type of atom or molecule in the source, or in the space between the source and the observer, can be identified. The atoms and molecules announce their presence by the unique wavelengths at which they either emit or absorb light. Hot matter, such as that in a star or surrounding a star, emits light. This light shows up as thin bright streaks, called emission lines, in the star's spectrum at the characteristic wavelengths of the atoms and molecules present in the hot matter. Cold matter, which is found in the broad expanses of space between stars, absorbs light, causing dark streaks—absorption lines—to appear in the spectrum.

Astronomers suspected that the diffuse interstellar absorption bands might be caused by molecules, rather than atoms. That seemed like a good bet because molecules generally produce

A massive star dubbed the Pistol Star, surrounded by a huge cloud of gas, blazes at 2 million to 15 million times the power of our sun in this Hubble Space Telescope image released in October 1997. Astronomers believe that the star, which is located near the center of the Milky Way Galaxy, may be one of the brightest stars in the galaxy.

more spectral lines than atoms, because there are many more kinds of molecules than atoms, and because all the lines formed by atoms were already known. But what kinds of molecules might they be? Most astronomers theorized that the molecules must be organic, because carbon is one of the most abundant elements in the universe and can form many types of chemical bonds.

To identify the molecules making up the diffuse bands, astronomers isolated molecules in the laboratory, looking for one that creates absorption lines at precisely the same wavelengths as those observed in the interstellar medium. Finally, Swiss researchers at the University of Basel led by chemist John P. Maier reported that they had found such a molecule, called C7–. C7– is a chain of seven carbon atoms with one extra electron, giving the molecule a negative electrical charge. C7– produced spectral lines in the laboratory that precisely matched those of five of the interstellar bands.

Thus, Maier and his group appeared to have made the first positive identification of an interstellar band. In the process, astronomers learned more about the large organic molecules in space that may have been the seeds of life on Earth—and perhaps on other planets.

Closing in on gamma-ray bursts. In mid-1997, astronomers made the first major breakthrough in pinpointing the source of gamma-ray bursts. Gamma rays are the most intense form of electromagnetic radiation, which also includes visible light, radio waves, infrared radiation, X rays, and ultraviolet light.

Astronomers have long believed that only nuclear processes, such as nuclear reactions at the centers of stars, could create the very high energies required to produce gamma rays. However, because the bursts usually last only 10 seconds or so, and because telescopes have been unable to accurately locate any gamma-ray source, astronomers have been unable to tell whether the bursts originate only in the Milky Way or come from faraway parts of the universe. A local origin would mean that the bursts are caused by nuclear events occurring at the present time. A cosmic origin in faraway galaxies would imply that the bursts were a phenomenon of the early universe, because of the length of time it would take for radiation to reach Earth from such a distance.

Gamma-ray bursts from space were first noted in the 1960's by U.S. defense satellites equipped to detect gamma rays emitted by nuclear bomb explosions on Earth. In 1991, NASA launched the Compton Observatory, an Earth-orbiting spacecraft with which astronomers discovered that short bursts of gamma radiation are quite common. However, neither the defense satellites nor the Compton Observatory could pinpoint the locations or distances of the sources precisely enough to identify them.

In 1996, the Italian and Dutch space agencies launched an orbiting observatory called BeppoSAX, equipped with both an instrument to detect gamma rays and an X-ray camera. From BeppoSAX, astronomers have learned that some gamma-ray bursts are quickly followed by a burst of X rays, which are much easier to pinpoint. Once a gamma-ray burst has been spotted, astronomers can use the X-ray camera to determine the point in space that the burst came from. They then know where to train their optical telescopes to look for a visible object.

Using BeppoSAX, astronomers discovered two gamma-ray bursts in mid-1997. Dutch astronomers led by Jan van Paradijs of the University of Amsterdam announced in April 1997 that the first of the two burst sources appeared to be in a very distant galaxy. But this object faded too quickly for astronomers to test whether it truly was a galaxy.

The second burst, detected in May 1997, was reobserved more promptly after its initial outburst. In June, two groups of astronomers at the California Institute of Technology in Pasadena, one led by Mark R. Metzger and the other by Stanislav G. Djorgovski, announced that the source of that burst appeared to be in a very distant, dim galaxy. The researchers calculated that the galaxy is 4 billion to 8 billion *light-years* from Earth. (A light-year is the distance light travels in one year, about 9.5 trillion kilometers [5.9 trillion miles].) In April 1998, Djorgovski and his team reported that they had calculated the distance to a burst spotted in December 1997 as well. That burst, which astronomers say was momentarily second in power only to the *big bang* (the theoretical explosion that gave birth to the universe), originated about 10 billion light-years from Earth. Both distances imply a cosmic origin for gamma-ray bursts.

All the light there ever was. Astronomers reported in January 1998 that they had mapped all of the energy ever produced by all of the stars that have existed since the beginning of time. The researchers, led by astronomer Michael Hauser of the Space Telescope Science Institute in Baltimore, produced the map by measuring electromagnetic radiation at far-infrared wavelengths, which are invisible to the human eye.

The astronomers became interested in calculating the amount of far-infrared radiation in the universe as part of an effort to understand how the universe originated and how it evolved. They realized that to calculate how much matter there is in the universe and when it was formed, they would need to measure all the radiation that was ever emitted into space by a variety of objects, primarily stars. To determine the number of stars in the universe now and in the past, the investigators needed to measure all the light that reaches Earth from space. But such an approach posed a problem: Much of the emitted light has been absorbed by grains of dust in interstellar space, so the astronomers had to take the absorbed light into account as well.

Fortunately, as interstellar dust absorbs light, it converts the light energy into heat radiation that glows at infrared wavelengths. Hauser and his team used this phenomenon to their advantage. The astronomers detected the infrared radiation in space with a special telescope aboard the Cosmic Background Explorer, a NASA-sponsored orbiting observatory. With the telescope, Hauser's group recorded radiation from sources at the far edge of the observable universe—emissions dating back to the origin of the first stars and galaxies. (The farther an object is from Earth, the longer its light takes to reach us, so we are seeing faraway objects as they were long ago.)

However, the team first had to subtract from the map radiation emissions coming from nearby objects—such as planets, stars, and dust clouds in the solar system and the Milky Way—because

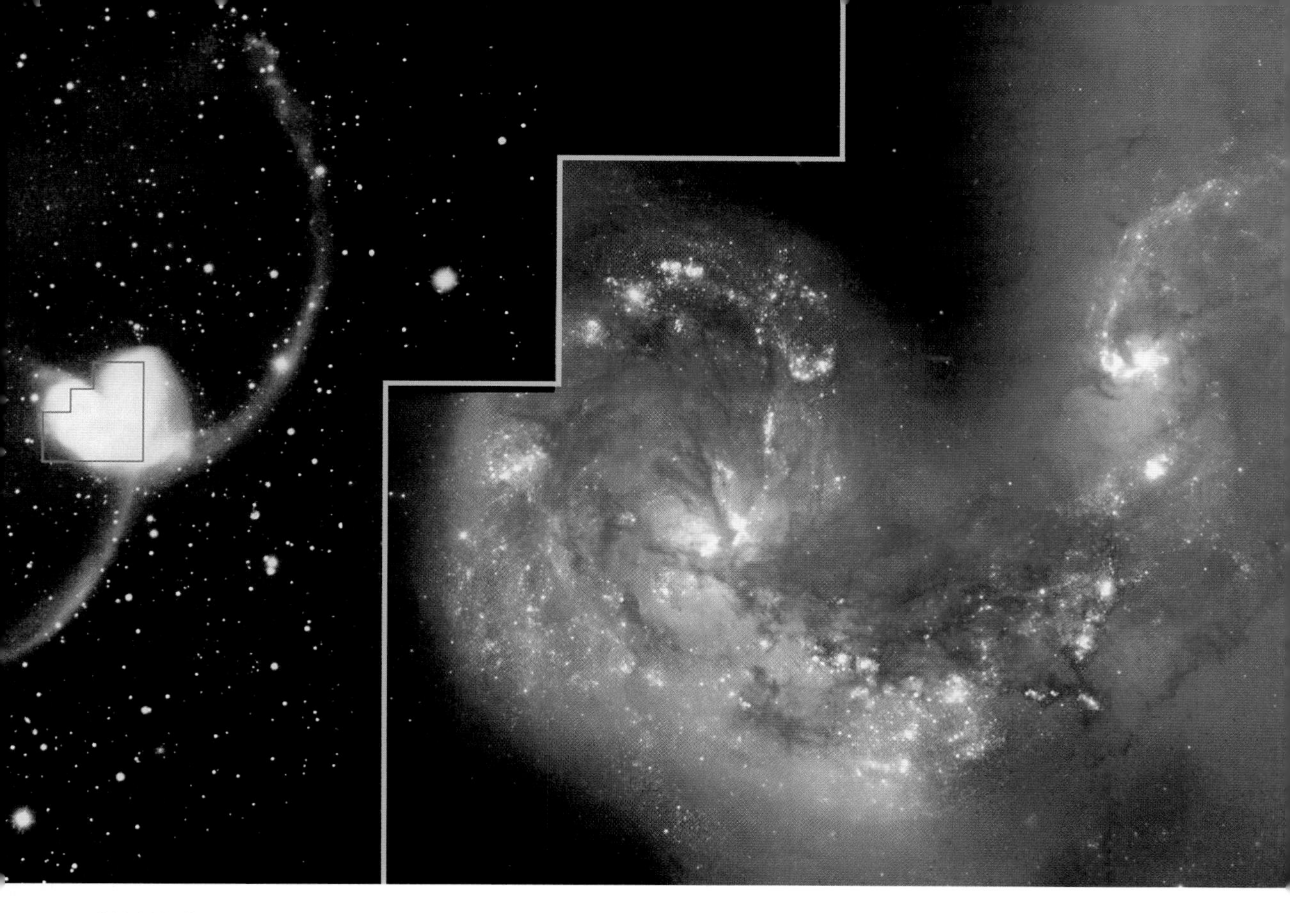

A Hubble Space Telescope image of two spiral galaxies colliding, *above right,* shows bright areas where millions of new stars are being born. The image, released in October 1997, gives astronomers a much clearer view of star formation than previous images taken with land-based telescopes, *above left.*

the bright light from these objects obscured the background infrared glow and would have biased the results. By January 1998, the team had completed the map, revealing a diffuse glow of far-infrared radiation across the entire sky.

The map showed that about 2.5 times more energy has been emitted since the universe began than astronomers had previously calculated by measuring visible starlight. The new map suggested that the universe is dustier than once thought, and that many stars—and probably entire galaxies—are obscured from our view. Such information will prove important to theorists studying the origin and evolution of the universe.

Extrasolar planets. Astronomers in 1998 offered more evidence for what may be planets outside our solar system. Possible extrasolar planets were first reported in 1996 and 1997, but they were detected by indirect means: measurements of small wobbles in the stars that the planets were orbiting, caused by the pull of the planets' gravity.

More-direct proof was offered in January 1998 by astronomer Alfred Schultz of the Space Telescope Science Institute in Baltimore and his colleagues. In images of Proxima Centauri, the closest star to our sun, taken with the Hubble Space Telescope (HST), Schultz and his team noted a dim object that seemed to be moving around the star. Similarly, in May, astronomer Susan Terebey of the Extrasolar Research Corporation in Pasadena, California, and NASA released an HST image that seems to show a planet that was ejected from a double-star system about 450 light-years from Earth. Some astronomers said the objects are probably brown dwarfs, gaseous bodies smaller in mass than a star but larger than a planet. More observations were needed to confirm that the objects are indeed planets.

On April 16, 1998, astronomers in Chile and Hawaii independently reported observing a disk of gas and dust around a star called HR4796, about 220 light-years from Earth. A hole in the middle of the disk, the astronomers said, may be a sign that the dusty material is being incorporated into newly forming planets.

Similarly, British and American astronomers using a telescope in Hawaii reported on April 23, 1998, that they had obtained detailed images of the already known dust disks around Vega, Fomalhaut, and Beta Pictoris—stars in the Milky Way. The disk around Fomalhaut also had a hole in it, suggesting that planets in the Milky Way may be more common than once believed.

Neutron star observed. The first visual observation of a neutron star was reported in September 1997 by astronomers Frederick M. Walter and Lynn D. Matthews of the State University of New York at Stony Brook. Neutron stars are highly compressed remnants of massive stars whose cores collapsed at the end of their lifetime. Compression changes the star's atoms to neutrons. Although neutron stars are more massive than the sun, their diameter is about 10 kilometers (6 miles), making them hard to see.

Neutron stars were only a theoretical possibility until the late 1960's, when astronomers discovered sources of radio emission that flickered on and off very rapidly, with perfect regularity. Astronomers ultimately attributed these repeating radio sources to *pulsars,* rapidly rotating neutron stars that emit beams of electromagnetic radiation. Other neutron stars were thought to be part of *X-ray binaries,* two-star systems in which gas drawn by gravity from a normal star to a neutron star companion is heated to the point of emitting X rays. But until the work of Walter and Matthews, no neutron star had ever been actually seen.

Walter first suspected that he had found a neutron star from faint X-ray emissions that he detected using the German ROSAT satellite. Matthews then focused the Hubble Space Telescope (HST) onto the same point in the sky. The images obtained with the HST showed a single, very dim object. Both the X-ray emissions and the faint visible light were properties that astronomers had expected from a neutron star. Investigations of this object were expected to help astronomers refine their theories about neutron stars. [Jonathan I. Lunine and Theodore P. Snow]

In the Special Reports section, see RETURN TO MARS.

Atmospheric Science

A powerful *El Nino*—a shift in global weather patterns triggered by the warming of surface waters in the equatorial Pacific Ocean—was a major atmospheric event of 1997-1998. But in the United States the spring of 1998 will also long be remembered for the devastating tornadoes that struck the Southeastern part of the country. Normally, only about 80 tornado-related deaths occur in the United States in an entire year. By late April 1998, however, there had already been more than 100 Americans killed by tornadoes.

Killer tornadoes. On the night of February 22 and the morning of February 23, 1998, several tornadoes ripped through central and eastern Florida. About 40 people were killed. In the afternoon and early evening of March 20, a series of tornadoes killed 14 people in Georgia and 2 in North Carolina. On the evening of April 8, tornadoes tore through central Alabama, killing at least 32 people, and several more people were killed by associated storms that same night in Mississippi and Georgia. The most violent of the Alabama twisters had estimated wind speeds in excess of 400 kilometers (250 miles) per hour. Few tornadoes are stronger than that.

Except for one death in Minnesota, every tornado-related fatality in the nation by April 1998 had occurred in states east of the Mississippi River and south of the Ohio River. Only Louisiana and South Carolina were spared. This pattern was highly unusual. Normally, the annual tornado season begins in the Southeastern United States in February, as winter begins to recede in that region. In March and April, the center of activity shifts north and west into the Midwest and the Great Plains. In 1998, however, the storm track remained over the Southeast throughout April and into early May. Scientists were puzzled by this lingering pattern. Some speculated that it may have been a residual effect of the receding El Nino phenomenon in the equatorial Pacific Ocean. El Ninos are known to disturb the globe-circling jet streams, which strongly affect weather patterns across the world.

By 1998, decades of research on tornadoes had enabled scientists to better

A satellite image of Earth's Southern Hemisphere, released in October 1997, reveals that the hole in the ozone layer (purple and black area) over Antarctica had reached an unprecedented height. The atmosphere's thick layer of ozone—a three-atom form of oxygen—protects Earth from much of the sun's harmful ultraviolet radiation. Since the late 1970's, the ozone hole, caused by certain chemical pollutants in the atmosphere, had extended from 14 to 19 kilometers (8.7 to 11.8 miles) in altitude, but measurements in 1997 showed that it had extended up to a height of 20 kilometers (12.4 miles). The hole reaches its maximum size each year in August, which is late winter in the Southern Hemisphere.

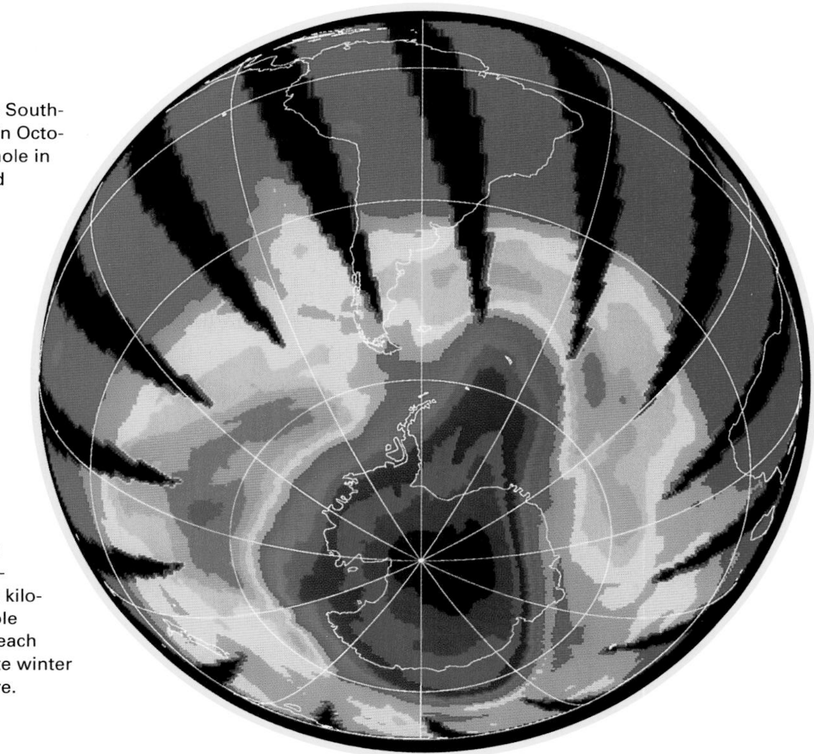

predict severe storms. Unfortunately, all this improved forecasting ability is of no use if the warnings do not reach the people in the path of the storm. This was the case with many of the tornadoes in spring 1998. The killer tornadoes in Florida and Georgia occurred in late evening, when many people had gone to bed. Many of those killed were apparently taken by surprise while they slept. Emergency-management officials across the United States were looking for better ways of warning unwary people of approaching tornadoes. One solution, an alarm-clock-sized radio that turns on automatically when a tornado warning is issued, was being used in some parts of the country in 1998, especially in schools.

While the tornadoes that struck Florida in late February were the most devastating of the early-1998 twisters, the most dramatic tornado was one that struck downtown Nashville, Tennessee, in the late afternoon of April 16. Pictures taken of this tornado showed that it did not have the classic funnel shape that is typical of almost all tornadoes. The Nashville tornado's core spun up masses of humid air that condensed to produce a broad, foglike column of cloud that extended from the thunderstorm that spawned it to the ground. This cloudiness hid the core of spinning, high-speed winds from view. For this reason, many people in Nashville failed to realize that they were witnessing a tornado until they saw the flying debris inside the cloud column. Fortunately, other people did realize what was happening in time to take shelter. The twister caused extensive damage to many of the high-rise buildings in downtown Nashville. No one was reported killed, though about 100 people were injured by flying debris.

Wind speed record refuted. When Typhoon Paka passed over the island of Guam on Dec. 16, 1997, wind-measuring equipment at Anderson Air Force Base reported a peak gust of 379 kilometers (236 miles) per hour, an apparent new world record for wind speed at the surface of the Earth. The long-standing record for surface wind speed was a gust of 372 kilometers (231 miles) per hour. It was recorded at the summit of Mount Washington, in New Hampshire, on April 12, 1934, during a spring snowstorm. U.S. officials sent teams of experts to investigate the 1997 winds in Guam.

Extreme surface winds usually occur at the top of a mountain or a high ridge, where there is a long, clear stretch of ground upwind. This was not the case in Guam, where the *anemometer* (an instrument for measuring wind speed) that reported Typhoon Paka's wind speed was located in the middle of a valley. The investigators concluded that the anemometer probably produced an erroneous reading due to a combination of heavy rain and powerful winds.

The investigators presented their findings to the National Climate Extremes Committee (NCEC), a government panel established in 1997 to evaluate national extremes of weather and climate. (Statistics on weather and climate extremes are important because they help government agencies establish standards for the construction of buildings and help scientists improve their understanding of Earth's climate.) The NCEC ruled in March 1998 that the reported wind gust from Typhoon Paka was not accurate. Consequently, the Mount Washington record stood.

The global warming debate continued in 1997 and 1998, especially over the issue of whether or not the observed increase in Earth's average temperature was being intensified by *anthropogenic* (caused by human activity) carbon dioxide emissions into the atmosphere. Such emissions contribute to the greenhouse effect, a warming of the atmosphere resulting from the trapping of *infrared radiation* ("heat rays") by certain atmospheric gases. These so-called greenhouse gases permit sunlight to pass through Earth's atmosphere to warm the surface of the planet, but trap a portion of the heat, slowing its reradiation back into space. This effect is essential to the maintenance of life on Earth; without the greenhouse effect, Earth's surface would be much colder.

Increasing the amount of carbon dioxide in the atmosphere by the burning of fossil fuels, such as gasoline in

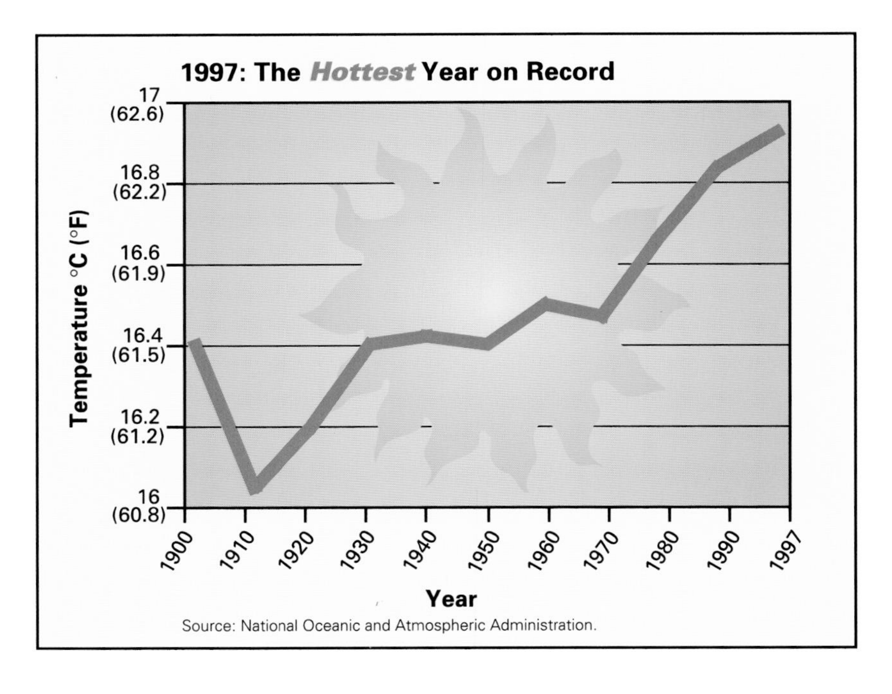

According to figures kept by the National Oceanic and Atmospheric Administration (NOAA), 1997 was the warmest year of the 1900's to date. Researchers at NOAA's National Climatic Data Center combined average surface-temperature measurements from all over the globe (both land and ocean) for the decades since 1900 and for 1997. The figure for 1997 was 16.92 °C (62.46 °F). The record-breaking warmth of 1997 continued a pattern of increasing global temperatures. Nine of the previous 11 years had also been the warmest on record.

automobiles, intensifies the greenhouse effect only slightly. By itself, this small enhancement would not be cause for concern. By 1998, however, many scientists believed that anthropogenic carbon dioxide emissions were just the first step in a larger process that was amplifying the warming significantly. These scientists argued that the slight warming caused by burning fossil fuels causes more water to evaporate from the oceans into the atmosphere. Water vapor is a very effective greenhouse gas, so the presence of more water vapor in the atmosphere significantly enhances the greenhouse effect.

While most scientists believed that global warming due to anthropogenic carbon dioxide emissions, possibly combined with a long-term natural warming trend, was almost certainly underway, many others believed that an anthropogenic warming effect had yet to be conclusively proven. Climatic conditions on Earth normally fluctuate, warming for a few decades, then cooling, then warming again, and so on. Some scientists, therefore, remained unconvinced that an intensified greenhouse effect had played any role at all in the global warming of the previous two decades. Nonetheless, new findings reported each year were making it apparent that Earth's climate was indeed warming slightly, either naturally or, more likely, from a combination of natural and human causes.

Weather forecasting in reverse. The global warming debate prompted scientists at the University of Massachusetts and the University of Arizona to reconstruct a detailed record of average annual global temperatures back to the year 1400. For their project, the scientists used such evidence as ancient tree rings, coral growth patterns, and ice cores, as well as historical records.

The data indicated that 1990, 1995, and 1997 were the warmest years since 1400. Either 1995 or 1997 could be considered the warmest years of all, depending on whether one compared temperatures over land, over the ocean, or both. When land and ocean temperatures were combined, both 1995 and 1997 ran well above the average for all of the 1900's.

The researchers also compared their estimated temperatures to trends in other influences on climate, such as variations in the sun's brightness or volcanic activity on Earth. They found that while these other factors appeared to have been important controlling influences in the past, greenhouse gases were apparently the dominant factor during the last few decades.

Scientists around the world called the study an early step in the process of describing a general rise in global temperatures and called for more such research. In addition, they noted that the findings were consistent with those of previous, less extensive studies.

El Nino and global warming. El Nino is essentially a regional phenomenon of the equatorial Pacific. However, in some years it grows so powerful that it alters weather patterns around the world. At a July 1997 meeting of meteorologists in Melbourne, Australia, much discussion centered on efforts to associate El Ninos and other worldwide climate events of the previous 20 years to the global warming phenomenon. Some experts argued that the effects of global climatic change, such as intense El Ninos, have apparently become more frequent and more intense in the last 30 years. For instance, the strongest El Nino on record prior to the 1997-1998 event was in 1982-1983. Furthermore, an unusually long-lasting El Nino persisted from 1990 through 1995.

Supporters of this view theorize that El Nino may be a fundamental mechanism by which Earth's climate deals with excess tropical heat. The sun continuously pours heat into the tropics. Normally, the patterns of flow in the ocean and the atmosphere work together to move this tropical heat toward the poles. From time to time, however, this ocean-atmosphere system is not sufficient to deal with all the excess heat. When this situation arises, an intense El Nino begins to form.

If El Nino really is a mechanism by which Earth's climate system compensates for a buildup of heat in the tropics, then any general warming of the planet would lead to more frequent occurrences of El Nino and other global climatic events. Statistics seem to verify that such events had indeed occurred with greater frequency in the previous two decades.

[John T. Snow]

See also OCEANOGRAPHY (Close-Up).

Surprising new findings about *circadian rhythms,* the daily cycles (such as waking and sleeping) that people and other animals go through, were reported in November 1997 by scientists at the Scripps Research Institute in La Jolla, California. The researchers, led by geneticist Steve Kay, discovered that one of the major proteins that regulates circadian rhythms in fruit flies functions throughout the fly's body—not just in the brain, as scientists had previously thought.

Since the 1970's, researchers have discovered a number of genes that play a role in regulating circadian rhythms. They have also isolated some of the proteins that these genes *code* for (direct the production of). In fruit flies, one of these proteins, called PER, is produced in varying amounts throughout the day to help control the fly's circadian rhythms. Although geneticists have not learned precisely how PER works, they had assumed that it functions only in the brain. They thought that different light levels perceived by the eyes and registering in the brain cause PER to adjust the body's "clock" to be synchronized with day and night.

To establish exactly where PER functions, the Scripps researchers linked a gene derived from jellyfish to the PER gene of fruit flies. The researchers knew that the jellyfish gene directs the production of a protein called green fluorescent protein (GFP), which generates the light seen in some jellyfish. The scientists reasoned that after the GFP gene was linked to the PER gene, the fruit flies would produce the fluorescent protein in all parts of the body where the PER protein normally functions.

When the scientists looked at the flies in the laboratory, they saw that GFP was glowing in every region of the body, including parts of the legs, wings, head, thorax, and abdomen. Moreover, when the fruit flies were cut up into individual parts and grown in laboratory dishes, the researchers found that each body part continued to glow—indicating that PER operates throughout the body independently of the brain.

In seeking to explain these results, the scientists said that although the brain's clock may oversee overall body rhythms (such as waking and sleeping), independent clocks may help different parts of the body respond to particular needs, such as the need for an insect's flight muscles to become active during the day.

A major unanswered question, according to the scientists, was whether PER-type proteins operate throughout the bodies of other animals, including humans.

Independent human clocks? An indication that humans too may have independent circadian-rhythm clocks operating throughout their bodies came from a study published in January 1998. Experimental psychologists Scott Campbell and Patricia Murphy of Cornell University's Medical College in White Plains, New York, reported that they altered the circadian rhythms of human volunteers by shining a light on the back of their knees.

The light, delivered by fiber-optic light pads during the night, changed the normal patterns of the daily cycles of temperature and hormone levels. For example, the body temperatures of some of the volunteers decreased in the afternoon—in contrast to the increase that normally occurs during this time of day.

The scientists said that if their results were confirmed by other researchers, the findings might lead to new treatments for jet lag, insomnia, and seasonal depression. People experiencing jet lag, for example, might be able to reset their daily rhythms to the new time zone by wearing light pads on their knees while sleeping.

Evolution in action. Midwestern farmers in the 1990's were witnessing a dramatic example of evolution in action that was limiting their ability to control corn rootworms, according to a January 1998 report. The report, by *entomologist* (insect expert) Joseph Spencer of the University of Illinois at Urbana-Champaign, described an alarming change in the behavior of rootworms, the *larvae* (immature form) of insects that feed on corn roots.

The adult form of the corn rootworm is a beetle that normally lays its eggs in a cornfield during the summer. If a farmer plants corn in the same field the next year when the eggs hatch, the young rootworms attack the corn roots, doing substantial damage to the crop. Farmers have minimized this problem by rotating corn and soybean crops from year

The world's second-smallest monkey, a dwarf marmoset, climbs up a tree in its home in the Brazilian rain forest. The monkey is shown at approximately its actual size—10 centimeters (4 inches) long. Ecologist Marc Van Roosmalen of the National Institute of Amazon Research in Manaus, Brazil, announced his discovery of the tiny primate in August 1997. The dwarf marmoset was the seventh new monkey species discovered in Brazil since 1990. Scientists said they expected several additional new species to be discovered in the area in upcoming years.

Discovering New Mammals

In mid-1998, biologists who had been part of a team exploring the tropical forests and isolated mountaintops of the Philippines published descriptions of some of their unexpected finds. Their discoveries included three new species of mammals, which were added to a growing list of creatures that scientists had never seen before. Two of the mammals were shrew-mice—small, burrowing animals that feed on insects and earthworms. The third was a hairy-tailed forest rat.

This same team of biologists had discovered other rodents in the Philippines, including the voracious, earthworm-eating Mt. Isarog striped rat, which the team found high on the forested slopes of an extinct volcano in 1988. In all, from the early 1980's to the late 1990's, scientists discovered more than 30 new species of mammals in the Philippines. Among them were fruit-eating bats, insect-eating mice, and large tree-climbing rats.

The discoveries in the Philippines were not unique. Throughout the world, more new species of mammals were found in the late 1900's than at any other time in the preceding 50 years. Between 1982 and 1998, approximately 250 new species of mammals were named and described.

This flurry of new discoveries did not match the peak reached between 1890 and 1920, the most active period for finding new mammals. During those years, natural history museums mounted the first major expeditions to vast, unexplored regions of the tropics. Zoologists discovered and described more than 1,300 mammal species—more than one-fourth of all the mammals ever found. Nonetheless, the number of discoveries made in the late 1900's was impressive.

Most of the new mammals, like those found before 1920, were discovered in the tropics, in such places as Madagascar, Southeast Asia, and Brazil. But some were found in countries with fewer wild areas, including the United States. Rodents were the most common of the newly discovered species. Biologists also found new bats, *ungulates* (mammals with hoofs), *insectivores* (mammals that feed chiefly on insects), *marsupials* (mammals that carry young in pouches), and *primates* (a group of animals that includes apes and monkeys). One primate newly discovered in the Brazilian rain forest is the second-smallest monkey ever found.

There are several reasons why so many new mammal species are being discovered. One reason is that biologists have developed new techniques for differentiating between species. Most new species can be identified and described on the basis of their *morphology* (physical structure), but sometimes morphology is unreliable. Now, researchers can analyze an animal's biochemical makeup, its *chromosomes* (structures in the cell nucleus that carry the genes), and even the molecular sequence of its DNA (deoxyribonucleic acid—the molecule genes are made of) to identify differences between mammals.

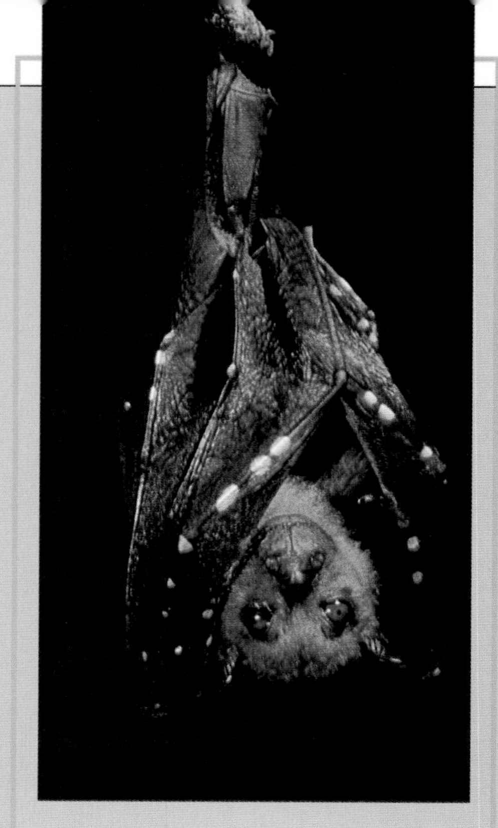

The tube-nosed fruit bat is one of more than 30 new species of mammals discovered in the Philippines in the 1980's and 1990's. Because its forest habitat is being destroyed by human population growth and economic activity, the bat is in danger of extinction.

With the new methods, biologists have learned that some mammals once thought to be a single species on the basis of their morphology are actually two or more distinct species. One such animal is Townsend's ground squirrel, found in a broad region of the Western United States. For more than 50 years, biologists had thought that all those squirrels belonged to one species. By the early 1990's, however, researchers had learned that there are three subgroups of Townsend's ground squirrel, each with a different number of chromosomes. With no evidence of interbreeding among these groups, researchers recognized the subgroups as three separate species.

Biologists have also discovered new species by

reexamining animal specimens stored for decades in natural history museums. By using routine measurements of morphology and various new methods, researchers found that some mammal specimens were undescribed species that had been misidentified as previously known species. In this way, the tiny Alaskan shrew *Sorex yukonicus* was discovered in 1993 in the collection of the University of Alaska Museum in Fairbanks. Altogether, more than 400 new species of mammals were uncovered between 1982 and 1998 through the reevaluation of misclassified specimens.

Nevertheless, as in the years between 1890 and 1920, many new species identified in the 1980's and 1990's were discovered in the wild, on expeditions to remote regions of the world. As expanding human populations and economic pressures cause destruction of tropical forests and other habitats, biologists have rushed to explore them.

In recent years, researchers have placed less emphasis on simply identifying new species than on testing specific theories of how species evolve, spread geographically, and function in natural communities. They have learned that when populations of a species of animal are geographically isolated, some populations may develop genetic differences. Over many generations, those differences can lead to the formation of new species. Thus, biologists were not surprised to find many new species in long-isolated places such as remote islands or mountaintops. Some researchers estimated that thousands of mammal species remained undiscovered.

Often, mammals discovered in isolated areas are *endemic* (found nowhere else). These animals live in very limited geographic areas and have specialized habits that require undisturbed habitats, such as mature tropical rain forests. Unfortunately, these characteristics place a species at especially high risk of extinction. By 1997, according to the World Wide Fund for Nature, an ecological group with headquarters near Geneva, more than two-thirds of the world's forests had been destroyed. As habitats are lost, an increasing number of species are threatened with extinction.

In some countries, the discovery of new species in the 1990's helped fuel conservation efforts. In the Philippines, after the discovery of two new species of mice on one of the islands, the government declared the mice's forest home a national park. Biologists hope that efforts to conserve what remains of the Earth's wild areas will enable future generations of researchers to continue discovering new species. They also hope that more and more people will come to understand that the biological diversity found in natural communities is a fundamental source of human health and well-being. [Eric Rickart]

See also BIOLOGY.

to year, so that when the rootworms emerge, they find only unappetizing soybean roots and die.

Spencer discovered, however, that some rootworm beetles have evolved to lay their eggs in soybean fields rather than cornfields. Due to this behavioral change, the rootworms hatch in cornfields the following year (if the farmer has followed the usual crop rotation) and have an abundant supply of food.

Spencer said that the altered behavior probably began when a genetic *mutation* (change) caused some adults to lay their eggs among soybean plants rather than corn plants. Because farmers planted corn the following year in the same fields, the young rootworms survived. They then passed on the genetic trait for laying eggs in soybean fields to the next generation.

The first damage to cornfields because of this phenomenon was observed in Illinois in 1993. By 1995, 9 counties in Illinois and 13 in northwestern Indiana had reported such damage. A survey in 1997 revealed that the problem was spreading rapidly to the north and east, as counties in Michigan and Ohio reported the problem.

This unfortunate development was expected to be costly for farmers, who now needed to purchase more chemical pesticides to apply to their fields. In a hopeful note for the future, some companies were working to develop a genetically engineered corn plant that would itself produce a pesticide fatal to rootworms.

Evolving fish. In another example of rapid evolution, marine biologists reported in September 1997 that increasingly murky water conditions in Lake Victoria in Africa were causing changes in the breeding habits and body color of fish called cichlids. The biologists, from the University of Leiden in the Netherlands, said that the murkiness, caused by soil erosion and by pollution, was also partly responsible for a severe reduction in the number of cichlid species in the lake.

Lake Victoria, which lies in parts of Kenya, Tanzania, and Uganda and has an area of 69,484 square kilometers (26,828 square miles), is the second largest freshwater lake in the world. It had originally been home to about 500 species of cichlids, brightly colored perchlike fish. However, as many as 60 percent of those species disappeared between the late 1970's and mid-1990's.

Biologists had previously known that one of the reasons for the decline was a population explosion of the Nile perch, a large fish that preys on cichlids. The Nile perch had been introduced into Lake Victoria in the early 1960's by British colonial authorities. Other known reasons for the decline included overfishing and the dumping of industrial wastes into the lake.

Widespread tree-felling around Lake Victoria since 1900 also led to increasing murkiness in the lake. The disappearance of forests meant that an extensive network of tree roots no longer existed to hold soil in place, so soil began to be washed into the lake. Adding to the lake water's cloudiness are agricultural fertilizers, which wash into the lake and promote the growth of great masses of algae.

The key discovery made by the Dutch researchers was that the loss of clarity in the lake had gotten so bad that female cichlids could no longer distinguish males of their own species from other males. Males of each species have adistinctive color that females recognize, but in the cloudy water, these colors could not be seen. This visual problem was causing males and females of different species to mate with one another. The different cichlid species are considered distinct by biologists because, under normal conditions of water clarity, they never breed with one another.

Mating between cichlid species, which are all closely related, results in healthy, fertile offspring. But the biologists found that most of the offspring from the different matches strongly resemble each other and that males' colorations are becoming increasingly dull. In other words, a formerly wide variety of brightly colored species are evolving into just a few dull-colored species. The biologists said this was a unique example of a loss of species diversity in the absence of any "classical" causes of extinction, such as a species dying out because of an inability to compete with another species.

Hot worm. The discovery of an unusual worm that lives at much higher temperatures than researchers had pre-

The locations of a protein that regulates "biological clocks" are indicated by a blue glow in various parts of a fruit fly's body. Biological clocks determine the timing of such daily cycles as sleeping and waking. Until California researchers demonstrated the presence of clock proteins throughout the fruit fly's body in 1997, scientists had thought that such substances functioned only in the brain. The new finding showed that, in fruit flies at least, there are independent biological clocks throughout the body.

viously thought possible for a multicelled organism was reported in February 1998 by molecular biologist Craig Cary of the University of Delaware in Newark. The Pompeii worm, as Cary named it, lives at the bottom of the ocean in tubes built on *hydrothermal vent chimneys.* These are chimneylike structures made of sulfur compounds that form over deep-sea hot springs from which "superheated" water and minerals spew forth. Temperatures in the worms' tubes reach as high as 79 °C (176 °F).

Biologists had known that certain single-celled microorganisms could survive in temperatures of as much as 112 °C (235 °F), but they thought that such high temperatures would destroy cellular components in more complex organisms. The previous multicelled record-holder for surviving great heat was a species of African ant in the Sahara that can withstand temperatures up to 54 °C (131 °F) for brief periods.

Cary found the Pompeii worm, which is about 6 centimeters (2.4 inches) long, by diving in an area south of Baja California in the famous research submarine *Alvin.* He placed a specially devised temperature probe inside several worm tubes, measuring an average temperature of 67 °C (154 °F)—with occasional surges of even greater heat. In contrast, the worm's head and gills, which stick out of the tube, are bathed in balmy water averaging only (22 °C) 72 °F.

Although Cary was not sure how the worms survived such conditions, he speculated that bacteria that form a fleecy pelt on the back of the Pompeii worm may somehow insulate the worm's body from the heat. Cary said additional research may reveal that the bacteria produce an enzyme that could be used for certain high-temperature industrial applications.

Early mammal. The discovery in Australia of a 115-million-year-old fossil animal jaw, reported in November 1997, threatened to change long-accepted views about the origin and history of mammals. Paleontologists from the Museum of Victoria in Melbourne, Australia, said that the jaw appeared to be the earliest known fossil from a *placental* mammal—a mammal that nourishes its unborn young within a *uterus* (womb). The animal, they said, resembled a shrew, a small insect-eating creature.

Scientists had long believed that the early history of Australia was dominated by *marsupials* (mammals that carry their developing young in external pouches) and *monotremes* (mammals that lay eggs). It was generally accepted that placental mammals did not appear in Australia until 5 million years ago, after originating in the Northern Hemisphere more than 95 million years earlier and slowly spreading around the rest of the world.

The scientists said that the jaw shows that the newly discovered animal, which was named *Ausktribosphenos nyktos,* had special cutting and crushing teeth called tribosphenic molars. Although these types of teeth are found in both placental mammals and marsupials, the fossil jaw lacks other features characteristic of marsupials. In addition, the five premolars and three molars in the jaw are typical of the number of teeth seen in placental mammals.

Paleontologists said that if *A. nyktos* was, in fact, a placental mammal, it could indicate that such animals originated in the Southern Hemisphere. Another possibility is that placental mammals originated much earlier than was thought—perhaps more than 180 million years ago, when all of the Earth's continents were combined into a single land mass called Pangaea.

Miniature monkey. A Brazilian researcher announced in August 1997 that he had discovered a new species of monkey with a body only 10 centimeters (4 inches) long. The monkey, called a dwarf marmoset, is just slightly larger than the smallest known monkey, the pygmy marmoset. That monkey grows to a length of 9 centimeters (3.5 inches).

Ecologist Marc Van Roosmalen of the National Institute of Amazon Research in Manaus, Brazil, said he discovered the tiny creature in a rain forest about 2,100 kilometers (1,300 miles) west of Rio de Janeiro. Based on its behavior and physical traits, Roosmalen said, the new-found monkey might be a link between the pygmy marmoset and larger species of marmosets.

The dwarf marmoset was the seventh new monkey species discovered in Brazil since 1990. Roosmalen said he was studying several other new species on which he expected to eventually publish reports. [Thomas H. Maugh II]

See also BIOLOGY (Close-Up).

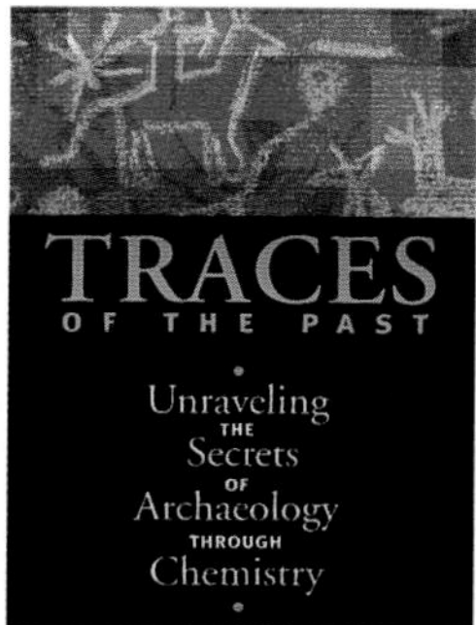

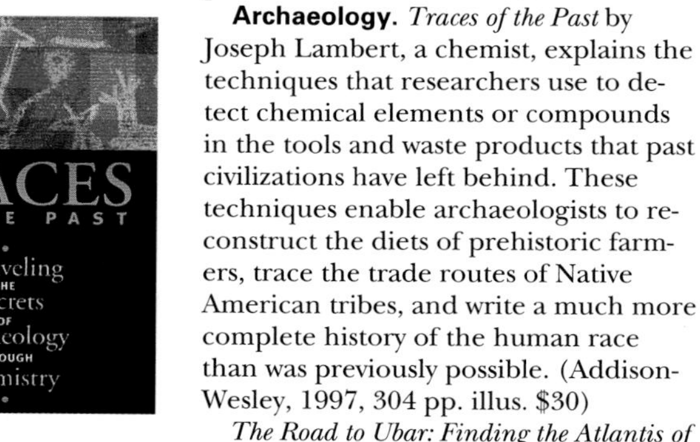

Here are 16 important new science books suitable for the general reader. They have been selected from books published in 1997 and 1998.

Archaeology. *Traces of the Past* by Joseph Lambert, a chemist, explains the techniques that researchers use to detect chemical elements or compounds in the tools and waste products that past civilizations have left behind. These techniques enable archaeologists to reconstruct the diets of prehistoric farmers, trace the trade routes of Native American tribes, and write a much more complete history of the human race than was previously possible. (Addison-Wesley, 1997, 304 pp. illus. $30)

The Road to Ubar: Finding the Atlantis of the Sands by Nicholas Clapp, a documentary filmmaker and amateur archaeologist, tells the story of the search for an ancient city 1,500 years after its disappearance. According to legend, Ubar was a prosperous stopping point for caravans carrying spices from southern Arabia to the Persian Gulf. Clapp, with the help of a team of explorers and scientists—and even astronauts aboard the U.S. space shuttle Challenger—rediscovered the city buried beneath drifting sands in present-day Oman. (Houghton Mifflin, 1998, 342 pp. illus. $24)

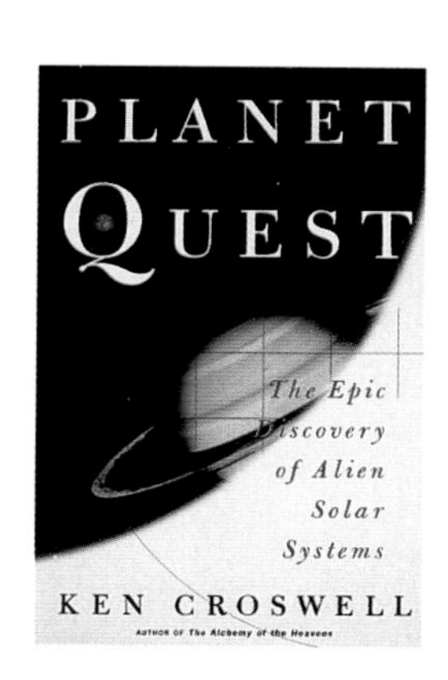

Astronomy. *Planet Quest* by Ken Croswell, a science writer, reviews the history of astronomers' attempts to discover planets orbiting stars other than our own sun. Many of the early discoveries could not be confirmed. However, since 1995, new astronomical techniques have led to the discovery of more than a half-dozen nearby stars that have planets. Croswell describes the controversies and the personalities behind the search. (Free Press, 1997, 324 pp. illus. $25)

Biology. *Cod* by Mark Kurlansky, a food and wine columnist, traces the natural and economic history of the cod and the cod fisheries from the time of the Vikings to the present day. The Vikings of the 900's discovered North America by following the range of the Atlantic cod. But by the late 1900's, declines in the cod population had forced a drastic reduction in the catch. Kurlansky shows how badly fishermen misjudged the ability of this hardy, productive—but vulnerable—fish to maintain its numbers in the sea. (Walker, 1997, 304 pp. illus. $21)

Huxley: From Devil's Disciple to Evolution's High Priest by Adrian Desmond, a historian, describes how one of the best-known scientists of the 1800's influenced, and was influenced by, the social and intellectual movements of his times. Desmond traces Thomas Henry Huxley's career from his work in classifying and interpreting fossils through his advocacy of education for the masses and his tireless popularization of science. He explains how Huxley's speeches and writings on evolution earned Huxley the nickname "Darwin's bulldog." (Addison-Wesley, 1997, 832 pp. illus. $37.50)

Cosmology. *Before the Beginning: Our Universe and Others* by Martin Rees, Great Britain's astronomer royal and a colleague of Stephen Hawking (who wrote the foreword to this book), explores the nature and origin of the universe. Rees addresses such questions as: How did the universe begin? How will it end? Why is it expanding? How much of the matter in the universe is invisible? Rees proposes no ready answers but points to some scientifically grounded speculation on the fate of our universe and its place among other possible universes. (Addison-Wesley, 1997, 291 pp. $25)

General Science. *Questioning the Millennium* by Stephen Jay Gould, a paleontologist, outlines the history of the present-day calendar. Gould explores the human tendency to impose arbitrary distinctions on the natural universe in order to find meaning and order in the chaos of events. He addresses such disagreements about the Third Millennium as when it will begin—January 1, 2000, or January 1, 2001. (Random House, 1997, 192 pp. illus. $17.95)

A Field Guide to the Invisible by Wayne Biddle, a science writer, examines some everyday things of which most people are only imperfectly aware. Biddle includes more than 50 short chapters, alphabetically arranged, about such topics as the origin of bad breath, burps, and body odor, the sources of the dust and pollen that make people sneeze, and the threat (or non-threat) of chemicals such as dioxin, freon, and sulfur dioxide. (Henry Holt and Company, 1998, 224 pp. illus. $25)

Mathematics. *The Number Sense* by Stanislas Dehaene, a mathematician and neuropsychologist, proposes that the ability of humans—and other mammals—to perform mathematical computations is as natural as life itself. Dehaene describes laboratory experiments in which rats were trained to count to 16, chimps were taught to add Arabic numerals, and newborn babies recognized the difference between two things and three. Though some students may regard mathematics as contrived and complex, Dehaene demonstrates that mammals have an innate sense of number, which varies in degree but reaches a surprising level of sophistication in many animals. (Oxford University Press, 1997, 288 pp. illus. $25)

Medicine. *A Commotion in the Blood: Life, Death, and the Immune System* by Stephen S. Hall, a science journalist, traces the development of our knowledge about the immune system over the past century. Hall demonstrates that doctors have triumphed over smallpox, diphtheria, and polio not because they have learned how to treat these illnesses but because vaccines were developed to stimulate the immune system to ward off the infections. Hall shows how doctors are trying to apply this knowledge to conquer cancer. (Henry Holt and Company, 1997, 448 pp. $30)

Oceanography. *Song for the Blue Ocean* by Carl Safina, a conservationist, explores the world's oceans, the people who live along their shores, and those who make their living from the sea. Safina explains the concerns of scientists and environmentalists, such as the fact that although the ocean today looks much as it always has, fishing and pollution have severely damaged one of the world's most precious natural resources. (Henry Holt and Company, 1997, 458 pp. $30)

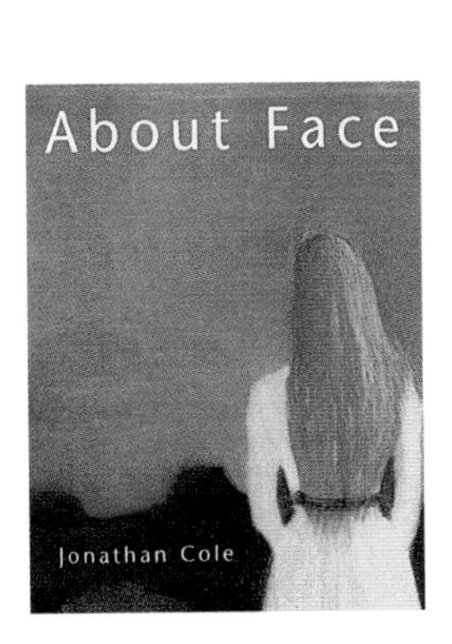

Paleontology. *Taking Wing: Archaeopteryx and the Evolution of Bird Flight* by Pat Shipman, an anthropologist, reviews the debates about how dinosaurs and birds may be related and about why and how animals developed the ability to fly. Scientists first began to suspect that dinosaurs were the direct ancestors of birds in the early 1860's, when workers discovered a fossil of *Archaeopteryx,* a winged reptile with feathers, in a German quarry. Shipman explains how the controversy is viewed today. (Simon and Schuster, 1998, 288 pp. illus. $25)

Psychology. *About Face* by Jonathan Cole, a physician, explores the role of the face in people's understanding of themselves. The face is a powerful organ of communication, able to convey in a single expression—such as doubt, terror, or passion—a message that might be impossible to relay in words. Cole's interviews with people who had lost either the ability to perceive faces (through blindness or neurological impairment) or to control their facial expressions (through such conditions as Parkinson disease) led him to conclude that the face strongly affects a person's emotional life. (MIT Press, 1997, 223 pp. $25)

How the Mind Works by Steven Pinker, a psychologist at the Massachusetts Institute of Technology, describes how and why the human mind works the way it does. Pinker reviews the latest findings of cognitive science and psychobiology to show how the various activities we call human—from art to religion to literature—arose as a result of the evolution of the brain. (W.W. Norton & Company, 1997, 660 pp. $29.95)

Technology. *Crystal Fire: The Birth of the Information Age* by Michael Riordan and Lillian Hoddeson, physicists, tells the story of the three scientists who shared the Nobel Prize in physics in 1956 for developing the transistor. By the late 1900's, this miniature building block of every electronic device from the digital watch to the laptop computer had become more common than the pencil. Riordan provides insight into the process of scientific discovery in the mid-1900's and the birth of the computer age. (W.W. Norton & Company, 1997, 352 pp. illus. $27.50)

Remaking the World: Adventures in Engineering by Henry Petroski, a professor of civil engineering and history at Duke University, explains the creative art—as well as the science—behind such projects as the Panama Canal and the Golden Gate Bridge. Petroski also discusses engineers whose lives are little known today but who made a great impact on the modern world. Among them are Karl Proteus Steinmetz, whose inventions ushered in the electrical power distribution system; Isambard Kingdom Brunel, who built the first large ocean-going steamships; and Karl Terzaghi, who founded the science of soil mechanics. (Knopf, 1998, 256 pp. illus. $24) [Laurence A. Marschall]

Two groups of chemists in England and the United States published independent reports in April 1998 describing new types of *catalysts* (compounds that increase the rate of chemical reactions) significant to the plastics industry. The scientists said that the new catalysts behave like metallocenes, catalysts used by manufacturers to create exceptionally strong, heat-resistant forms of plastics such as polyethylene and polypropylene. However, unlike metallocenes, which are mostly compounds of zirconium—a rare and expensive metal—the new catalysts are compounds of either iron or cobalt, which are both relatively cheap and plentiful. Such catalysts, according to the scientists, might lead to a wide range of stronger, more durable, and less expensive plastic consumer goods.

The English group that reported the catalysts was led by Vernon Gibson of Imperial College in London and included researchers from BP Chemicals Ltd., a division of the oil giant British Petroleum Company PLC. Maurice Brookhart of the University of North Carolina in Chapel Hill led the American group, which also included chemists from E.I. du Pont de Nemours & Company in Wilmington, Delaware.

Both groups of scientists manufactured polyethylene by passing the gas ethylene through iron and cobalt catalysts. The ethylene molecules collected on the surfaces of the catalysts, where chemical reactions caused them to link up and form polyethylene. The chemists said the catalysts had to be mixed with an aluminum-containing compound called methylalumoxane in order for the chemical reactions to proceed.

The scientists reported that in their experiments, ethylene molecules reacted at only one particular site on the catalysts. By contrast, ethylene reacts at widely scattered sites on conventional catalysts. The single reaction site resulted in the formation of polyethylene *polymers* (long-chain molecules) with almost identical sizes and shapes instead of the jumble of different sizes and shapes produced by ordinary catalysts. This uniformity made the polymers much stronger than ordinary polyethylene.

The researchers at BP Chemicals said they hoped to use the iron and cobalt catalysts to make advanced plastics with useful new properties. And the DuPont scientists said they wanted to take advantage of the catalysts' ability to convert ethylene into very pure alpha-olefins, key ingredients in plastics.

Silicon chemical sensor. In October 1997, two California chemists separately reported that they had developed a new type of *chemical sensor* (a device to detect the presence of chemicals in a solution) that was both more versatile and less expensive than existing sensors. M. Reza Ghadiri of the Scripps Research Institute in La Jolla, California, and Michael Sailor of the University of California at San Diego said they each based their sensor on silicon wafers, the same material used to make computer chips.

The silicon chip developed by the researchers contains millions of microscopic pillars on its surface. Attached to the pillars are chemicals that bind to *organic* (carbon-based) molecules in the solution being tested. When this binding takes place, it changes the *refractive index* of the pillars—the speed at which light passes through them. As a result, light rays that are beamed through the pillars and simultaneously reflected from the base of the chip combine to create an *interference pattern* (a series of dark stripes caused by the intersection of light waves) that is unique to the particular organic chemical being analyzed. The interference pattern, besides altering the color of the chip, is projected onto a light detector, where it is converted into an electronic signal. This signal is then fed into a computer, which identifies the chemical.

This detection system, besides being up to 100 times as sensitive as previous chemical sensing devices, is also extremely versatile because it can detect a wide range of substances. Among these substances are particular proteins and certain stretches of deoxyribonucleic acid (DNA), the double-stranded molecule that makes up genes.

In 1998, the California researchers were trying to modify their sensor so that it could not only identify chemicals in solution but also determine their amounts. To accomplish this, the scientists planned to build arrays of sensors, with each sensor designed to respond to a different chemical concentration.

Gel chemical sensor. A second chemical sensor, also described as versatile and inexpensive, was reported in Octo-

Optical amplifier
Chemists at the University of California at San Diego demonstrate two laser beams crossing paths inside of a new material they developed. The material is a type of organic polymer composite—a stack of thin films consisting of large carbon-based molecules. Energy from one laser beam is being transferred to the other inside the composite. The beam gaining energy (upper right) appears thicker than the beam losing energy (lower right). The chemists reported in July 1997 that the composite had potential applications in *optical amplification* (the intensification of light signals), a technology very important in telecommunications.

ber 1997 by chemists John Holtz and Sanford Asher at the University of Pittsburgh in Pennsylvania. This sensor consists of microscopic beads of the common plastic polystyrene embedded in a *gel* (a jellylike substance).

The gel is designed to respond to a particular chemical in a solution being tested. As the gel absorbs the chemical, it swells. The swelling changes the spacing between the polystyrene beads—causing the beads to *refract* (bend) light in a certain way. The result is a change in the color of light passed through the beads. The color indicates the concentration of the chemical in the solution.

According to the Pittsburgh scientists, the gel sensor can be tailored to measure a wide range of substances. The researchers hoped that their sensor would eventually find commercial applications in simple testing kits, such as ones to measure lead in home drinking water.

DNA shrinks electric circuits. Researchers in Israel reported in February 1998 that they had used DNA as a sort of scaffolding to build electrically conducting silver wires between two gold electrodes. The technique, reported by scientists with the Technion-Israel Institute of Technology in Haifa, produced wires less than half the width of the thinnest conducting wires made by conventional methods. The Israelis said that their technique might lead to the production of computer chips that would be much smaller and faster than any currently made.

To create the silver wires, chemist Yoav Eichen and physicists Uri Sivan and Erez Braun first bonded a short stretch of DNA called an oligonucleotide to each of two gold electrodes, which were spaced 12 microns apart—about 1/10 the width of a human hair. The oligonucleotides provided a foundation upon which the scientists attached longer strands of DNA to form a bridge between the electrodes.

The scientists then immersed the assembly into a bath containing positively charged silver *ions*—silver atoms that had lost electrons. The ions adhered to the DNA, which had been given a negative charge by the scientists. Next, the researchers added chemical agents that transformed the ions into grains of silver metal that linked up into long, narrow strands between the electrodes.

The investigators found that these silver strands, about 1,000 times thinner than a human hair, could carry a small electric current between the two electrodes. This observation suggested that similar wires might someday be used in tiny electronic devices.

The Technion group said it believed that its technique would be able to yield even narrower silver wires. But the researchers noted that many technical hurdles had to be cleared before their DNA method could be used to make a fully functioning computer chip.

Mimicking photosynthesis. In a March 1998 account, a group of scientists from the United States and France described their success in getting artificial membranes to mimic photosynthesis, the process by which plants convert the sun's light energy into food. While the system was only intended to provide a laboratory model of how photosynthesis works, its developers said they believed that it might also have a number of practical applications. Among the applications cited by the scientists was the use of sunlight to power chemical reactions creating drugs or other commercially important chemicals inside artificial cell-like structures.

In order to mimic photosynthesis, chemist Thomas Moore and his colleagues at Arizona State University in Tempe and biochemist Jean-Louis Rigaud of the Curie Institute in Paris, first prepared membranes consisting of *liposomes* (hollow spheres made of fatlike molecules called lipids). Inside the liposome walls they inserted a chemical complex that included *porphyrin* (a molecule found in chlorophyll, the green pigment in plant cells) and *quinone* (a type of catalyst).

When the scientists suspended the membranes in water and exposed them to sunlight, the chemical complex absorbed the light energy and transferred hydrogen ions from the water to the interior of the membranes. When the ion concentration reached a certain level, it activated an *enzyme* (a protein catalyst)

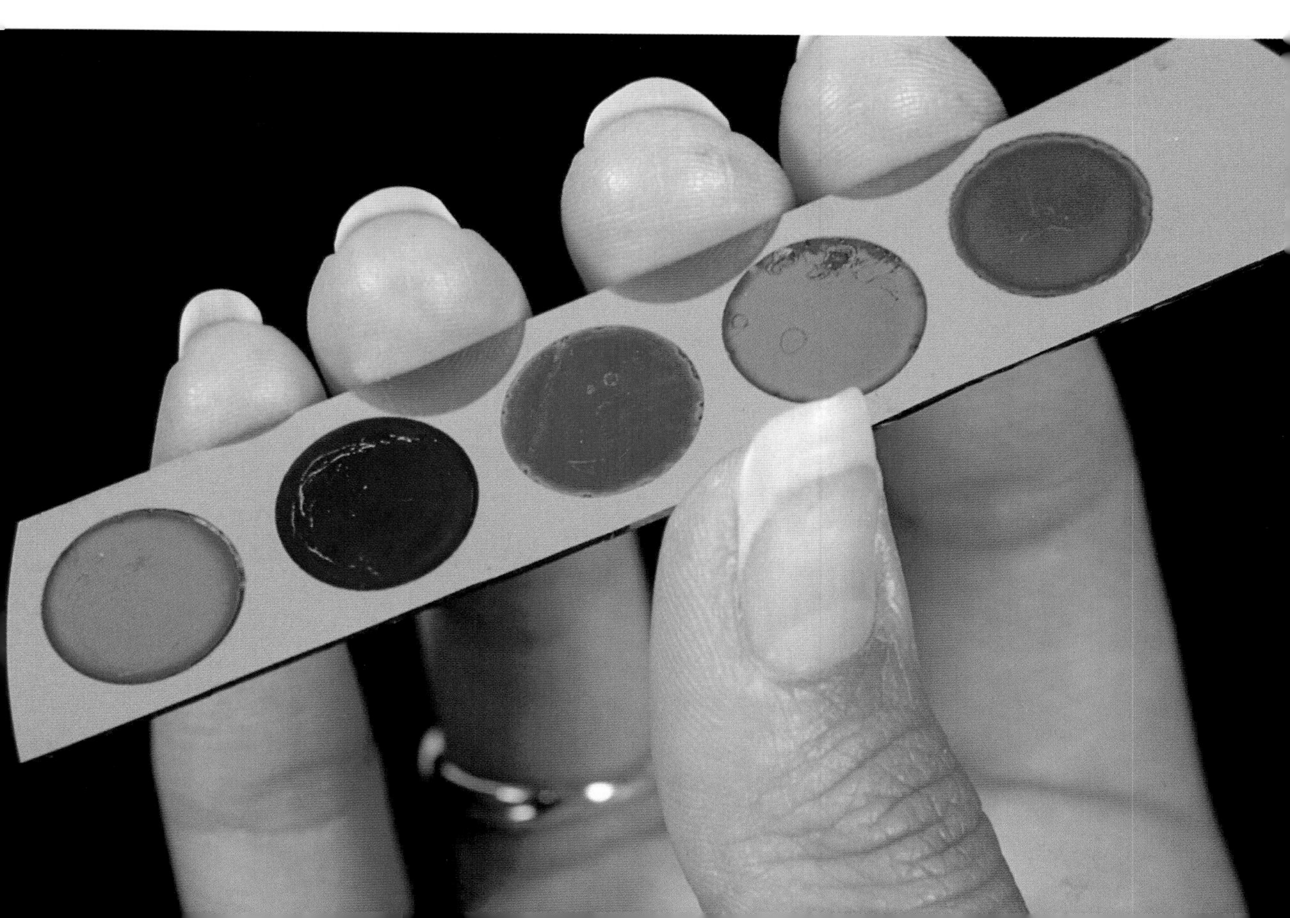

Colorful sensors
Silicon chips, each consisting of millions of microscopic pillars, display different colors. In October 1997, California chemists said they had developed the chips, similar to those used in computers, to identify unknown chemicals in solutions. Light waves beamed through the pillars and simultaneously reflected from the base of the chip combine to produce a certain pattern, which is dependent on the chemical in the solution. A computer then analyzes the pattern to identify the chemical. The color that the wafers display is also indicative of the chemical.

called ATP synthase. This enzyme then manufactured a high-energy molecule called ATP (adenosine triphosphate) from materials in the membrane. This process duplicated what happens inside the leaves of plants, where ATP is produced to store energy from the sun until the plant needs it for growth.

The researchers acknowledged that practical applications of the new membrane system are probably far off. However, they said that their use of sunlight to make ATP was an important step toward the goal of using sunlight to drive chemical manufacturing processes.

Artificial enzyme. In January 1998, chemists at Stanford University in Stanford, California, announced that they had, for the first time, made an artificial compound that is similar in both structure and behavior to an enzyme. The achievement, by a team of researchers led by chemists T. Daniel Stack and Keith Hodgson, held great promise for the drug and chemical industries, which hoped that artificial enzymes might be able to withstand extreme industrial conditions better than natural enzymes. Enzymes that are obtained from nature are very delicate and often break down when exposed to heat or to strong acids or bases.

The compound prepared by the Stanford group was intended to mimic a natural enzyme called galactose oxidase, which is commonly used to manufacture drugs and perfumes. After designing the compound to have a pyramidlike arrangement of atoms similar to the natural enzyme, the researchers discovered that the artificial enzyme behaved much the same way as natural galactose oxidase. Both compounds transformed one type of organic compound called an alcohol into another type called an aldehyde under similar conditions of temperature and pressure.

The Stanford scientists said their artificial compound promoted the alcohol-to-aldehyde reaction too slowly to make it commercially useful. Nonetheless, many chemists said that this research could be the first step in the development of a new class of industrial catalysts modeled on natural enzymes.

[Gordon Graff]

Computers and Electronics

Among the happiest trends for consumers in the personal computer (PC) market in 1997 and 1998 was the continuing decline in prices. And while prices promised to drop even further, the performance and capabilities of the machines continued to improve. Other major stories included huge mergers among telecommunications and computer companies and the growth of digital broadcasting technology.

PC prices dropping. By the beginning of 1998, prices for well-equipped PC's had dipped below the $1,000 mark, with further reductions almost certain to follow. Indeed, some computer industry analysts felt that prices for similar PC's might drop below $600 by the end of 1998. Ironically, even as prices for personal computers dropped, the number of households purchasing a new PC also declined. While industry estimates showed that as many as 50 million American households had at least one personal computer, the number of households acquiring their first computer failed to rise at expected rates. This indicated that much of the unit sales growth in the consumer PC industry was occurring in households that were purchasing a second or even a third computer for home use.

New chip breakthrough. The size of computer circuits took another step downward in September 1997, when International Business Machines (IBM) of Armonk, New York, announced the development of circuits that were more than 500 times thinner than a strand of human hair. The technology, called CMOS 7S, used copper circuitry, which is far more efficient in conducting an electric current than the aluminum circuits traditionally used in computer chips. Additionally, CMOS 7S could fit up to 200 million transistors on a single chip, further shrinking the amount of space required for high-level computer functions. Because the new technology packed more circuitry into a smaller space, industry analysts expected it to speed the incorporation of more sophisticated computer functions into smaller, lighter but more powerful electronic devices, including portable telephones and computers, and small appliances.

Wireless technology. The U.S. Federal Communications Commission (FCC) began auctioning off the rights to a formerly unused spectrum of broadcast frequencies in February 1998. The frequencies were to be used to broadcast digital signals. Companies who purchased these rights planned to offer such services as wireless Internet access (at speeds up to 1,500 times faster than access over traditional telephone lines), local telephone calls, private computer networks, and high-definition video broadcasting.

The new technology, called "broadband wireless" or "local multipoint distribution system" (LMDS), uses high-frequency microwave signals instead of wires or cables. LMDS networks are less expensive to install than wire-based systems, and have a much larger data capacity. Data is carried to, and between, users over a network of microwave antennas owned by the local provider. Analysts expected the first customers for LMDS service would be businesses. Residential customers could use the same systems, but the cost of LMDS service was expected to be too high for individual consumers for several years.

World Wide Web. The phenomenal growth in the popularity of the Internet, particularly the World Wide Web (the Web), continued in 1997 and 1998. The number of registered *Web sites* (individual groupings of electronic pages at one address, featuring various assortments of text, images, video, and sound) continued to multiply at a tremendous rate. Some estimates placed the number of pages published on the Web as high as 300 million—divided among tens of millions of Web sites—a figure that was expected to double before the year 2000.

Indexing services that collected and sorted the contents of Web pages so users could access them more simply with Web *browsers* (Web-navigating programs) had become a big industry by 1998. According to the U.S. Commerce Department, the Internet was growing at a faster rate than any other technology in history, expanding from barely 3 million users in 1994 to more than 100 million worldwide in 1998.

Legal troubles for Microsoft. The Microsoft Corporation of Redmond, Washington, the world's largest *software* (computer programs) company, faced serious legal problems throughout much of 1997 and 1998. Its difficulties escalated in May 1998, when the United States Department of Justice, 20 individual states, and the District of Columbia filed suit against Microsoft, charging that the company had an unfair advantage when marketing its other software products against competitors. Such an advantage could be a violation of federal antitrust laws, which are intended to prevent any one company from dominating a market or group of markets.

Microsoft's Windows 95 software was the world's most popular *operating system* (the master program that controls all of a computer's basic functions), and was thus the most essential of all computer software. Microsoft's competitors in the Web-browser market accused Microsoft of using the immense popularity of Windows 95 to force computer makers to install Microsoft's Internet Explorer Web-browsing software on every new computer that used Windows 95. Competing companies argued that such a requirement effectively limited their ability to compete for customers.

In December 1997, a federal judge ordered Microsoft to end this practice and to provide computer manufacturers with versions of Windows 95 that did not include Internet Explorer. Microsoft appealed, seeking to overturn the decision. The situation was all the more critical for Microsoft because the company feared that the government would try to block the intended release of a new version of the Windows operating system, Windows 98, to consumers in June 1998.

With Windows 98, Microsoft had made Internet Explorer an integral part of the operating system, using the browser as the primary means of accessing files on the computer as well as on the Web. Microsoft argued that this integration would make every aspect of the computing environment—desktop machine and software as well as Internet and telecommunications—simpler for the user. But its competitors argued that this blending of operating system and Web browser would discourage consumers from purchasing Web browsers sold by other companies.

Internet2. In April 1998, U.S. Vice President Al Gore announced the debut of Internet2 (I2), a new computer-based

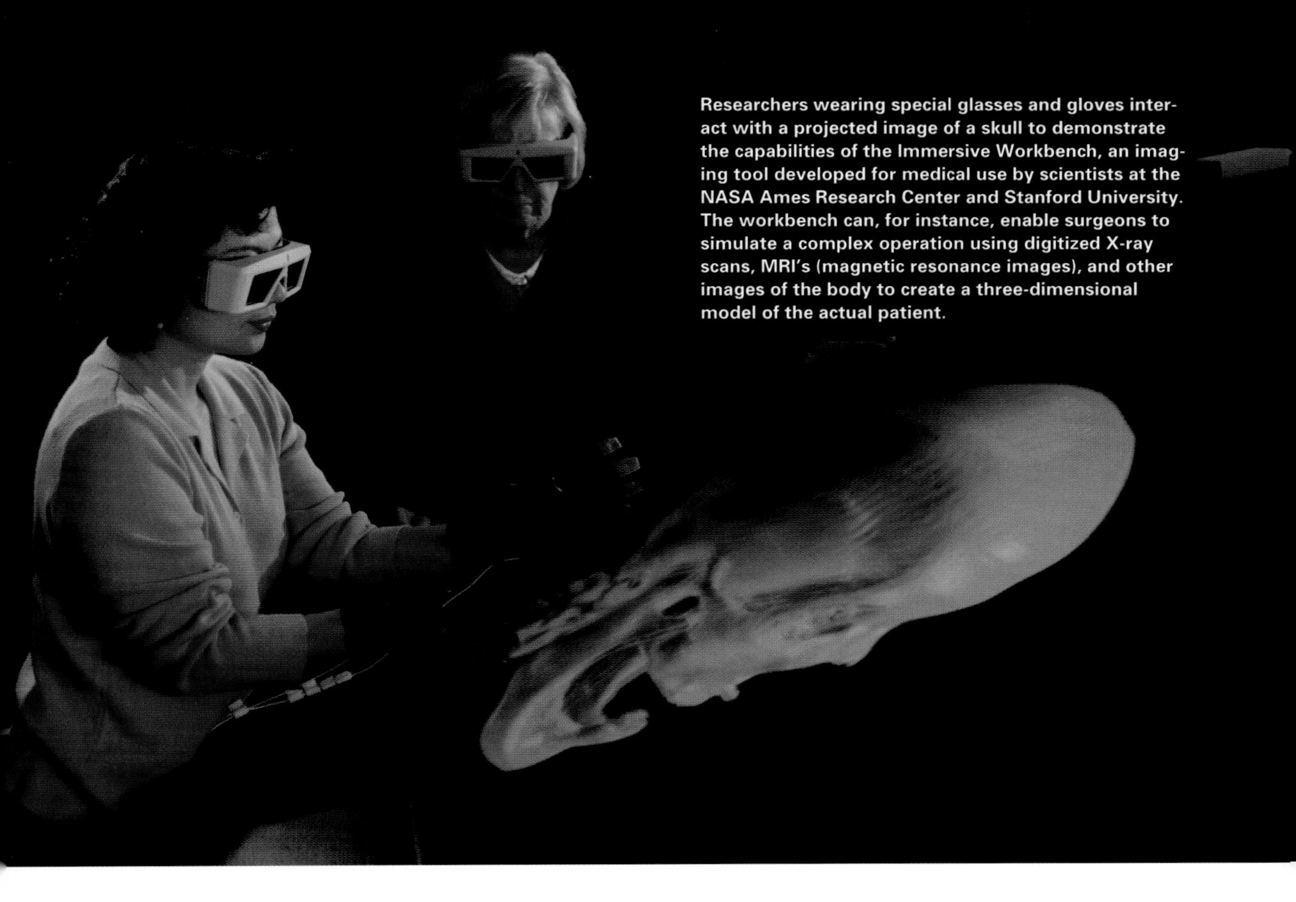

Researchers wearing special glasses and gloves interact with a projected image of a skull to demonstrate the capabilities of the Immersive Workbench, an imaging tool developed for medical use by scientists at the NASA Ames Research Center and Stanford University. The workbench can, for instance, enable surgeons to simulate a complex operation using digitized X-ray scans, MRI's (magnetic resonance images), and other images of the body to create a three-dimensional model of the actual patient.

telecommunications network that was planned to connect more than 100 universities and other institutions worldwide, beginning in fall 1998. The first Internet had been developed by a similar collaboration among the government, academic institutions, and industry over the previous three decades.

The growing popularity of the Web led to increasingly frequent "traffic jams," particularly at popular, highly visited sites. As growing numbers of people used the Web, the capacity of traditional phone lines became strained. The more information that moved over the lines, the slower the system became. It was clear that existing telephone systems would soon be unable to meet the demand for fast, reliable Internet access. While a variety of new strategies were taking shape to face this challenge, including the use of cable television lines for Internet traffic, most solutions, such as I2, were only in the introductory or planning stages by mid-1998.

Data transfers over I2, which will use state-of-the-art telecommunications networks, were expected to be up to 1,000 times faster than is possible with the existing Internet. At such high speeds, I2 users could download all 22 volumes of *The World Book Encyclopedia* in one second or less.

The U.S. government's determination to increase the speed and efficiency of the Internet was made clear in January 1998, when President Bill Clinton made that goal part of his State of the Union address. Although I2 was still in the planning stages in mid-1998, it was widely expected that the technologies developed for its implementation would ultimately reach the broader Internet and World Wide Web.

WebTV a success. The continued growth and popularity of the World Wide Web led some manufacturers to develop simple, stripped-down computers known as "set-top PC's." These devices were designed to access the Internet using a traditional television, a TV-style remote control, and a telephone line. The first device of this type, introduced in 1996 by a small California company called WebTV Networks, did not find a large market.

In August 1997, WebTV was purchased by Microsoft, and in December 1997, the company introduced a newer design, WebTV Plus. Manufactured by several major consumer electronics companies and priced below $300, the WebTV Plus set was better equipped, coming with a modem, its own television tuner, a large hard disk for data storage, and a keyboard. WebTV Plus seemed, by early 1998, to be more warmly embraced by consumers than earlier versions. Many in the consumer electronics industry hoped that such devices would attract consumers who were reluctant to purchase a complete desktop PC system, thus helping to offset the slowing growth in that market.

Industry consolidation. Several computer, electronics, and telecommunications companies merged in 1997 and early 1998, seeking to increase their ability to compete vigorously in the expanding global marketplace. These corporate consolidations generally resulted in much larger businesses, which hoped to offer a wider array of products and services to their customers.

The groundwork for a major industry consolidation was laid in January 1998, when Compaq Computer Corporation of Houston announced its intention to buy Digital Equipment Corporation of Boston for $9.6 billion. By merging with Digital, Compaq, a leading manufacturer of PC's, sought to become a major competitor in the market for larger business computers and computer networks, a huge market dominated by IBM.

Consolidation also affected the on-line services industry in September 1997 when America Online Inc. (AOL) of Vienna, Virginia, bought out one of its competitors, CompuServe Corporation of Columbus, Ohio. Customers connect to on-line services with a personal computer that is connected to telephone lines by a modem. Typical on-line services provide e-mail and Internet access as well as special content areas such as electronic magazines and chat rooms. The sale solidified AOL's position as the world's largest on-line service. By the end of 1997, AOL had more than 10 million customers. AOL officials announced that they would continue operating the far smaller CompuServe as a separate business.

The CompuServe deal was actually part of a three-way agreement. CompuServe was initially purchased by WorldCom, Inc., of Jackson, Mississippi, a fast-growing telecommunications company. By 1997, WorldCom had become one of the world's major providers of high-capacity Internet lines and long-distance services. In the deal, WorldCom kept Compu-Serve's data communications division but sold the on-line service component to AOL. In exchange, WorldCom received AOL's network communications division.

WorldCom made headlines again in October 1997 by announcing one of the largest corporate takeovers in history. WorldCom offered more than $30 billion to purchase MCI Telecommunications Inc. of Washington, D.C., the second-largest long-distance company in North America. The proposed merger was being studied by federal antitrust and regulatory officials. In any case, by mid-1998, the parties had not reached agreement on the merger.

The WorldCom-MCI deal was itself dwarfed in May 1998, when Southwestern Bell Communications (SBC) of San Antonio announced a $62-billion merger with Ameritech Corporation of Chicago. The two "Baby Bell" companies, along with five others, were created in 1984, when federal antitrust actions required that the American Telephone & Telegraph Company (AT&T) divest itself of its local telephone service. If permitted to go forward by federal officials, the merger would be the second-largest corporate merger in history and would make SBC and Ameritech one of the most powerful telecommunications companies in the United States.

Satellite glitch. The failure of a communications satellite disrupted service to pager networks and some broadcast networks in the United States in May 1998. It was the worst outage in the 35 years since communication satellites were first launched. Tens of millions of people rely on pagers, portable devices that enable them to receive electronic messages via radio signals.

The satellite, named Galaxy 4, spun out of position on May 19, and lost contact with ground-based receivers. Galaxy 4 carried as much as 90 percent of the pager traffic in the United States as well as TV and radio programming.

A 3-D experience for the home
ChequeMate's C-3D imaging system, *below,* is a VCR-sized device that digitally manipulates standard video images to create a three-dimensional effect. It then blends the manipulated picture with the original one. Special glasses, with liquid-crystal "shutters" that open and close many times each second, trick the brain into seeing the two images as a single 3-D picture. The system works equally well with either broadcast TV or recorded video.

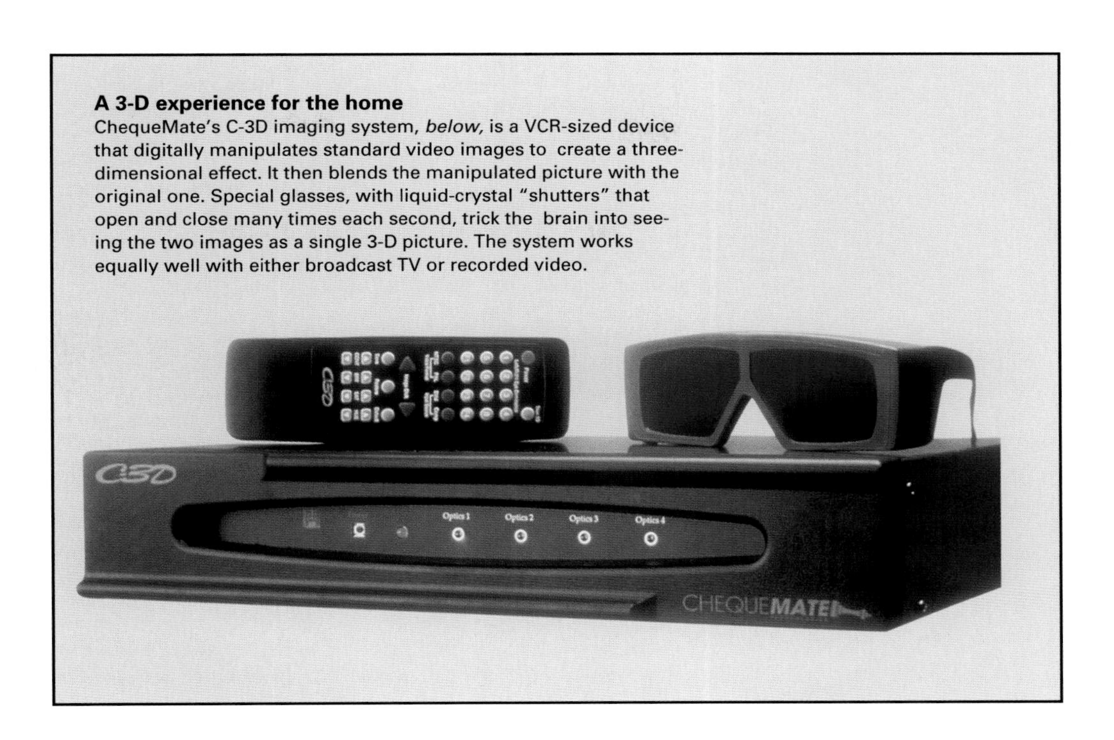

The reason for Galaxy 4's failure was not immediately known, and it was not expected to recover. The satellite's owner, PanAmSat of Greenwich, Connecticut, rerouted Galaxy 4's traffic to other systems while it moved another satellite into position to take Galaxy 4's place, a process that took several days.

Reinventing the Apple. Steve Jobs returned to the role of chief executive of Apple Computer, Inc., of Cupertino, California, in September 1997. Jobs, who cofounded Apple in 1976, was ousted as the company's chief executive in 1985, but returned as an adviser in late 1996. He became interim leader of the company at the request of Apple's board of directors in July 1997.

Sales of the company's Macintosh computer, which used Apple's own operating system, were plummeting. By 1997, the Windows operating system was running on the vast majority of the world's desktop computers. Meanwhile, Apple's share of the global desktop computer market had declined from close to 10 percent in 1994 to less than 3 percent in 1997.

Jobs shocked the industry in August 1997 by accepting a $150-million investment in the company from its former archrival, Microsoft. In September, Apple rolled out a line of desktop Macintoshes that used a speedy new version of the PowerPC *microprocessor* (the chip that executes a computer program's instructions) jointly developed by Apple, IBM, and Motorola Corporation of Schaumburg, Illinois. Called the G3, the new chip was designed to compete against the popular Pentium II microprocessor developed by Intel Corporation of Santa Clara, California, which was used in Windows-based PC's.

In April 1998, Apple announced that it had turned a profit for the second financial quarter in a row under Jobs's tenure—after eight quarters in a row of losses. In May 1998, Apple introduced the iMac, a sleek desktop machine optimized for accessing the Internet. In a statement introducing the new product, Jobs described the iMac as "next year's computer for $1,299; not last year's computer for $999." The iMac was due out in August 1998. [Keith Ferrell]

During 1998, the governments of the United States and Canada considered proposals to drastically reduce the ballooning numbers of lesser snow geese in order to prevent the birds from destroying their Arctic nesting habitat. The proposals—which included extending the length of the hunting season and increasing hunting in and around refuges—were made in June 1997 by the Arctic Goose Habitat Working Group. The 17-member group included scientists from government conservation agencies, private conservation organizations, and universities.

Biologists estimated that the number of lesser snow geese that bred in the Canadian Arctic and wintered along the Gulf Coast of the United States was between 4.5 million and 6 million birds in 1997—a 300 percent increase since 1969. This huge population is a major problem in the birds' breeding range. As the birds forage for food, including the roots and leaves of many plants, they strip the tundra of vegetation. Lacking the insulation provided by the blanket of vegetation, the underlying *permafrost* (layer of frozen soil) melts, forming a barren wasteland of mudflats that is of no value to the geese or any other wildlife. Once stripped, the tundra may never recover.

The phenomenal growth of the lesser snow goose population is related to the rise of modern agricultural practices and the creation of numerous wildlife refuges in the Southern United States. The mechanical equipment used to harvest corn and rice on farms leaves a large amount of grain lying in the fields. This waste grain is a good source of food for the geese during the winter—increasing the survival of the birds. Wildlife refuges nurture the geese by providing even more food and limiting hunting.

The proposals made by the Arctic Goose Habitat Working Group were designed to double or triple the number of snow geese that hunters are allowed to kill each year. Hunters killed between 300,000 and 500,000 snow geese annually in the 1990's. The higher limit would be maintained until the goose population was reduced to 50 percent of its 1997 level. Some of the proposals, if enacted, would require changes in the Migratory Bird Treaty between Canada and the United States, which has protected North American waterfowl since 1916.

Flies versus ants. In July 1997, scientists with the U.S. Department of Agriculture (USDA) released several thousand phorid flies near Gainesville, Florida, in the hope that the parasitic insects would attack and destroy fire ants—one of the most annoying pests in the Southern United States. Since escaping from ships docked at Mobile, Alabama, in the 1920's, the stinging South American ants have spread to 11 Southern states as well as to Puerto Rico. Conservationists hoped that control of the fire ants might protect wildlife that suffer from the ants' stings and allow populations of harmless native ants to rebound.

Fire ants aggressively attack any person or animal that disturbs their basketball-sized nest mounds. Each ant stings repeatedly, producing multiple pus-filled bumps on the skin of its victims. Conservationists said that chemical insecticides, the most common weapon used to control fire ants, pollute the environment and have not always been effective against the ants.

A female phorid fly, which is the size of a pinhead, lays a single egg inside the body of a fire ant. Each female can lay more than 100 eggs, and thus can implant a biological time bomb in more than 100 ants. The tiny *larva* (immature fly) that hatches from an egg eventually crawls into the ant's head, which soon falls off. After the fly completes its development, it emerges from the mouth of the severed head.

In addition to the Florida release, USDA scientists released phorid flies in Texas in early 1998 and planned to eventually release the flies in other states. Although phorid flies were not expected to eliminate fire ants in the United States, scientists said the flies could at least make the ants much less troublesome to humans, livestock, and wildlife.

Lemur release. In November 1997, five black-and-white ruffed lemurs born and raised at the Duke University Primate Center in Durham, North Carolina, became the first lemurs reared in captivity to be released in their native habitat—the rain forests of Madagascar. Lemurs are cat-sized primates, the group of animals that also includes

North America's rich but imperiled "ecoregions"

Scientists with the World Wildlife Fund, an international conservation organization, reported in September 1997 that 13 of 116 "ecoregions" in North America contain just as much natural diversity as the Everglades in Florida, one of America's richest natural areas. An ecoregion is a large area that is relatively uniform in species, climate, and geology. The report said that the 13 ecoregions cited (listed and mapped below) are imperiled by development, pollution, and other factors.

1. Hawaiian Moist Forests
2. Hawaiian Dry Forests
3. Appalachian Mixed *Mesophytic* (moderate moisture) Forests
4. Southeastern Mixed Forests
5. Northern California Coastal Forests
6. Southeastern *Conifer* (cone-bearing tree) Forests
7. Florida Sand Pine *Scrub* (stunted tree or shrub)
8. British Columbia Mainland Coastal Forests
9. Central Pacific Coastal Forests
10. Klamath-Siskiyou Forests (containing many conifers)
11. Sierra Nevada Forests
12. Central Tall Grasslands
13. California Coastal Sage and *Chaparral* (an area of shrubs and small trees)

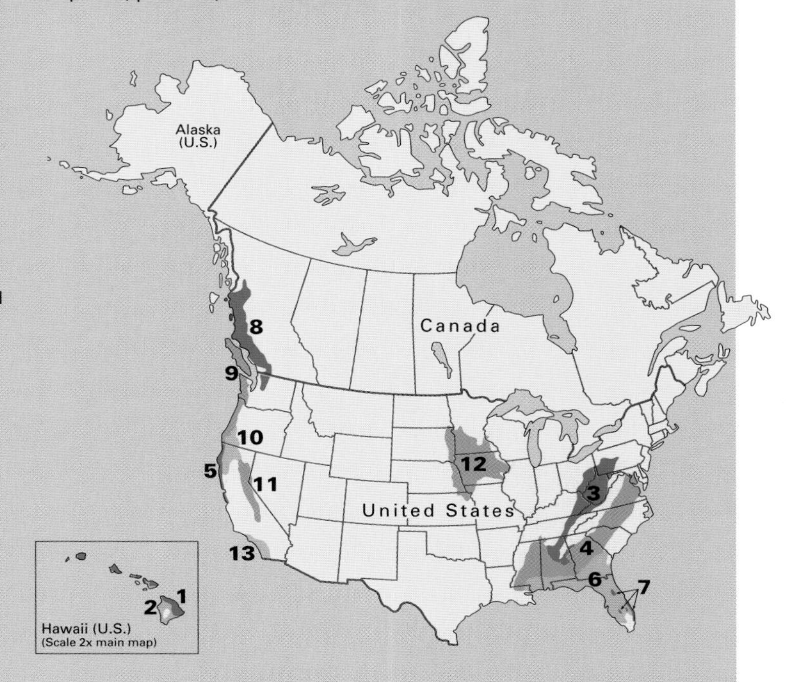

A marbled murrelet enjoys a meal of fish, *above.* This threatened sea bird nests in the temperate coastal rain forests of Siuslaw National Forest in Oregon, *right.* Its home lies within the Central Pacific Coastal Forest ecoregion, which has been logged so intensively that only small pockets of rain forest remain.

monkeys, apes, and humans. Many species of lemurs are among the world's most endangered animals. Their forest homes have been cleared for timber and farmland, and the island's native people have long hunted lemurs for food.

The five animals from Duke were released in Madagascar's Betampona Natural Reserve, a tropical rain forest of more than 2,025 hectares (5,000 acres), where the dangers to lemurs are not as great as in other areas. Between 30 and 35 black-and-white ruffed lemurs were already living in the reserve, but biologists said the reserve could support at least 300 of the primates.

Conservationists with the lemur-release program, an international project involving many zoos and conservation organizations, planned to keep track of the released animals for several months to monitor their progress. They also planned to release about 15 additional black-and-white ruffed lemurs in Madagascar over a three-year period.

New threat to whales? One of the major concerns of conservationists in 1998 was the proposed construction of the world's largest salt production facility on the Pacific coast of Baja California in Mexico. A number of U.S. and Mexican conservation groups argued that the salt facility, planned by a company jointly owned by the Mexican government and the Japanese industrial giant Mitsubishi International Corporation, would pose a threat to gray whales. The whales give birth to and nurse their young in the nearby San Ignacio Lagoon for three months every year.

Gray whales represent one of the greatest success stories in the history of conservation. Since being hunted nearly to extinction in the 1800's and early 1900's, gray whales had rebounded to their original, prehunting population of about 21,000 by the 1990's. In 1994, the whales were removed from the endangered species list. Conservationists feared that the proposed salt factory might suddenly reverse this progress.

The salt factory would produce more than 7 million metric tons (7.7 million tons) of salt each year for export. To achieve such an enormous output, Exportadora de Sal, S.A. (ESSA), the company planning the facility, would build 300 square kilometers (115 square miles) of evaporation ponds—ponds in which seawater is evaporated, leaving only salt—on land near the lagoon. About 22,700 liters (6,000 gallons) of salt water would be pumped from the lagoon into the ponds every second—an amount that some conservationists believed could alter the lagoon's ecosystem and harm whale calves.

Also of concern was a proposed concrete pier that would be about 1.6 kilometers (1 mile) long. Conservationists feared that the pier might interrupt the path of gray whales migrating along the coastline. In addition, ships moored at the pier would present a risk of oil spills.

ESSA countered these concerns by claiming that the risks posed by the factory would be minimal. For example, the company said that supports for the pier would be widely spaced to allow the whales free movement along the coastline, and that the lagoon water pumped into the evaporation ponds would be replenished by seawater from the Pacific. During 1998, ESSA was working on studies to better determine the environmental impact of the salt facility.

Wolf program illegal. A federal judge in Wyoming ruled in December 1997 that the U.S. Fish and Wildlife Service's program to return gray wolves to Yellowstone National Park and central Idaho was illegal. Wolves had been exterminated from these areas by a federal program in the 1930's. The judge, acting on a lawsuit brought by ranchers and farmers, ordered the removal of dozens of wolves released by scientists in the areas since 1995. Also to be removed were all the offspring of those wolves. The judge, however, delayed the enforcement of his ruling until the Fish and Wildlife Service and conservation groups had a chance to appeal.

The judge based his ruling on the fact that because the released animals were designated an exception to the Endangered Species Act, they were not protected from being shot if they preyed on livestock. By contrast, gray wolves that have moved naturally from Canada into the same areas are fully protected as an endangered species and cannot be legally shot. The judge decided that because there was no way for a farmer or rancher to tell the difference between the two groups of wolves, migratory wolves might be mistaken for released animals

Newly freed lemur
A black-and-white ruffed lemur peers from a branch after being released into the rain forest in Madagascar in November 1997. It was one of five lemurs raised in captivity at Duke University in Durham, North Carolina, that were released by an international group of conservationists who are trying to rebuild the natural population of these endangered primates. The group planned to release 15 additional lemurs over a three-year period.

and shot. That, he said, would undermine the Endangered Species Act.

About 160 wolves were affected by the removal decision as of mid-1998. But biologists said that number could double by the time a decision was reached on the appeal.

Wolves in Arizona. Federal biologists released 11 Mexican gray wolves into the White Mountains of eastern Arizona in March 1998. Scientists believed that wolves had been absent from the Southwestern United States since the 1970's, when they were eradicated. Plans called for an eventual release of some 100 wolves in the Apache and Gila national forests in the region.

Endangered species list. Gray wolves, bald eagles, and more than two dozen other species were to be either removed from the endangered species list or have their status improved from "endangered" to "threatened," according to a May 1998 announcement by the U.S. Department of Interior. Department Secretary Bruce Babbitt said the decision reflected the fact that the Endangered Species Act, criticized by some observers, had worked to protect those species. Before any species could be removed from the list, however, a lengthy process of public commentary on the move was required.

Galapagos Islands protection. In March 1998, the government of Ecuador approved measures to increase protection of the unique plants and animals on the Galapagos Islands, a group of volcanic islands in the Pacific Ocean that are a province of Ecuador. Conservationists had been concerned that the islands were becoming increasingly threatened by overfishing, overpopulation, and alien plants and animals. British naturalist Charles Darwin, who proposed the theory of evolution, studied wildlife on the islands in the 1830's.

Among the protective steps taken were an expansion of the no-fishing area around the islands from 24 kilometers (15 miles) to 64 kilometers (40 miles), a reduction in the number of people allowed to live on the islands, and the institution of a system of inspection and quarantine for foreign species brought to the islands. [Eric G. Bolen]

Deaths of Scientists

Jacques Cousteau

Elizabeth B. Keller

Mary Sears

Notable scientists and engineers who died between June 1, 1997, and June 1, 1998, are listed below. Those listed were Americans unless otherwise indicated.

Barton, Derek H. R. (1918–March 16, 1998), British scientist who shared the 1969 Nobel Prize in chemistry for adding a third dimension to the analysis of chemical compounds. In 1950, Barton published "The Conformation of the Steroid Nucleus," which showed that organic molecules are three dimensional and that chemical properties of an organic molecule can be inferred from its form. Three-dimensional, or conformational, analysis altered human understanding of organic molecules and how they interact in life processes. It provided a theoretical framework for computer programs used to design drugs, making Barton's work central to chemical research and development.

Cousteau, Jacques (1910–June 25, 1997), French oceanographer who developed underwater cameras; designed a submarine for ocean research; and built a series of underwater laboratories to prove that humans could work and live on the ocean floor. He coinvented the Aqua-Lung, the first scuba device, which allowed skindivers to reach much greater depths than had been previously possible. In 1953, Cousteau published *The Silent World,* a first-person account of the development of scuba diving, which sold millions of copies and was made into a documentary that won a 1957 Academy Award. To finance an underwater laboratory, Cousteau produced a National Geographic television special, broadcast in 1966, which spawned a series of similar programs. Although Cousteau, who held no scientific degree, was attacked by oceanographers questioning the scientific value of his work, he incontestably explored more of the Earth and opened more of its surface to human endeavor than anyone else in history.

Esau, Katherine (1898–June 4, 1997), international figure who dominated the study of plant anatomy and physiology during much of her career, which stretched from czarist Russia to the 1990 publication of her final scientific paper. Esau is credited with laying the foundation for much current research on plant physiology. In the process of researching the effects of viruses on plant tissues, she noticed that virus particles travel from cell to cell along structures—called plasmodesmata—that are actually smaller in diameter than the viruses they carry. Esau is credited with widening the scope of molecular biology by pointing out this phenomenon to fellow scientists and suggesting that research into virus transportation might unlock secrets of cellular function. Her own work expanded from the creation of plant hybrids in agriculture to light-microscope studies of plant anatomy, including how certain tissues transport food from plant leaves to roots. In 1989, Esau became the first botanist to be awarded the National Medal of Science.

Hitchings, George H. (1905–Feb. 27, 1998), biochemist who, with colleague Gertrude Elion, shared the 1988 Nobel Prize for physiology or medicine for research that led to the creation of drugs for the treatment of gout, leukemia, malaria, and meningitis as well as for septicemia and other bacterial infections. Hitchings and Elion's research also produced drugs to treat human immune system disorders. Those drugs included Imuran, which eventually made organ transplants possible, and AZT, used in the treatment of AIDS. Hitchings also developed a biochemical approach to chemotherapy, used to treat cancer. The approach is based on his theory that the growth of cancerous cells can be slowed or checked with chemical compounds that selectively kill rapidly dividing cells. Scientists have estimated that the research Hitchings and Elion carried out eventually produced drugs that saved more than 1 million lives in Hitchings's own lifetime.

Keller, Elizabeth B. (1917–Dec. 27, 1997), biochemist who uncovered how proteins are made by cells and who devised a model to describe transfer RNA, the molecule that assembles proteins from building blocks called amino acids. Using paper, pipe cleaners, and pieces of Velcro, Keller fashioned models of the subunits of a particular type of transfer RNA to explain the role it plays in attracting certain amino acids and linking them into proteins. The model was published by Robert W. Holley, who included it with his own work on deciphering the genetic code of RNA, for which he was awarded a share of the 1968 Nobel Prize for physiology or medicine.

David N. Schramm

Robert Serber

Eugene Shoemaker

Kendrew, John C. (1917–August 23, 1997), British molecular biochemist who shared the 1962 Nobel Prize for chemistry with Max Perutz for discovering the structure of some major proteins. Using X rays, Kendrew unraveled the complexities of myoglobin—the red protein in muscles—while Perutz concentrated on hemoglobin, the oxygen-carrying protein in red blood cells. Their work helped establish molecular biology as a new field of study and furthered knowledge of protein function.

Matthews, Drummond (1931–July 20, 1997), geophysicist whose theory that the ocean floor is spreading provided much of the foundation on which the theory of plate tectonics was based. The Earth's crust, according to this theory, consists of a collection of rigid plates moving in relation to each other. Matthews advanced his ideas on sea-floor spreading to explain the existence of alternating positive and negative magnetic bands on the ocean bottom.

Schramm, David N. (1945–Dec. 19, 1997), leading astrophysicist who conducted pioneering studies on the formation of the universe and demonstrated how the light elements—hydrogen, helium, and lithium—were produced. Schramm championed the big bang theory of the origin of the universe and prodded the scientific world into taking it seriously. He is credited with playing a major role in melding particle physics with astrophysics and nuclear physics to gain insights into the early universe. He calculated that ordinary matter makes up little of the total mass of the universe, and he insisted, in the face of considerable scientific argument, that only three families of elementary particles exist in the universe, a prediction proven correct in 1989 through the use of particle accelerators.

Sears, Mary (1905–Sept. 2, 1997), researcher with the Woods Hole Oceanographic Institution, in Massachusetts, and author and editor of marine publications. Sears began at Woods Hole in 1932 as a research assistant and remained for more than 60 years, first as a planktonologist and then as a senior scientist in the biology department. In the years when Woods Hole was a summer operation, Sears taught at Wellesley College, in Wellesley, Massachusetts, and conducted research at Harvard University. She worked with naval intelligence during World War II (1939– 1945), leading the Oceanographic Unit of the Navy Hydrographic Office and coordinating naval oceanographic research at Woods Hole. In 1959, she chaired the First International Congress on Oceanography at the United Nations. Sears edited several journals, including *Deep-Sea Research,* which she founded in 1953, as well as *Oceanography,* published by the American Association for the Advancement of Science. In 1980, *Deep-Sea Research* dedicated an entire issue to Sears, who was characterized as having done more to advance oceanography than any other woman in the field.

Serber, Robert (1909–June 1, 1997), theoretical physicist with the Manhattan Project between 1941 and 1945 who, with physicist J. Robert Oppenheimer, laid the theoretical foundation for the creation of the atomic bomb. Colleagues described Serber, who was able to thoroughly absorb and then communicate ideas, as an "ideal bridge" between theoretical and practical scientists working on the project. After the bomb was tested, Serber advised the military on its use and was with the first group of Americans to enter the atom-bombed Japanese cities of Hiroshima and Nagasaki to assess the damage. After World War II, Serber lectured widely to experimental physicists in the hope that nuclear energy could be channeled into peaceful uses.

Shoemaker, Eugene (1928–July 18, 1997), astronomer who, with his wife Carolyn, is credited with the record discovery of 32 comets and 1,125 asteroids. The couple's most famous find—Comet Shoemaker-Levy, discovered with David Levy in March 1993—actually consisted of fragments from a comet that had been ripped apart by the gravitational pull of Jupiter. Shoemaker calculated that these fragments would strike Jupiter, producing a tremendous spectacle. In July 1994, more than 20 comet pieces smashed into the planet, igniting fireballs visible on Earth. A planetary geologist with the U.S. Geological Survey, Shoemaker was well known for his conviction that the impact of a comet or asteroid caused the extinction of the dinosaurs and that comets and asteroids continue to threaten life on Earth.

[Scott Thomas]

The marketing of any new drug in the United States requires formal approval by the federal Food and Drug Administration (FDA). In 1997 and early 1998, the FDA approved some 40 new drugs. Among the new medicines getting the agency's nod were drugs for the treatment of impotence, schizophrenia, Parkinson disease, and obesity.

Impotence remedy. The FDA in March 1998 approved sildenafil citrate (trade name Viagra) as the first drug designed to treat impotence. Between 10 million and 20 million men, particularly those who are middle aged or elderly, suffer from impotence. Impotence is characterized by the inability to achieve or maintain an erection.

In clinical trials involving more than 3,000 men with varying degrees of impotence associated with physical conditions such as diabetes, spinal-cord injury, and prostate ailments, the drug was up to 70 percent effective. Up to 89 percent of men whose impotence was due to psychological causes said they were helped by the drug.

Viagra is taken by mouth about one hour before sexual intercourse. The drug does not directly cause penile erection but rather prolongs the effects of a naturally occurring chemical in the penis that causes muscles in the spongy, erectile tissues to relax, allowing the arteries to expand and fill with blood, resulting in an erection.

Men participating in the clinical study reported mild side effects from the drug, including headaches, flushed skin, and indigestion. About 3 percent of the patients also reported temporary vision problems, ranging from blurred vision to a green or blue halo effect. Also, because Viagra can cause a sudden drop in blood pressure, experts warned that the drug poses a health risk to men who are taking certain medications to control high blood pressure.

A new therapy for schizophrenia was approved by the FDA in September 1997. Clinical tests on quetiapine, marketed under the name Seroquel, showed the drug to be highly effective in managing the symptoms of schizophrenia in a number of patients. The most commonly reported side effects of quetiapine included sleepiness, nasal stuffiness, and minor weight gain.

Schizophrenia is a mental illness that causes severe functional impairment in patients. The condition usually occurs before the age of 45, with 75 percent of patients developing the disease between the ages of 16 and 25.

Researchers believe that there is excessive chemical activity in the brains of schizophrenic patients. Nerve cells secrete chemical substances called neurotransmitters to communicate with other nerve cells. However, in schizophrenic patients, experts believe, certain regions of the brain secrete an abundance of a neurotransmitter called dopamine. Quetiapine is one of a new class of drugs that are called atypical antipsychotic drugs. Researchers theorize that these medications work both by blocking the excessive dopamine activity in the brain and by altering the activity of other brain chemicals.

Treatments for Parkinson disease. Two new drugs for the management of Parkinson disease—ropinirole (sold as Requip) and pramipexole (sold as Mirapex)—were made available in the United States in 1997. Both drugs are used to treat the symptoms of Parkinson disease but not the cause, which involves the death of brain cells that produce dopamine.

Ropinirole and pramipexole improved the symptoms of Parkinson patients by mimicking dopamine in the brain, researchers who tested the drugs reported. The drugs worked specifically in the parts of the brain that are most affected by Parkinson disease.

Parkinson disease affects between 500,000 and 1 million Americans. The symptoms of the disorder, which include tremors, rigidity, slowed and delayed muscle movement, and the inability to stand straight, develop slowly but continue to worsen over time. Because Parkinson patients suffer from insufficient dopamine activity in the brain, physicians usually treat the condition with a drug called levadopa, or L-dopa, which reinforces the action of the brain's remaining dopamine. While the use of L-dopa is effective in controlling the symptoms of the disease in many patients, many of the patients may develop severe side effects. In addition, they slowly become less responsive to the therapeutic action of the drug.

Although both ropinirole and pramipexole caused side effects, including

New drug in the battle against baldness
A man with male pattern baldness, *below, left,* later exhibits moderate hair regrowth, *below, right,* in clinical studies of the antibaldness prescription drug finasteride, sold under the name Propecia. The U.S. Food and Drug Administration in December 1997 approved the use of Propecia. The drug promoted hair growth or stopped hair loss in 83 percent of the men who participated in research studies.

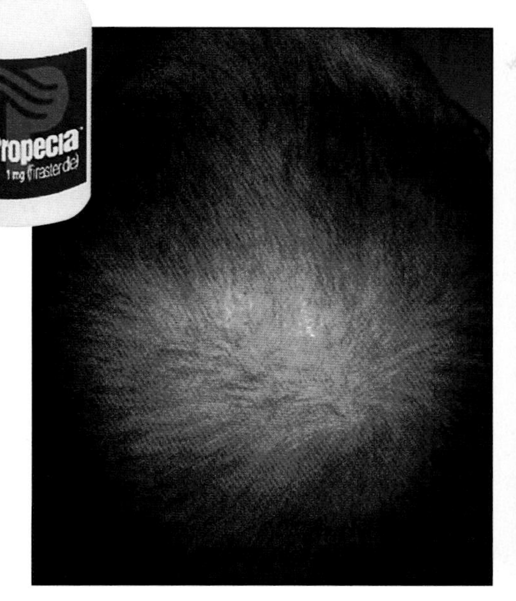

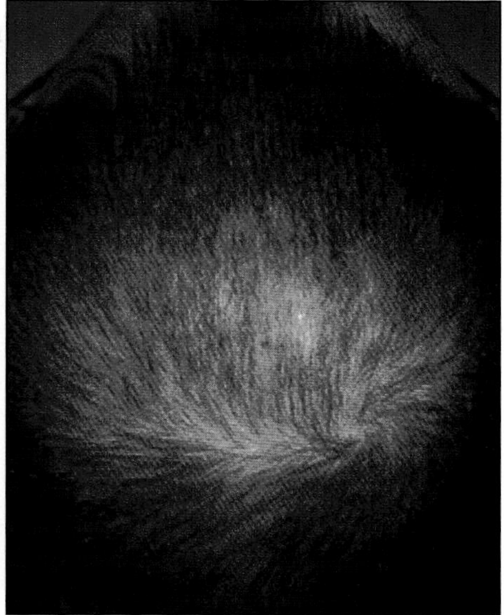

nausea and dizziness, in some Parkinson patients, researchers said the new drugs offered several advantages. For example, the drugs may make it possible to delay the use of L-dopa until later stages of the disease. In clinical studies, the two new drugs also improved patients' response to L-dopa when used in combination with it, and allowed the researchers to administer lower dosages of L-dopa.

A new treatment for obesity was made available in November 1997 when the FDA approved the use of sibutramine, sold under the brand name Meridia. Clinical studies of sibutramine showed that the drug diminished appetite by increasing the levels of two brain chemicals—serotonin and norepinephrine—and slowing the breakdown of the serotonin produced naturally by the body. Research on obesity has indicated that appetite control in humans is regulated by the levels of serotonin and norepinephrine in the brain. In the clinical trials, people who used sibutramine for one year, in addition to exercising and following a prescribed diet, lost an average of about 4.3 kilograms (10 pounds) more than participants who had received a *placebo* (an inactive substance).

However, the FDA cautioned that sibutramine carries several risks, including elevated blood pressure and increased pulse rate. Thus, the drug is not recommended for patients with high blood pressure or an irregular heartbeat or for people who have survived a stroke. Other side effects may include constipation, difficulty sleeping, and agitation. The FDA said that the drug should be used only in combination with a reduced-calorie diet and only by seriously obese patients.

Meridia became the first drug approved by the FDA for combating obesity after two popular weight-loss drugs were barred from use. In September 1997, the FDA removed the drugs fenfluramine, which had been sold as Pondimin, and dexfenfluramine, which had been sold as Redux, after physicians linked their use to potentially life-threatening heart valve damage.

[Thomas N. Riley]

Different species of trees exchange large amounts of carbohydrates through networks of fungi connecting their roots, according to an August 1997 report by ecologists led by Suzanne Simard of the British Columbia Ministry of Forests in Canada. Carbohydrates are energy-rich carbon compounds that are produced by plants through the process of photosynthesis, in which carbon dioxide and water are chemically combined using the energy of sunlight. This study indicated that, rather than being competitors for energy resources as ecologists had thought, trees may, in effect, help each other by sharing energy resources.

Simard had shown in 1995 that fungi called ectomycorrhizal fungi cover up to 90 percent of the root tips of *coniferous* (cone-bearing) trees such as the paper birch and Douglas fir. (Similar fungi also live in or on the roots of other land plants.) This relationship enables the fungi to obtain carbohydrates and other nutrients from the tree roots, while the trees obtain various minerals, such as nitrogen and phosphorus, from the fungi. Fungal strands called hyphae form an extensive underground network that connects different trees to one another.

To learn what amount of carbohydrates was being exchanged between trees through this network, Simard in her 1997 study sealed young paper birches and young Douglas firs, all growing in close proximity, in airtight plastic bags and exposed them to carbon dioxide gas. The gas to which the birches were exposed contained an *isotope* (form) of carbon different from the one in the gas that was delivered to the firs.

After allowing the plants to process the carbon for several days, Simard chemically analyzed the foliage and roots of each tree. She found not only that a large amount of carbon had been passed among the trees but also that the carbon transfer occurred in both directions through the fungal hyphae.

In addition, Simard found that when a tree was shaded from sunlight (thereby making it unable to perform photosynthesis and produce carbohydrates), there was an increase in the amount of carbohydrates that the fungal network delivered to that tree's roots from adja-

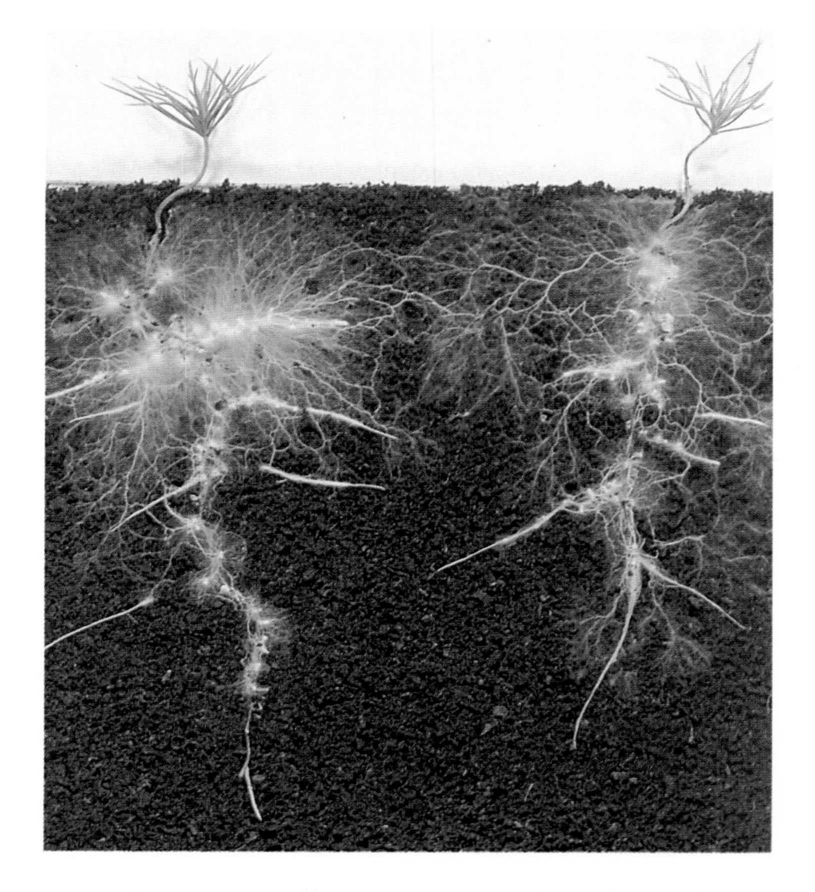

An extensive network of whitish fungi covers and connects the roots of two *saplings* (young trees) of different conifer species. Canadian ecologists reported in August 1997 the surprising finding that large amounts of energy-bearing carbohydrates flow between trees of different species through such fungal networks. This finding implied that trees in a forest share, rather than compete for, energy resources.

Two ants visit the "food bodies" on an acacia leaf. The leaves of acacia trees in eastern Africa produce these fat and protein sources to attract certain species of aggressive ants, which protect the trees from other hungry insects. In July 1997, zoologists explained how these trees get their flowers pollinated by bees despite being patrolled by the ants. The zoologists, at the University of St. Andrews in Scotland and Oxford University in England, found that acacia flowers produce a chemical substance that repels the ants but not bees. After pollination, the flowers stop producing the chemicals and the ants return to their guard duty.

cent trees. This finding indicated that tree seedlings growing up in the shade of mature trees are able to obtain nutrients from surrounding trees, even ones of different species.

Because of the surprising results of this study, ecologists expected new areas of research to open up into the nature of forest ecosystems. Such research should help clarify the role of fungal networks in maintaining diversity and stability in forest ecosystems.

Plants defenses against insects. New findings on the methods that plants use to protect themselves from insects were reported in February 1998 by *entomologist* (insect specialist) Anurag Agrawal of the University of California at Davis. Agrawal found that *induced resistance,* a type of defense that a plant develops after an animal starts to eat it, works both as a major deterrent to insects and a stimulant to increased seed and flower production. In contrast to induced resistance, some plants protect themselves with *constitutive resistance,* defenses that are maintained at all times, even in the absence of insect attacks.

Agrawal provoked induced resistance in wild radish plants early in the growing season by placing caterpillars of the cabbage white butterfly on the plants' leaves. In response to the leaf damage caused by the hungry caterpillars, the plants increased their secretion of certain toxic substances. The plants also grew greater numbers of bristly hairs, called trichomes, on their leaf surfaces.

Agrawal found that these defenses, besides deterring further attacks by the caterpillars and other insects, resulted in a 60 percent increase in seed production by the plants. This increase was in comparison with plants that had not been provoked into induced resistance early in the season and were later attacked. Agrawal also found that the plants he had provoked produced more flowers than the other plants.

This study did not address why a plant would evolve an induced mechanism of deterrence rather than a constitutive one. However, Agrawal suggested some reasons, including the possibility that plants with induced resistance are less likely to be attacked by insects that have developed adaptations to constitutive resistance.

Ecological "dead zone." In August 1997, the White House Office of Science and Technology launched an 18-month investigation into how to stop a worsening ecological disaster called the "dead zone" off the coast of Louisiana. This zone is an area of stagnant, oxygen-depleted water that cannot sustain fish or other marine life. The zone appears every summer and since 1993 it has grown to at least 17,600 square kilometers (6,800 square miles) in size and has had a devastating impact on both sport and commercial fishing.

Oceanographer Nancy Rabalais of the Louisiana Universities Marine Consortium and coastal ecologist Eugene Turner of Louisiana State University in Baton Rouge had linked the dead zone to nitrogen-based fertilizers that drain from farms in the Mississippi River Basin and are carried to the Gulf of Mexico. These fertilizers cause *blooms* (sudden, uncontrolled growths) of algae that accumulate when the Gulf currents decrease each summer. As masses of the algae die, they are decomposed by large numbers of bacteria, which use up the oxygen in the water.

Turner discovered the dead zone in 1974, when it was just a small area of water. The phenomenal growth of the zone began with the massive floods that hit the Midwestern United States in 1993. Turner and Rabalais theorized that the floods may have caused the sudden spread of the zone by pouring huge amounts of fertilizer into the Gulf. The researchers said they were uncertain, however, why the dead zone continued to grow to such an enormous size each year after 1993.

Some ecologists speculated that the dead zone is part of a larger problem in which excess nitrogen is overwhelming a number of coastal areas and other ecosystems around the world. Besides depleting coastal waters of oxygen and fish, excess nitrogen also weakens trees, making them more susceptible to disease. The nitrogen comes not only from fertilizers but also from nitrogen-oxide emissions by automobiles and factories.

[Robert H. Tamarin]

Energy

Fuel cells continued to move into more areas of commercial use in 1998 because of their high efficiency and reliability, low environmental impact, and adaptability to various fuels. A fuel cell is a device that converts chemical energy to electrical energy by combining a fuel and an *oxidizer* (a substance that removes electrons in a chemical reaction).

Emission-free bus. In September 1997, the Chicago Transit Authority (CTA) added the world's first zero-emission fuel-cell buses to its fleet of vehicles. The three experimental buses were fitted with fuel cells designed by Ballard Power Systems of Vancouver, Canada.

The fuel cells used hydrogen, produced from natural gas, to power a series of chemical reactions. The reactions generated an electric current, which was used to drive an electric motor. The only emission produced by the buses was water vapor.

The CTA began a two-year study of the fuel-cell technology in 1995 and began testing the first zero-emissions bus on city streets in 1996. The 1997 study marked the first time fuel-cell buses were used as part of a regular service route. Researchers planned on using data from the study and the regular use of the buses to create "the next generation" of clean transportation systems in North America.

Activated-radical combustion. Motivated by an increasing interest in fuel economy, reliability, and reduction of pollution emissions, researchers at Sonex Research, Incorporated, in Annapolis, Maryland, announced in 1998 that they were making progress on a new type of combustion process called activated-radical combustion.

Exhaust released from an engine usually includes smoke (traces of unburned fuel and soot), carbon dioxide, carbon monoxide, and water vapor. However, the exhaust also contains extremely small, harmless amounts of chemicals called activated radicals. Researchers at Sonex Research developed a method to trap these chemicals in exhaust fumes and then use them as a *catalyst* (a substance that speeds up a chemical reaction) to promote combustion rather than using a spark plug. The presence

The Chicago Transit Authority began operating three experimental fuel-cell-powered buses on regular routes in September 1997. The buses use chemical reactions involving hydrogen and oxygen to create electricity, which is used to power an electric motor. The only emission produced by the buses is water vapor.

of activated radicals makes it possible for engine ignitions to occur at the lower compression levels typically found in gasoline engines.

This process differs from the way standard engines operate. There are two main types of internal combustion engines: spark-ignition engines, such as those that power most automobiles, and diesel engines, the kind associated with trucks and locomotives.

The engines differ in the ignition and combustion process they use to produce power. A spark-ignition engine uses electricity and spark plugs to ignite the fuel and air mixture in the engine's cylinders. A diesel engine, however, is a compression-ignition engine, which greatly compresses the air in the cylinders, causing the temperature of the air to rise. Fuel injected into the hot, compressed air then immediately ignites.

Sonex's technology includes specially shaped piston crowns containing extremely small chambers called microchambers. The microchambers retain the activated radicals so that they are available for the next ignition.

Some car manufacturers were experimenting with using activated-radical ignition in an automobile engine, and planned to introduce such a model by the end of 1998. A similar approach to the one developed by Sonex had already succeeded by 1998 in an off-road racing motorcycle developed by the Honda Motor Company of Tokyo. Honda's technique relied on injecting some of the residue containing the activated radicals into a combustion chamber to continually enhance combustion.

Solar-powered aircraft. A remotely piloted, solar-powered aircraft established a new altitude record on July 7, 1997, after it flew at 21,802 meters (71,530 feet)—more than 1,370 meters (4,500 feet) higher than any other propeller-driven aircraft. The aircraft, called Pathfinder, was tested at the Pacific Missile Range Facility in Kauai, Hawaii. The plane was built by AeroVironment of Simi Valley, California.

Pathfinder had a 30-meter (100-foot) wingspan and weighed 216 kilograms (475 pounds). Six propellers were driven by electric motors powered by a thin

Switchable windows
A type of window glass that can be switched from transparent to mirrored was successfully tested in December 1997 by scientists at Philips Research Labs in the Netherlands. The glass is coated with a thin film of a metallic alloy and sealed in a cell. Hydrogen gas pumped into or out of the cell causes the metallic film to become reflective, *top,* transparent, *middle,* or opaque, *bottom.* The researchers said windows based on this technology could be used for energy conservation.

Reflective

Transparent

Opaque

film of solar cells attached to the upper surface of the wing. The solar cells used approximately 15 percent of the solar energy striking them. Most of the cells on the transparent wing surfaces were double-faced solar cells, which allowed them to also use sunlight reflected by the Earth's surface. The aircraft carried a back-up battery as well.

Pathfinder was part of the National Aeronautics and Space Administration's (NASA) Environmental Research Aircraft and Sensor Technology (ERAST) program. The program's goal is to develop a remotely piloted aircraft capable of flying at altitudes of 18,000 to 19,500 meters (60,000 to 65,000 feet) and at speeds as slow as 25 kilometers (15 miles) per hour or as fast as 95 kilometers (60 miles) per hour for weeks or months. NASA researchers said that such unpiloted aircraft could be used to carry sensors and other monitoring equipment to study environmental threats such as pollution and volcanoes.

Aircraft designers in 1998 were developing other solar-powered planes scheduled for flights by the year 2003. Among the remotely piloted aircraft was Helios, a lightweight plane with a 68-meter (222-foot) wingspan capable of carrying payloads as heavy as 90 kilograms (200 pounds) for months at a time.

More efficient electric motors. New federal standards for minimum efficiency levels of most electric induction motors—those motors operating on alternating current (AC)—took effect in October 1997. The regulations, which were part of the U.S. Energy Policy Act of 1992, affected general-purpose motors of between 1 and 200 horsepower. The regulations require the motors to be between 75.5 percent and 95 percent efficient. According to experts, the new regulations should help conserve energy. Since up to 65 percent of electrical energy in the United States is consumed by motors, even small gains in efficiency can produce significant energy savings.

Manufacturers expected the motors to be 8 to 25 percent more expensive than earlier models. The increased cost stems from additional materials needed to boost the motors' efficiency. Experts said that temperatures in the new motors will rise more slowly than in earlier models, which should enable the motors to last longer. [Pasquale M. Sforza]

The development of a new protective coating for use with rolling, sliding, or rotating machine parts that is smoother and more durable than coatings previously available was announced in October 1997. The coating was developed by scientists at Argonne National Laboratory in Argonne, Illinois.

The material, called near-frictionless carbon (NFC), is produced by converting gas containing carbon into a *plasma,* an extremely hot gas consisting of atomic nuclei and electrons. An object that is to be coated with NFC is placed into a specialized chamber, and the material drifts down from the plasma onto the object to cover it. In experiments, NFC had a *coefficient of friction* (CF)—the ratio between the force needed to move an object and the force pressing the object against a surface—of less than 0.001, much lower than that of any other coating. The lower the CF, the easier an object slides against another object. For example, the CF of wood sliding on wood is between 0.25 and 0.50. Metal sliding on metal has a CF between 0.15 and 0.20. Teflon, a material used to coat many kitchen utensils, has a CF of 0.04.

The researchers reported that NFC has many potential benefits. For instance, it can be produced in large quantities in a relatively short time, adheres to many surfaces, including plastic, and is very durable. In lab experiments, Argonne researchers tested the material's wear resistance in a machine that slid a sapphire ball in a circle on the coating's surface while being subjected to a downward pressure of 145,000 pounds per square inch (psi), or nearly 10,000 times atmospheric pressure. After 5 million cycles, the material had been worn down by just 1 micron (0.00004 inch).

Ali Erdemir, the project's team leader, reported that the coating's wear resistance showed promise for use in coating automobile engine parts such as turbocharger rotors and fuel injector elements. The material also had other potential applications, including for oilless bearings, spacecraft mechanisms, and rolling and sliding gear systems.

Exploring Chernobyl. A team of researchers from Ukraine and the United States were finalizing plans in the spring of 1998 to field-test a small, remotely controlled robot by sending it into the Chernobyl nuclear power plant. The plant, located in Ukraine, once part of the Soviet Union, in 1986 was the site of the world's worst nuclear accident. Some areas at the site were still so radioactive in 1998 that just a few seconds of exposure would have caused a person to absorb a lethal dose of radiation.

The researchers, including teams from Carnegie Mellon University in Pittsburgh, RedZone Robotics in Pittsburgh, the U.S. Department of Energy (DOE), the National Aeronautics and Space Administration (NASA), and several other institutions, planned to send the remote-controlled robot, called Pioneer, into the most radioactive parts of the plant in the fall of 1998. Pioneer was designed to test the structural integrity of a concrete and steel structure built around the site to contain radiation. Experts had reported in the late 1990's that the protective structure was beginning to deteriorate, and engineers worried that it might collapse.

Pioneer, which was about 1.2 meters (4 feet) long and moved on treads, carried equipment that would enable it to photograph the plant and create a three-dimensional virtual reality model of the building interior. An onboard drill was to take samples from concrete walls, which the researchers planned to examine for clues about the extent of the radiation damage. Pioneer also contained a variety of other radiation and environmental sensors. Information gathered by Pioneer was expected to help the researchers develop methods to clean up the former power plant.

Roving robot. Scientists from Carnegie Mellon University and NASA successfully tested a remote-controlled robotic rover in June and July of 1997. The vehicle, called Nomad, traveled more than 200 kilometers (125 miles) in Chile's Atacama Desert as engineers and scientists in the United States remotely piloted it with radio signals relayed by satellite. The researchers said they hoped that in the future, Nomad could be used to explore the moon or Mars in much the same way that Sojourner Rover, in the Mars Pathfinder mission, was used in July 1997. The researchers chose the Atacama Desert because its barren landscape of craters, rocks, and loose sands without any vegetation is

Highway of the Future?

Being a motorist in today's world is often less than pleasant and is sometimes harrowing. Creeping along in traffic jams; skidding on wet pavement; narrowly avoiding a collision on a crowded expressway. These are experiences nearly every driver is familiar with.

They are also ones that technology may someday eliminate. Engineers in 1998 were working to develop, or perfect, various automated systems aimed at making driving experiences safe and hassle-free. These innovations, all of which use computer technology, are called Intelligent Transportation Systems (ITS's). Among ITS's that were already in use in the late 1990's were highway signs that warn motorists of traffic jams ahead. ITS's being planned for the future included highly advanced automated highway systems designed to take over control of a vehicle from the driver for a sustained period. Such a system was successfully tested in 1997.

The simplest form of ITS is called an Advanced Traffic Management System (ATMS). These systems, which were already in use in 1998, use sensors placed in the pavement, closed-circuit cameras, and even human observers in helicopters to monitor traffic congestion. The traffic information is posted on electronic highway signs, Internet Web sites, and special radio stations that broadcast traffic advisories to alert motorists of potential problems. Units for receiving and displaying this information were available in 1998 as an option on some more-expensive automobiles, and experts predicted that within 10 years, navigation systems providing up-to-the-minute traffic information would be standard equipment on most vehicles.

Using modern technology to make driving easier is not a new idea. Some vehicle-control systems tested in the late 1970's became common features on automobiles in the 1990's. One such system is antilock brakes. When a driver has to stop quickly, the natural reaction is to simply step hard on the brake pedal. But that can cause the wheels to lock, or stop rotating—especially on slick pavement—and send the car into a skid. Antilock brakes employ sensors at each wheel to monitor wheel speed. When the system detects a locked wheel, a small on-board computer adjusts the wheel's braking force so it keeps rotating. With the wheels rotating, the driver can steer the car while still quickly reducing speed.

In the late 1990's, a more sophisticated vehicle control system, dynamic skid control, was being offered on some vehicles in the upper price range. In dynamic skid control, sensors monitor a car's *yaw* (rotation around its center) and the distance and direction that the steering wheel has been turned. If the system detects that the vehicle is about to go into a spin, an on-board computer determines which wheels need to be braked to keep the car on a straight path, and it applies braking pressure to only those wheels.

Another system that was updated for the 1990's was cruise control. Since its introduction in the 1960's, cruise control has enabled a driver to set a desired speed on a panel on or near the steering wheel. The driver can then remove his or her foot from the accelerator, and the system automatically maintains that speed.

A newer technology, called adaptive cruise control, was being offered on some foreign-model cars in 1998 and was expected to become widely available by the year 2000. Adaptive cruise control uses a sensor mounted behind the front grill to monitor the car's distance from other vehicles. The driver programs the system's computer with a particular distance that must be maintained between the vehicles. If the driver's automobile gets too close to the car ahead of it, the system automatically reduces speed to maintain the required distance.

Experimental versions of other intelligent vehicle systems, which offered even more automated assistance to drivers, were being tested in 1998. Engineering and automotive experts predicted that much of this new technology, known as Advanced Vehicle Control Systems (AVCS's), would be available as options on most vehicles by the early 2000's. Such systems included obstacle- and collision-avoidance systems, which use cameras, computers, and sensors to detect and avoid obstacles or dangerous situations in the road ahead. If the computer determines that a crash is going to occur, it applies the brakes and steers the vehicle to avoid a collision.

Experts predicted that various elements of AVCS technology would eventually be incorporated into automated highways. When entering the on-ramp of one of these advanced roadways, a driver would enter destination information into an on-board computer. Further information from the driver would be unnecessary. The computer would take control of the vehicle, leaving the motorist free to read, eat, or even sleep.

Because all the cars on an automated highway would be computer-driven, the vehicles could be more tightly spaced than ones traveling on conventional roads. Some engineers have calculated that an automated freeway could handle 6,000 vehicles an hour, three times the number that could be accommodated by the typical expressway in 1998.

Motorists traveling on an automated highway would be able to relax, work on a portable computer, even take a nap, while their vehicle was guided to its destination by advanced technology. Completely automated driving systems were still up to 20 years in the future in 1998, but many new accident-avoidance systems were expected to become standard on vehicles in the meantime.

Automated highways, however, need specialized equipment to work. Car-guidance magnets must be embedded in the pavement in the center of each lane, spaced about 1 meter (3 feet) apart. (An alternative guidance-system proposal would use radar-reflecting tape fastened to the road surface.) And sensors must be planted in the pavement or mounted at the side of the road to monitor traffic and relay the information to a central controlling station. The station would regulate the flow of traffic similar to the way an air-traffic control tower monitors the movement of airplanes.

Vehicles, too, would have to be specially equipped. Instruments called *magnetometers,* which measure the intensity and direction of magnetic fields, would be mounted under the front and rear bumpers of a car and keep the vehicle centered in its lane on a roadway using embedded magnets. (An on-board radar device would be used for a reflecting-tape guidance system.) Behind the front grill, a distance-measuring sensor, similar to those used with adaptive cruise-control systems, would measure the distance to the next car in the lane and detect obstacles in the road. Other sensors would detect vehicles to the rear and in adjoining lanes. Electronic motors would operate the car's throttle, steering wheel, and brakes. A radio receiver would receive instructions from the controlling station telling the vehicle what speed to maintain. All the car's systems would be operated by an on-board computer.

As of 1998, no guidelines had been established on whether the speed limit on automated highways would be greater than that on standard highways. But there was no hurry to decide that question, because many experts predicted that it might be 20 years before fully automated highways were in regular use.

Nonetheless, an experimental version of an intelligent highway system was successfully tested in August 1997. The project was sponsored by the National Automated Highway System Consortium (NAHSC), a group of auto manufacturers, government transit departments, and university research centers. The experiment was mandated by the 1991 Intermodal Surface Transportation Efficiency Act. The test was carried out on a 12.2-kilometer (7.6-mile) stretch of Interstate 15 north of San Diego, equipped with both embedded magnets and radar-reflecting tape to demonstrate the two competing systems.

Some cars were equipped with sensors that detected the buried magnets, while others used a small radar transmitter and receiver to follow the tape. Steering, acceleration, and braking were all done automatically, under the control of on-board computers.

The federal government in 1998, however, considered intelligent highways to be a long-term vision and not a short-term goal. Opponents questioned the reliability and safety of completely automated driving systems. Another concern was the high cost of installing the technology in cars. Detractors also claimed that automated systems would create tangled liability issues if an accident occurred. The need for new laws and new forms of insurance to sort out such issues led some critics to question whether motorists would want automated highways.

As research into intelligent vehicle systems continued in the late 1990's, engineers expected many elements of the technology to become standard on most new vehicles within a few years. And some experts, despite the doubts of critics, predicted that the day when people would be traveling on fully automated highways was not that far down the road. [Larry Webster]

similar to the landscapes on the moon and Mars.

Nomad weighs 720 kilograms (1,600 pounds) and is about the size of a compact automobile. It moves on four aluminum wheels with cleats, which provide traction in soft sand. Nomad was operated by an electric motor powered by a gasoline generator that enabled the robot to travel at speeds up to 50 centimeters (20 inches) per second. However, the scientists were exploring ways to power the vehicle using solar energy.

Miniature turbine. A research team at the Massachusetts Institute of Technology Gas Turbine Laboratory in Cambridge announced in January 1998 that it was making progress in experiments aimed at creating a miniature turbine engine by the year 2000. The engine would be about the same size as a shirt button, weigh less than 1 gram (0.035 ounce), and be capable of producing 10 to 20 watts of electricity—about the same amount of electricity consumed by an outdoor porch light.

The miniature turbine would be similar in design to a conventional jet engine. It would include three key components: a turbine wheel, a combustion chamber, and a compressor wheel. Fuel burning in the combustion chamber would send exhaust gases through the blades of the turbine wheel, causing it to rotate, which in turn would drive the compressor wheel via a central shaft.

By early 1998, the research team, led by Gas Turbine Laboratory director Alan Epstein, had been able to construct a tiny silicon turbine wheel 4 millimeters (0.16 inch) in diameter and a combustion chamber 2 millimeters (0.08 inch) long. The third component, the compressor wheel, was still under development.

In order to produce power, the investigators said, the turbine wheel would have to spin at 2.5 million revolutions per minute. The researchers were testing low-friction air bearings in 1998 that would permit the wheel to spin at that speed. However, since operation at such a high speed would generate considerable heat, the scientists were also experimenting with constructing the microturbines out of a material called silicon

A turbine wheel just 4 millimeters (0.16 inch) in diameter was tested in mid-1997 by engineers at the Massachusetts Institute of Technology's Gas Turbine Laboratory in Cambridge. The wheel, which was constructed out of silicon, was part of a project to create a miniature turbine engine. The researchers said they hoped to have the tiny engine ready for a demonstration by the year 2000.

Work was completed in November 1997 on the first phase of the Three Gorges Dam, the largest flood-control and hydroelectric power project in the world. The project is designed to generate 18,200 megawatts of electricity by taming China's Yangtze River. The dam, which will be 2,300 meters (7,550 feet) long and 180 meters (590 feet) high, was scheduled to be completed in the year 2009.

carbide, which is more durable and temperature-resistant than silicon. Such miniature engines could be used to power various devices, including communications devices for the military.

Basilica repairs. Engineers, architects, and art restorers in 1998 worked to re-create and reinforce the vaulted ceilings of the Basilica of St. Francis in Assisi, Italy, following a series of earthquakes in September and October 1997. The ceiling was covered with *frescoes* (wall and ceiling paintings) attributed to Giovanni Cimabue and Giotto, artists of the late 1200's and early 1300's.

Using modern construction materials and techniques, restorers were working from photographs and archival drawings to complete the repairs to the basilica by the year 2000. Workers filled cracks in the vaulted ceiling's unseen portions with a special salt-free mortar and applied fiber adhesive strips to cracks to prevent them from widening further. The restorers also used reinforcing rods to connect the top of the ceiling to the roof. The rods were specially designed to swing back and forth and to absorb shock from future quakes.

Repairs slated for late 1998 included the restoration of damaged arches and ribs with thick carbon-fiber elements. Repair crews planned to attach the ribbing to the top of the restored ceiling vault and then connect it to the roof with springs. Restoration experts were also sifting through numbered boxes containing preserved fragments of the collapsed frescoes in an effort to reconstruct the damaged artworks. The building's destroyed exterior portions were to be rebuilt with stone mined from the same quarry used to provide material for the original structure.

Germ-detecting sensors. The development of a prototype sensor capable of detecting *pathogens* (disease-causing microorganisms) at very low levels was announced in February 1998 by researchers at Virginia Polytechnic Institute and State University in Blacksburg. The sensor can detect pathogens in parts per trillion. Other such devices available in 1998 allowed measurements no lower than parts per billion. The Virginia Tech researchers said that the need for

The Sky Tower, the highest structure in the Southern Hemisphere, opened in Auckland, New Zealand, in July 1997. The tower stands 328 meters (1,076 feet) tall and was designed to resist earthquake damage and withstand wind gusts of up to 200 kilometers (125 miles) per hour. The tower features a revolving restaurant and four observation decks.

such technology stemmed from the threat of biological warfare that had confronted the allied forces during the Persian Gulf War (1991).

William Velander, a biochemical engineer, adapted a technology he had invented to purify certain drugs that are present at trace levels in blood plasma. Kent Murphy, an electrical engineer, combined that process with an optical-fiber sensing device. In tests the sensor yielded results that were 20 times more sensitive than from previous sensing devices, and gave researchers results within seconds, rather than the one hour typically needed with other existing sensors.

The prototype sensor does have certain drawbacks, according to the Virginia Tech scientists. For example, the device is too large to be portable. However, the researchers in late 1998 were working on the development of a battery-powered belt-pack device that someday could be worn on a battlefield.

Three Gorges Dam. Engineers in China in January 1998 began construction on the second phase of the Three Gorges Dam on the Yangtze River, the largest flood-control and hydroelectric power project in the world. During the second phase, work crews planned to install generators at the site. About 550,000 people were expected to be forced to move from their homes along the river during the second phase. Some 1.2 million people were to be displaced once the $24.5-billion project was completed in the year 2009.

The dam was being designed to generate 18,200 megawatts of electricity for economic development in central China by taming the Yangtze River. In the first phase of construction, which was completed in November 1997, workers cut a channel in one of the river's banks to temporarily redirect the flow of water. On the other bank, workers began moving about 58 million cubic meters (2.05 billion cubic feet) of granite to build a navigation lock for river traffic.

When completed, the Three Gorges Dam will contain almost 15 million cubic meters (530 million cubic feet) of concrete and will be 2,300 meters (7,550 feet) wide and 180 meters (590 feet) high. The dam's reservoir, which will extend for 560 kilometers (350 miles), will entirely or partly submerge several cities and towns. [Andrew G. Wright]

A large area of Southeast Asia was covered by a thick blanket of smoke from jungle fires in September and October 1997. Many of the fires were set by corporations clearing land for commercial crop production. Because of unusually dry weather conditions caused by *El Nino* (a warm Pacific Ocean current that occasionally upsets weather patterns across the world), the fires soon raged out of control. In September, an estimated 7,700 square kilometers (3,000 square miles) of jungle were on fire, and smoke stretched some 3,200 kilometers (2,000 miles) across six nations.

Smoke from the fires combined with industrial emissions and car exhausts to create potentially hazardous smog in several major cities in Southeast Asia. Children in Indonesian cities were warned to stay indoors after officials canceled school. An Indonesian official estimated that 20 million people suffered respiratory problems as a result of the pollution. Medical experts predicted that breathing the polluted air would inflict long-term health damage on thousands of people.

The fires also posed a threat to plants and animals in the biologically diverse rain forests of the region. Experts reported that orangutans in Borneo, a large island belonging mostly to Indonesia, were greatly threatened as a result of the fires because flames had gutted their habitats. Orangutans depend on an intact rain forest for food and shelter.

The EPA buys a neighborhood. The U.S. Environmental Protection Agency (EPA) in November 1997 launched a project to purchase 358 houses in Pensacola, Florida, after scientists discovered unhealthy levels of various chemical contaminants, including dioxin, a poisonous substance, in the ground. The $24-million project was scheduled to be completed in the year 2000.

The low-income neighborhood targeted by the EPA had become surrounded by industrial plants that emitted large quantities of pollutants. According to the EPA, contaminants deposited in the soil from the facilities had reached potentially toxic levels and posed a health risk to people living in the area. The

Work crews spray oil-dispersing chemicals into Tokyo Bay in Japan in July 1997 after a tanker ran aground on a shallow reef. The accident ruptured 2 of the ship's 14 oil tanks, spilling 1.5 million liters (400,000 gallons) of light crude oil into the bay. Government officials originally believed the spill to be 10 times greater than it was, leading some environmentalists to charge that more chemicals were used to disperse the spill than was necessary.

The Eiffel Tower in Paris is obscured by heavy smog that shrouded the city in early October 1997. Weak winds and unseasonably warm weather led to a dangerous build-up of nitrogen dioxide across the city. The severe pollution prompted French officials to enforce new clean-air regulations, including restrictions on the operation of private vehicles. Many Parisians relied on public transportation to reach their destinations.

EPA said that after all of the residents had left the area, it would clean up the site and designate it for industrial use.

Maternal lead and cavities. A mother's exposure to lead prior to the birth of her child or during breast-feeding may increase the child's risk of developing cavities, researchers at the University of Rochester in New York reported in September 1997. The investigators, led by William H. Bowen, a dentist and microbiologist, studied the transport of lead from female rats to their fetuses and newborns.

Bowen and his colleagues fed pregnant female rats water that contained high levels of lead. They found that the lead crossed the placenta, the structure that supplies nutrients to a fetus and helps eliminate wastes from its blood, and entered the bodies of the rat fetuses. Lead was also transported to the newborn rat pups in breast milk. The researchers found that pups exposed to lead through their mother developed 40 percent more cavities and produced 30 percent less saliva than pups whose mothers had not been exposed to lead.

Because lead shares some chemical properties with calcium, the body normally stores lead in bone. During pregnancy and *lactation* (milk production), the body breaks down bone to release calcium to provide nutrients for the offspring. At the same time, however, lead stored in the bones is also released, to be transferred to the offspring, in whose body it gets incorporated into both bones and teeth.

The researchers theorized that the lead deposits in the baby rats' teeth created structural weaknesses that may have contributed to the development of cavities. They further speculated that the reduction in saliva may have added to the problem, because saliva cleanses the teeth by washing away food particles. Saliva also contains naturally occurring antibiotics that kill bacteria in the mouth responsible for cavities. In addition, saliva replenishes the minerals in teeth.

Health experts believe that the high lead content found in some people's bodies probably comes from a number of sources, including air pollution and lead-tainted drinking water. The latter source is especially common in older homes with lead pipes. Lead-based paints can also contribute to lead exposure, especially when children eat paint chips that flake off of interior surfaces.

Inaccurate pollution reports. A research team at the University of Southern California in Los Angeles, headed by environmental engineer Ronald C. Henry, reported in June 1997 that oil refineries and other industrial companies make significant errors in reporting the amounts of certain pollutants emitted from their facilities. Reports of estimated emissions are required by all American companies that annually emit 100 tons or more of smog-forming pollutants known as volatile organic compounds (VOC's).

The researchers used automated air-sampling monitors to make hourly measurements of 54 *organic* (carbon-based) compounds in an industrial area of Houston over a six-month period. The measurements were then compared to manufacturers' reports filed with the state of Texas, which had commissioned the study.

Henry's team found that industry emission reports inaccurately reported both the amount of chemicals and the type of chemicals released into the atmosphere. The inaccuracies included both underestimates and overestimates of emissions. However, the researchers and other environmental pollution experts agreed that the inaccurate reports were most likely unintentional errors and not fraud.

Annual air pollution inventories are used by government officials to monitor and adjust pollution-control strategies. When VOC emissions are underreported, the researchers said, pollution-control strategies may later prove to be inadequate and air quality may endanger human health. When such emissions are overestimated, the figures can result in the installation of costly but unnecessary pollution-control devices. Experts believed the study could stimulate new regulations requiring companies to make actual measurements of VOC emissions rather than just estimate such emissions.

Second-hand smoke dangerous. Researchers reported in January 1998 that second-hand smoke may cause atherosclerosis—the formation of fatty deposits on the inside of arteries—in nonsmokers who are exposed to at least one hour of smoke per week. The study was conducted by teams of researchers from Wake Forest University in Winston-Salem, North Carolina; Johns Hopkins University in Baltimore; the University of Minnesota in Minneapolis; and the University of Bergen in Norway.

From 1987 to 1989, the researchers used *ultrasound* (high-frequency sound) imaging to examine the carotid arteries, which deliver blood to the brain, of nearly 11,000 people between the ages of 45 and 65. The patients were reexamined three years later. The researchers reported that arterial blockages built up 50 percent faster in smokers than people who had not been exposed to smoke. They also found that the rate of deposit formation was 20 percent higher in nonsmokers who had been exposed to second-hand smoke than in nonsmokers who were not exposed to smoke.

Mysterious fish killer. Nearly 12,000 fish were found dead in Maryland and Virginia waterways in August 1997. The fish, which had developed red or white rotting, ulcerated lesions, resembled dead fish that had been discovered in the coastal waters of North Carolina since the 1980's.

Environmental scientists blamed the fish deaths on a microscopic toxin-producing organism called *Pfiesteria piscicida* that can affect not only marine life but also people. Workers and scientists who had been exposed to the organism while removing infected fish from the waterways or examining the dead fish complained of shortness of breath, nausea, skin lesions, and loss of memory.

Although scientists in 1998 were still debating the reasons for the 1997 outbreak, many experts blamed increased water pollution. Some scientists said that the outbreaks, called blooms, of some microorganisms appeared to be the result of increasing levels of nitrogen pollution in the water. Experts said that nitrogen comes from a variety of sources, including agricultural fertilizers, human sewage, livestock waste, and air pollution. Scientists and fishermen in the spring of 1998 were waiting to see whether more fish would become infected with *Pfiesteria piscicida.*

[Dan Chiras]

One of the greatest crises to affect the history of animal life in the seas was the mass extinction that ocurred about 250 million years ago during the late Permian Period and early Triassic Period. Paleontologists have argued that this event affected organisms in a random way—that is, the species that survived were simply lucky. In November 1997, however, paleontologists Richard K. Bambach at Virginia Polytechnic Institute and State University in Blacksburg and Andrew Knoll of Harvard University in Cambridge, Massachusetts, provided evidence that survival was not based on luck but on certain anatomical features.

In 1995, Bambach, Knoll, and their colleagues had argued that the Permo-Triassic extinction may have resulted from the introduction of carbon dioxide (CO_2) into shallow ocean waters. Abundant levels of CO_2 would have accumulated in the deep-ocean waters from the decay of dead organisms during a time when deep-ocean circulation stagnated. When global climatic changes cooled the ocean's surface, the warmer, CO_2-laden water at the bottom moved up.

Bambach and Knoll followed up this hypothesis in 1997 by comparing the survival rate of organisms that could protect themselves from high levels of CO_2 and those that could not. Animals that are resistant to the gas are called CO_2-buffered organisms. These include *mollusks* (shellfish and related organisms), marine worms, and some *arthropods* (joint-legged animals). CO_2-buffered organisms have either gas-exchange organs, such as gills, or closed blood-circulation systems, or both. These features would help filter out harmful gases in the seawater and protect internal organs. The nonbuffered organisms, including corals and most *echinoderms* (the group that includes starfish, sea urchins, and sea lilies), have no such screening systems, so their internal organs would have been directly exposed to the high CO_2 levels.

By compiling statistics on the organisms that were wiped out and those that survived, the researchers discovered that the buffered group seemed to have an advantage. Before the Permo-Triassic extinction, there were 748 *genera* (the plural of genus) of nonbuffered organisms and 274 buffered organisms. But the extinction wiped out about 91 percent of nonbuffered genera and only about 50 percent of buffered genera.

Bambach and Knoll's new data supported the CO_2 link to the Permo-Triassic extinction and provided a reasonable explanation for the changes in various populations of organisms. Most researchers said, however, that the CO_2 theory would require further testing.

New dates for land dwellers. Two reports in 1997 challenged traditional views on when sea-dwelling *vertebrates* (animals with backbones) first invaded the land. In August, paleomagnetist Ebbe Hartz and his colleagues at the University of Oslo in Norway argued that fossil beds in east Greenland, which were believed to be from the end of the Devonian Period (410 million to 360 million years ago), are actually from the middle of the Mississippian Period (360 million to 300 million years ago). The east Greenland fossil beds are important because they contain some of the oldest known *tetrapods* (four-legged animals).

Hartz's team based the new dates partly on the *paleomagnetism* (ancient magnetic orientation) of rocks. When molten rock solidifies, the grains of the rock align with the Earth's magnetic field. The direction of the grains, therefore, acts as a permanent record of a rock formation's geographical position before the land mass it is situated on moves with the shifting of continental plates over hundreds of millions of years. By using other geological evidence that acts as a timeline of the continental shifts, the researchers determined that the east Greenland rocks probably formed about 340 million years ago in the middle of the Mississippian Period.

This date places the evolution of tetrapods about 20 million years later than the traditionally accepted dates. Some scientists were skeptical of the new dates, but other researchers agreed that the study required a reassessment of the early stages of tetrapod evolution.

Another study of early vertebrates was reported in December 1997 by paleontologists John Bolt of the Field Museum of Natural History in Chicago and Eric Lombard of the University of Chicago. The researchers dated fossils of animals called microsaurs found in Kansas to about 332 million years ago, 10 million years earlier than other microsaur speci-

Baby dinosaur from Italy
The discovery of the fossilized remains of a baby dinosaur—the first dinosaur ever found in Italy—was reported by paleontologists Cristiano dal Sasso of the Museo Civico di Storia in Milan, Italy, and Marco Signore of the Universita degli Studi di Napoli in Naples. The fossil, measuring about 24 centimeters (9 inches) long and dated at about 113 million years old, shows razor-sharp teeth resembling those of much larger predatory dinosaurs such as *Tyrannosaurus rex.* The fossil also revealed the best-preserved internal organs of a dinosaur ever found, including a windpipe, muscle tissue, intestines, and a liver. The researchers believed the dinosaur died shortly after it hatched from its egg.

mens. Microsaurs were tiny primitive amphibians measuring about 15 centimeters (6 inches) long.

The new date for the microsaurs moves these animals, which show evidence of thorough adaptation to a land environment, closer to the emergence of land-dwelling vertebrates. The researchers believed further studies of microsaurs and other early animals would help them better understand the transition of animal life from water to land.

The color of ancient animals. An armor-plated fish from approximately 370 million years ago was probably red and silver, according to an August 1997 report by paleontologist Andrew Parker of the Australian Museum in Sydney. Parker's analysis of the fossil fish suggested for the first time a way to determine the colors of extinct organisms.

Most attempts to determine the colors of extinct animals are based on the colors of closely related living species and on speculation. Parker's approach was to study the microscopic structures that give color to living organisms. He then searched for the remains of similar structures in fossilized organisms.

Parker's conclusions were based partly on tiny preserved structures called *diffraction gratings,* which split white light into a rainbow of colors. The gratings create a silvery appearance, much like the hologram on a credit card. These structures were located on the underside of the fish. Parker also identified red *pigment cells* (color-containing cells) that left impressions in the skeletal plates on the top of the fish.

Besides satisfying our curiosity about the appearance of prehistoric organisms, the abililty to identify an animal's color offers clues to the habitat it lived in. Because color is often a camouflage for animals, scientists can draw conclusions about the conditions of a local environment. Parker concluded that the fish may have lived in a well-lit environment, because the silvery bottom of the fish would have been difficult to see from below. The red on the fish's back may have helped it blend in with reddish foliage or mud when the fish was seen from above. Parker hoped that other fossilized pigments would be found in

FOSSIL STUDIES: A Science Year Close-Up

A Slow Explosion of Life

An event that paleontologists refer to as the "Cambrian explosion" was one of the most dramatic episodes in the history of life on Earth. Until recently, the standard explanation of the event went as follows: In a relatively brief interval of geologic time—about 10 million to 30 million years—almost all *phyla* (major body forms of animals), including those with skeletons, suddenly evolved in the shallow seas of the Earth after nearly 3 billion years of dominance by single-celled organisms. This rapid evolution, according to the fossil record, occurred during the Cambrian Period, which lasted from about 545 million to 500 million years ago.

But scientists were always puzzled by what could have caused this event. In 1997 and 1998, several researchers reported on new findings that led paleontologists to rethink the accepted scenario. The studies suggested that the Cambrian explosion may not have been quite so explosive after all and that complex life forms may have appeared much earlier than 545 million years ago.

A key consideration in this puzzle is the interpretation of a group of unusual fossil imprints that are found in sandstone formations from the Vendian Period (about 560 million to 545 million years ago) but not in deposits from the later Cambrian Period. This group of specimens, called Ediacaran (*ee dee ACK a run*) fossils, was named for the Ediacara Hills north of Sydney, Australia, where the first of them were found. Various kinds of Ediacaran fossils have subsequently been discovered in many other parts of the world in rocks of about the same age. Researchers have hypothesized that the organisms that made these impressions disappeared without leaving any descendants. Most paleontologists have assumed that Ediacaran organisms were much simpler animals than the animals that followed and may not have been animals at all. It is this interpretation, to some extent, that makes the appearance of so many complex animals during the Cambrian Period seem so explosive.

In August 1997, paleontologists Mikhail Fedonkin at the Russian Academy of Sciences in Moscow and Benjamin Waggoner, then at the University of California at Berkeley, offered evidence to change that interpretation. They discovered new and better-preserved specimens of a previously known Ediacaran organism called *Kimberella* in 550-million-year-old deposits on coastal cliffs along the White Sea in northern Russia. The 35 specimens ranged in size from 3 to 105 millimeters (0.1 inch to 4 inches). Most scientists had believed *Kimberella* was similar to a jellyfish or some other simple animal. The new fossils reveal, however, that it was more complex.

The Russian fossils suggest that *Kimberella* had *bilateral symmetry*—similar right and left sides. In contrast, *radially symmetrical* animals, such as jellyfish, have body parts arranged circularly around a central point. Fedonkin and Waggoner argued that *Kimberella* was a wormlike ancestor of mollusks, a group of animals that includes such modern organisms as snails, clams, and octopuses.

The fossil specimens revealed that *Kimberella* was shaped rather like a slug, with a mouth in the forward part of its body and possibly a complete digestive system. It also seems to have had a soft protective shell. The edges of the organism had a rufflelike tissue resembling a flap of "skin" called the mantle in mollusks, which creates either a shell or tough protective skin on the organism.

Therefore, at least one fairly common Ediacaran organism appears to represent a link to a major group of animals that became more diverse during the Cambrian Period. The researchers also offered a new interpretation for the sudden disappearance of *Kimberella* and other Ediacaran organisms from the fossil record in the Cambrian Period. These soft-bodied organisms, according to the researchers, would have stood little chance of being preserved in later deposits because burrowing and scavenging animals, which were abundant in the Cambrian Period, would have exposed the Ediacaran remains to destructive bacteria. The *Kimberella* research thus offered a new explanation of early complex life forms and their relationship to later, more diversified species.

A breakthrough in fossil research reported in September 1997 led to additional studies that some paleontologists thought were even more significant to the interpretation of the animal life before the Cambrian Period. In the 1970's, Chinese scientists had reported the discovery of tiny spherical fossils from Cambrian deposits in China and Siberia, but the fossils were always a bit of a mystery. Paleontologists Stefan Bengston at the Swedish Museum of Natural History in Stockholm and Yue Zhao at the Institute of Geology in Beijing resumed research on these fossils. By examining them with a scanning electron microscope, they discovered that they were embryos of jellyfishlike animals in various stages of development. Their 1997 report prompted other researchers to reexamine other small fossils.

In February 1998, two separate research teams reported their findings using microscopic analysis of tiny fossils from south-central China. The fossils were preserved in 570-million-year-old deposits of calcium phosphate.

A well-preserved 550-million-year-old fossilized impression of a *Kimberella* organism, *right,* from the northern coast of Russia provides evidence of complex animal life at a time much earlier than paleontologists had previously thought. The organism, illustrated *below,* had a soft shell and a rufflelike flap that resembles a mantle, a physical feature of modern mollusks, the group of animals that includes snails, clams, and octopuses.

The first team, led by paleontologist Chia-Wei Li of the National Tsing Hua University in Hsinchu, Taiwan, reported that some of the specimens they examined were juvenile sponges. The fossils were, therefore, the oldest known representatives of a modern phylum. They identified other specimens as embryos that showed characteristics of bilaterally symmetrical organisms.

The second group, led by paleontologist Shuhia Xiao at Harvard University in Cambridge, Massachusetts, examined embryos in various stages of cell division, including specimens with only two, four, and eight cells. The reseachers concluded that the patterns of cell division clearly belonged to bilaterally symmetrical animals such as worms. The examination of other embryos revealed unusual patterns of cell division that suggested a relationship to modern crustaceans, the tremendously diverse group of marine and freshwater *arthropods* (joint-legged animals) that include barnacles, lobsters, crab, and shrimp. These embryos, therefore, demonstrated that animals more complex than sponges or jellyfish existed at least 570 million years ago. Indeed, all of these studies—perhaps the most important paleontological discoveries of the century—provided compelling evidence for animal life at a time much earlier than previously assumed and earlier even than the Ediacarian organisms.

Another discovery, reported in December 1997, could extend the record of complex animal life still further back in time. The study was based on *trace fossils,* evidence of animal activity rather than the remains of bodies or embryos. Paleontologists Martin Brasier of Oxford University and Duncan McIlroy of the University of Liverpool, both in England, found small strings of fossil *feces* (bodily waste) produced by a sediment-feeding organism in 600-million-year-old sandstone on the Scottish island of Islay. The significance of the fossil feces is that they represent the activities of an ancient organism that had a mouth at the front of its body and an anus at the back of it. In other words, it was a wormlike animal with a complete digestive tract. Although these fossils offer only indirect evidence, they imply the existence of rather sophisticated, bilaterally symmetrical animals at a time well before the Vendian or Cambrian periods.

The vast diversification of life that appears in the fossil record from the Cambrian Period still warranted further explanation in 1998. Nonetheless, the new research removed some of the mystery of the "Cambrian explosion" scenario by showing that several groups of complex organisms existed at least 570 million years ago, thereby extending the length of the "explosion" by at least 25 million years.

[Carlton E. Brett]

Fossilized embryos

Microscopic analysis of tiny spherical fossils revealed the early stages of cell division in 540-million-year-old embryos. Paleontologists Stefan Bengston at the Swedish Museum of Natural History in Stockholm and Yue Zhao at the Institute of Geology in Beijing reported their findings in September 1997.

The magnified image of a fossilized embryo less than 1 millimeter (0.04 inch) in diameter reveals the developmental stage of an unidentified organism when it consisted of 256 cells.

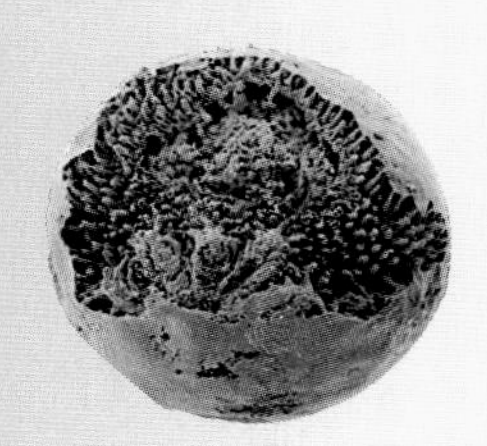

The fossilized embryo of a jellyfishlike animal, *left, top,* shows an early stage of development. The same kind of embryo fossil at a later point of development, *middle,* shows the shape beginning to change. The spiral-shaped fossil, *below,* is a hatched organism of that species. The study of these fossils created a valuable method for learning about early soft-bodied animals that were not well preserved in the fossil record.

the bony skeletal plates of other groups of organisms, including dinosaurs.

Mistaken identity. Several nests of dinosaur eggs that had been attributed to a plant-eating dinosaur were actually the nests of meat-eating animals, according to a report in September 1997 by paleontologist Jack Horner of the Museum of the Rockies in Bozeman, Montana. In 1979, Horner had found several dinosaur nests in ancient tidal flats in Montana. He concluded that the nests had belonged to plant-eating dinosaurs of a type called *Orodromeus* because fossils of that species were found nearby. But a better-preserved nest found in 1996 led to a different conclusion.

The new nest contained the remains of a *Troodon,* a meat-eating dinosaur. Horner, therefore, decided to evaluate all of the nests more carefully. He found well-preserved embryos with features that clearly linked the nests to *Troodon.* Horner concluded that the *Orodromeus* remains were prey that a *Troodon* parent had killed.

An ongoing debate among paleontologists about whether birds evolved from dinosaurs continued in 1997 and 1998, with researchers providing evidence to both defend and refute the hypothesis. Some innovative—though far from decisive—research by two groups supported the claim that there is no link.

Some obvious similarities exist between the skeletons of small, predatory dinosaurs called theropods and *Archaeopteryx,* a prehistoric bird that lived about 150 million years ago. The debate hinges on the question of whether those similarities are the result of a direct evolutionary link or the result of separate lineages of animals evolving characteristics that only appear to be linked.

Sinosauropteryx fossils, found in fossil beds 120 million to 140 million years old in China in 1996, had offered support to the theory of a dinosaur origin for birds. The specimens have what appear to be feathers—or structures that could have evolved into feathers—along the dinosaurs' spines, but the actual nature of those structures remained unresolved in 1998. The fossils also revealed well-preserved internal organs.

In November 1997, physiologist John A. Ruben of Oregon State University in Corvallis reported that fossilized lungs from the theropod *Sinosauropteryx* more closely resemble the lungs of crocodiles than the breathing mechanisms of modern birds. He concluded, therefore, that *Sinosauropteryx* did not represent an evolutionary link to birds.

A crocodile has a diaphragm that acts as a kind of bellows to draw air into the lungs and force it out. A bird, in contrast, possesses a more complicated system of air pockets and spaces, including air sacs within hollow bones. In general, the fossil lungs indicated a bellowslike breathing system like that of crocodiles. Although the research cast doubt on the bird-dinosaur link, most researchers cautioned that the specimens were too crushed to draw solid conclusions about how *Sinosauropteryx* breathed.

In October 1997, ornithologists Alan Feduccia and Ann C. Burke of the University of North Carolina in Chapel Hill argued that similarities between the three *digits* (fingerlike appendages) in birds' wings and the three digits of theropods' front claws are the result of *convergence,* the phenomenon of two unrelated evolutionary lineages independently developing similar characteristics.

The ancestors of both birds and theropods had five digits, but two of those digits were "lost" as the organisms evolved. According to fossil evidence, theropods lost the last two digits—the equivalent of a human's "ring" finger and little finger. The scientists reasoned that a test of the link between birds and dinosaurs, then, would be to determine if birds had also lost the same two digits.

Feduccia and Burke compared the embryonic development of modern birds to that of alligators, the closest living relatives of birds. (They have a common ancestor that predated dinosaurs.) Alligators have five fully developed digits on each front limb. In an alligator embryo, the fourth digit—the ring finger—develops before the others. Assuming that the stages of development are the same in bird embryos, the scientists concluded that birds had lost the first and fifth digits—the thumb and little finger—as they evolved. Therefore, the three-digit similarities between birds and theropods may be an example of convergence rather than evidence of evolutionary ties. [Carlton E. Brett]

Genetics

The complete *genome* (total amount of genetic information) of the common bacterium *Escherichia coli* was published in September 1997 by researchers at the University of Wisconsin at Madison, led by geneticist Fred Blattner. This report represented the conclusion of a 15-year-long race between Blattner's team and a research group at the University of Japan, which completed its work just days after the Wisconsin scientists.

Both teams of scientists concluded that the DNA (deoxyribonucleic acid—the molecule that makes up genes) of *E. coli* consists of over 4.6 million *base pairs* (pairs of individual DNA subunits). The base pairs are organized into more than 4,200 genes, many of which had not been previously known. (By contrast, the human genome consists of some 3 billion base pairs arranged into as many as 100,000 genes.)

Most scientists said that *E. coli* is the most important of several microorganisms that have had their genomes *sequenced* (decoded) since 1995. *E. coli* is widely studied in research laboratories and has a number of industrial applications. In addition, the role that some strains of *E. coli* play in causing food poisoning has helped make the bacterium's name a household word. For example, many Americans learned of the potential danger of *E. coli* in August 1997, when Hudson Foods, Incorporated, of Rogers, Arkansas, recalled 11 million kilograms (25 million pounds) of ground beef thought to be contaminated with a deadly strain of the bacterium.

Scientists hoped that knowledge of the microbe's genome would lead to a better understanding of what makes some strains of *E. coli* dangerous to people and other strains harmless. Such knowledge could lead to new drugs to fight food poisoning.

DNA chips. The use of a *DNA chip* (a tiny array of DNA) to find a disease-causing gene was reported in March 1998 by researchers led by neurologist Vivian Cheung of the Children's Hospital of Philadelphia. Although scientists had speculated that DNA chips, also called DNA microarrays, could be used to find the locations of specific genes in human *chromosomes* (structures that car-

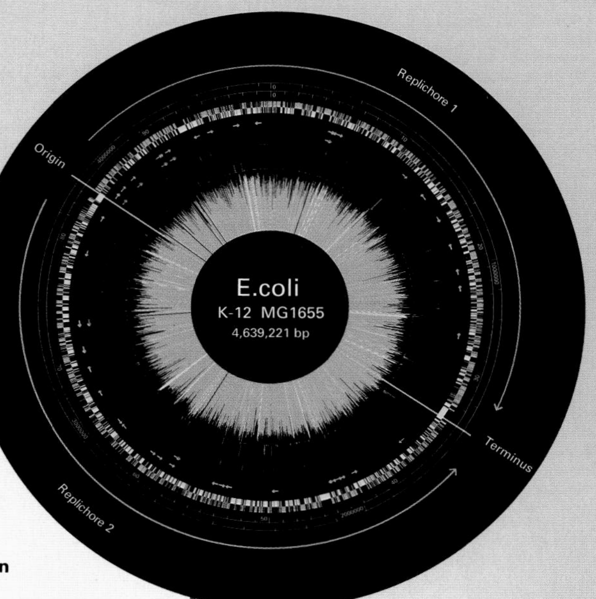

Mapping the genes of microbes

Since 1995, biologists have *sequenced*, or decoded, the *genomes* (complete set of genetic material) of several microorganisms. Knowledge of these genomes enables scientists to better understand not only how the microbes function but also how they have evolved. In addition, the genomes have a number of commercial applications, including in the design of new antibacterial drugs.

The genomes of more than a dozen microorganisms, including the ones listed below, had been sequenced as of mid-1998. Approximately 50 other microbial genome projects were underway during 1998. Unless stated otherwise, each microbe listed is a bacterium.

Microbe	Approx. number of genes	Description
Saccharomyces cerevisiae	6,000	brewer's yeast
Escherichia coli	4,300	lives in animal guts; some strains cause infections; used in research and industry
Bacillus subtilis	4,100	lives in soil; used in research and industry
Mycobacterium tuberculosis	4,000-4,500	causes tuberculosis
Synechocystis sp	3,200	blue-green alga
Archaeoglobus fulgidus	2,400	archaeon (bacterialike organism) found in oil wells
Methanobacterium thermoautotrophicum	1,900	archaeon found in sewage
Haemophilus influenzae	1,700	causes ear infections, meningitis, and pneumonia
Methanococcus jannaschii	1,700	archaeon found in deep-sea vents
Helicobacter pylor	1,600	causes peptic ulcers
Borrelia burgdorferi	850	causes Lyme disease
Mycoplasma pneumoniae	680	causes pneumonia
Mycoplasma genitalium	470	causes genital and respiratory infections

The structure of the genome of an *Escherichia coli* strain called K-12 MG1655 was diagrammed, *above,* by researchers at the University of Wisconsin. The bacterium consists of 4,639,221 *base pairs* (molecular units in genes). The two circular strands of *E. coli's* single chromosome, which contains the genes, are shown as the outer orange and yellow circles. The blocks in the circles represent the genes on the strands. The central orange "sunburst" is a predictor of gene expression. (Some genes are expressed, or activated, more often than other genes.) A longer ray of the sunburst suggests that a gene is expressed less often compared with others. Green lines on the diagram split the chromosome into two halves, called replichores, which *replicate* (reproduce) separately when the bacterium divides. The replication of each replichore begins at the origin point (left) and ends at the terminus (right). Other elements in the diagram include numerical scales, which are used to identify the locations of particular genes. A microscope photograph of *E.coli, left,* shows the bacterium's long tail.

ry the genes), these results were the first to show how this could be done.

A DNA chip is like a computer chip in the respect that enormous amounts of information can be packed together in a small area. In the case of DNA chips, however, the information is in the form of DNA molecules. The goal of using a DNA chip is to identify one particular piece of DNA in the array. Finding that piece of DNA can mean that a particular gene has also been located.

A DNA chip is used to find a disease-causing gene by comparing pieces of DNA from two unrelated people who have the same disease. This procedure is based on the likelihood that these people will share few, if any, genes with the same base-pair sequence other than the gene responsible for the disease.

In the first step of the procedure, DNA fragments from one of the person's chromosomes are placed on a glass slide. Then, DNA fragments from the other person's chromosomes are added to the slide. In places where the DNA fragments from the two people are similar, DNA subunits called nucleotides will bind to each other. In places where the fragments are different, the nucleotides will not bind. The unbound fragments are removed from the array by special proteins added by the researchers. The DNA pieces that remain in the array contain the gene (or genes) responsible for the disease.

The gene located by Cheung's researchers was a gene responsible for a form of hyperinsulinism, a disorder characterized by high levels of the protein insulin and low levels of the sugar glucose. This gene had previously been found using other analytical methods, but Cheung wanted to demonstrate that a DNA chip could be used to find the same gene more quickly and efficiently. Although most scientists agreed that she did in fact demonstrate this, they said that DNA-chip technology would need additional refinements before it could be widely used in genetics work.

Parkinson disease. The discovery of a gene that plays a role in Parkinson disease, a degenerative nerve disorder in which certain brain cells die, was announced by an international team of scientists in June 1997. The team included geneticists from the National Human Genome Research Group in Bethesda, Maryland; the Robert Wood Johnson Medical School in Piscataway, New Jersey; and institutions in Italy and Greece. The study helped shed light on some little-understood factors underlying this disabling disease, which causes such symptoms as trembling hands, rigid muscles, and slow movement.

To find the gene, the researchers collected blood samples from Italian and Greek families that had a history of Parkinson disease. Comparisons of the DNA in the blood specimens revealed that all family members with the disease carried a defective copy of a gene that directs the production of alpha-synuclein, a brain protein of unknown function. None of the healthy family members carried the defective gene.

The scientists were not sure how a defective alpha-synuclein gene might lead to the death of brain cells, but they speculated that it might cause the production of a protein unable to fold up normally. Proteins unable to fold in the right way do not function correctly, resulting in various problems in the body.

Two genes for epilepsy. Defective genes cause at least one form of epilepsy, a brain disorder characterized by seizures, according to a study published in January 1998 by a team of researchers led by geneticist Mark Leppert of the University of Utah in Salt Lake City. The scientists said they had identified two genes that are defective in people with benign familial neonatal convulsions (BFNC), a type of epilepsy that primarily strikes newborns and later usually fades away.

Epilepsy affects 20 million to 40 million people worldwide. Some forms of the disorder are known to be caused by brain damage resulting from injuries, tumors, or infections. Other cases, however, are passed from generation to generation within families, implying that genetic factors are involved. BFNC is among the types of inherited epilepsies.

To find the genes behind BFNC, the researchers studied the chromosomes of several families with a history of the disease. They found that every person who had suffered from BFNC as an infant had the same abnormal section of a certain chromosome. Additional research allowed the scientists to zero in on two *mutated* (altered) genes, named KCNQ2 and KCNQ3, in this section.

How Life Depends on Death

The human body can be thought of as an extremely complicated machine consisting of over 100 trillion cells organized into several hundred different cell types. Building such a machine and keeping it functioning for the length of a human life span is no easy task. Biologists have long known that *cell proliferation* (the division and multiplication of cells) is essential for the development and maintenance of a body. However, since the 1970's, scientists have begun to realize that a normal, healthy life also requires that old and damaged cells be killed. In other words, many individual cells must die for an organism to live.

The death of defective cells is brought about through the activation of a genetic suicide program called apoptosis, or programmed cell death. Scientists have determined that all normal cells in a person's body carry such a genetically encoded self-destruction plan. Although this system usually lies dormant within a cell, it can be activated at a moment's notice when needed. For example, apoptosis works during a baby's development in the womb to guide the growing *embryo* (unborn baby) into the proper shape and to get rid of excess cells. Apoptosis is also called upon when a cell develops a damaging *mutation* (genetic alteration) or is infected by a virus. If such a cell's suicide program is functioning, the cell will die before it causes any harm to the rest of the body.

Unfortunately, devastating diseases can occur when a cell has faulty genes that prevent apoptosis from functioning. In some situations, a cell that has sustained genetic damage, such as from excess ultraviolet radiation from the sun's rays, fails to commit suicide. The cell may then go on to divide and give rise to a cancerous tumor. Blocked apoptosis can also lead to disorders of the immune system. In other cases, malfunctioning apoptosis programs cause cells to commit suicide when they should not. This out-of-control cell death contributes to *neurodegenerative* (destructive to nerve cells) diseases such as amyotrophic lateral sclerosis (Lou Gehrig's disease) and increases the damage from strokes and heart attacks.

Scientists in the 1980's and 1990's invested a large amount of energy in trying to better understand the mechanisms of programmed cell death. They hoped that this understanding would lead to new and improved treatments or cures for the many disorders traced to aberrant apoptosis.

By 1996, several groups of biologists had identified many components of the death machinery and also some of the biochemical factors that interfere with them. Among the latter is a family of proteins, including one called the Bcl-2 oncoprotein. Components of the system include a family of *enzymes* (proteins that promote chemical reactions) known as caspases, which are the foot soldiers of apoptosis. By cutting apart certain critical proteins in a cell, caspases are responsible for most, if not all, of the cellular destruction that occurs during apoptosis.

Caspases are present in cells most of the time as inactive compounds called procaspases. One of the main events at the beginning of apoptosis is the development of procaspases into active, death-promoting caspases. Scientists have discovered two distinct mechanisms involved in the activation of caspases.

The first mechanism has been gradually uncovered since the late 1980's by cancer researchers including Peter Krammer of the German Cancer Research Institute in Heidelberg and Vishva Dixit. Dixit was at the University of Michigan Medical School in Ann Arbor until 1997, when he moved to Genentech, Incorporated, in South San Francisco, California. The researchers have found that immune cells known as cytotoxic T lymphocytes (CTL's) continuously patrol the body for cells that show signs of being cancerous or infected by a virus. When a CTL identifies such a cell, it immediately activates a protein, called a Fas receptor, on the surface of the cell. The Fas receptor then begins attracting a number of procaspases to the cell. The concentration of a large number of procaspases on a tiny area on the cell membrane somehow promotes the development of the procaspases into caspases, which then go about breaking down the cell.

The second mechanism of caspase activation was described in a series of papers during the 1990's by biochemist Xiaodong Wang and his colleagues at the University of Texas Southwestern Medical Center in Dallas. Wang set out to purify the ingredients involved in caspase activation from cells grown in culture. He succeeded in isolating two previously unknown proteins, which he called apoptotic protease activating factors (Apafs). His research showed that the two Apafs—Apaf-1 and Apaf-2—work together to coax a certain procaspase, procaspase-9, into its active form.

Scientists noted that Apaf-1 is similar to an apoptosis protein that had previously been identified in the roundworm *Caenorhabditis elegans,* a tiny organism with fewer than 1,000 cells. This worm is such an uncomplicated animal that the apoptosis of its cells is much easier to understand than that of human cells. In the early 1980's, researchers at the Massachusetts Institute of Technology in Cambridge discovered two proteins—

CED-3 and CED-4—that are essential for apoptosis in *C. elegans.* The MIT team showed that CED-3 is the worm's equivalent of the caspases in humans, and that CED-4 is required for CED-3 activation. Following Wang's findings, apoptosis researchers realized that CED-4 has the exact same role in the roundworm that Apaf-1 has in humans.

The fact that roundworms and humans have such similar apoptosis mechanisms underscores how fundamental programmed cell death is to animal life. Scientists believe that apoptosis must have arisen very early in the evolution of animals.

As Wang continued his research on the Apafs, he made a stunning discovery that took many scientists by surprise. He said he found that Apaf-2 is actually the same substance as cytochrome c, a protein that had long been known by biologists for its role in cellular respiration, the process by which cells generate most of their energy. Biologists had always assumed that cytochrome c was present only in the mitochondria, the cell's power plants. However, Wang discovered that for a cell to die, cytochrome c must move outside a mitochondrion and into the *cytoplasm* (the main body of the cell), where it works with Apaf-1 to activate procaspase-9. As of mid-1998, scientists had not determined how cytochrome c escapes from the interior of the mitochondria.

These revelations about apoptosis have important potential applications in the treatment of diseases—particularly cancer, which is characterized by malfunctioning apoptosis and the unchecked growth of abnormal cells. In 1998, cancer treatments under development included various drugs designed to reactivate the stalled apoptosis program in cancer cells. One of these drugs was safingol, which had shown promise in helping certain anticancer drugs fight tumors in studies at Memorial Sloan-Kettering Cancer Center in New York City. Another substance undergoing research was ceramide, which biologists at Duke University Medical Center in Durham, North Carolina, had shown removes the block on apoptosis in tumor cells grown in the laboratory. In contrast to these substances, other drugs were being developed to block apoptosis in order to prevent the death of cells that occurs in neurodegenerative diseases, strokes, and heart attacks.

In the late 1990's, researchers were armed with much new knowledge about the critical role that cell death plays in preserving health. Continued research was expected to lead to an array of new drugs to regulate apoptosis. If these expectations are fulfilled, apoptosis will have developed in a short time from a mostly ignored process to a fundamental cornerstone of efforts to combat disease and keep the machinery of the body running smoothly. [Michael O. Hengartner]

Cell death versus tumor growth

Apoptosis, or programmed cell death, is the system a cell uses to destroy itself when it has sustained serious damage. Sometimes, however, the apoptosis system of a cell fails to function.

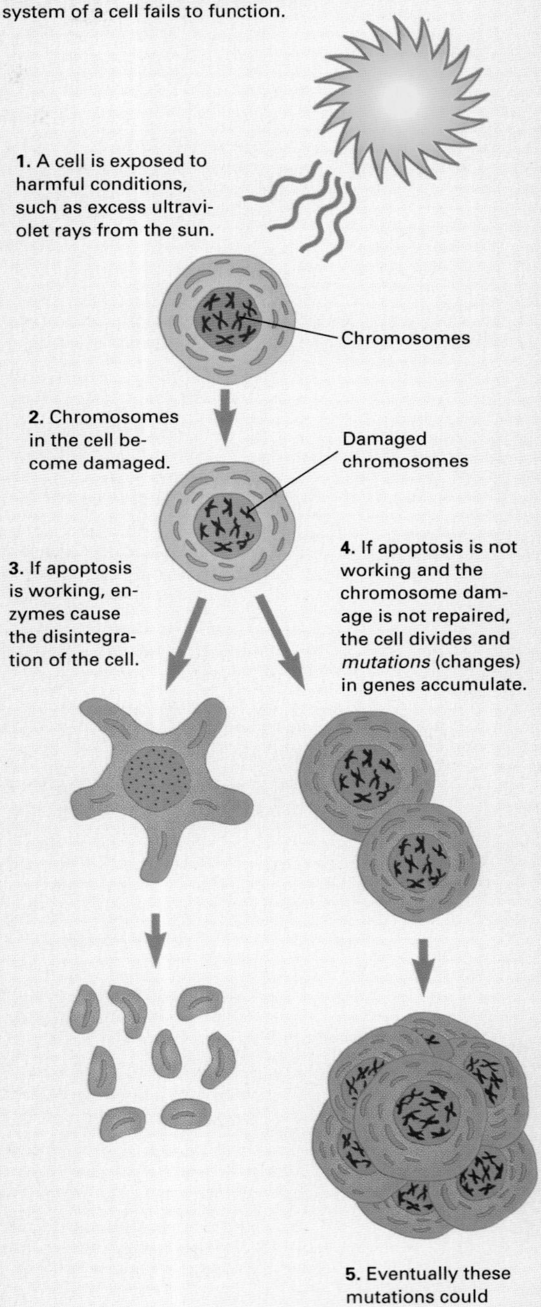

1. A cell is exposed to harmful conditions, such as excess ultraviolet rays from the sun.

Chromosomes

2. Chromosomes in the cell become damaged.

Damaged chromosomes

3. If apoptosis is working, enzymes cause the disintegration of the cell.

4. If apoptosis is not working and the chromosome damage is not repaired, the cell divides and *mutations* (changes) in genes accumulate.

5. Eventually these mutations could lead to the growth of a tumor.

The scientists determined that the normal forms of the genes help control the flow of substances such as potassium and sodium into and out of cells. Cells of the nervous system need to be able to control this flow in order to communicate with one another. The scientists said that defective KCNQ2 and KCNQ3 genes would result in the inability of cells to regulate the movement of chemicals. They speculated that this inability could cause brain cells to fire off nerve messages like a machine gun, overloading the nervous system and leading to seizures.

The scientists said that their findings might lead to the discovery of yet other genes that play a role in the development of epilepsy. Researchers hoped that the discovery of such genes and the understanding of their functions would help them develop improved treatments for epilepsy.

Genetic clue to baldness. A gene responsible for a type of *alopecia* (hair loss) was described in January 1998 by a group of geneticists led by Angela Christiano of Columbia University in New York City and Wasim Ahmad of Quaid-I-Azam University in Pakistan. The gene, when mutated, causes an extreme form of alopecia called alopecia universalis, which results in the total absence of hair on the head and body.

The gene, named "hairless" by the researchers, was discovered by studying several members of a large family in Pakistan that suffers from alopecia universalis. Extensive tests of the family's DNA located the general chromosomal region where the gene was, but the great number of genes in this region made it difficult to narrow the search down to the hairless gene itself. Fortunately, the scientists had also been studying a hair-loss gene in mice, and they speculated that the mouse gene might be similar enough to the human gene to help them in their search.

Using the known location of the mouse gene as a guide, the researchers tracked down what they suspected was the equivalent gene in humans. Their suspicions were confirmed when they found that this gene was mutated in all family members with alopecia.

The scientists explained that the normal form of the gene promotes hair growth, probably by directing the production of a type of protein called a transcription factor. Transcription factors work to activate various genes in a cell, many of which are required to construct complex structures such as hairs. If the hairless gene is mutated, a certain transcription factor is missing, and hairs cannot be made.

Christiano said that even though alopecia universalis is very different from male pattern baldness, the discovery of the hairless gene provided researchers with important clues to an understanding of both hair growth and hair loss. The researchers hoped that their study would lead to the discovery of genes responsible for other types of hair loss. Knowledge of how such genes work might eventually lead to new treatments for baldness.

Cholesterol gene. The location of a gene that helps cells regulate cholesterol levels was announced in July 1997 by cell biologist Peter Pentchev and his colleagues at the National Institute of Neurological Disorders and Stroke in Bethesda. The gene, when mutated, causes a rare childhood disorder called Niemann-Pick type C disease. Children with this disease have cells that are unable to process cholesterol in the normal way. Instead of breaking cholesterol down as a waste product or incorporating it into their makeup, cells become glutted with these fatlike molecules and die off. Most children who suffer from this disorder die before they reach adulthood.

Although rare, Niemann-Pick type C disease came to national attention in the mid-1990's when three grandchildren of Ara Parseghian were diagnosed with the disorder. Parseghian is known to many Americans as the football coach at the University of Notre Dame in South Bend, Indiana, from 1964 to 1975.

In its search for the gene, Pentchev's team took advantage of the fact that previous research had narrowed down the gene's location to a small region on a certain chromosome. To find exactly where in this region the gene was, the scientists took very small pieces from the region and placed them, one at a time, into a laboratory culture of cells from a person with the disorder. The researchers hoped that if a normal copy of the gene was contained in one of the pieces

Geneticist Angela Christiano of Columbia University in New York City announces in January 1998 that her team has discovered the gene responsible for a severe type of baldness called alopecia universalis. Behind her is a photograph of a man suffering from the disorder, which is characterized by a total lack of hair on the head and body. Christiano said the gene's discovery could lead to a better understanding of more common forms of baldness.

placed in the cells, it would direct the production of a certain protein that would restore the ability of the cells to process cholesterol.

When one of the chromosome pieces did restore the cells' cholesterol-processing ability, the scientists were able to isolate the gene, which they named NPC1. The scientists went on to show that the gene was mutated in patients with the disease.

In a related study also reported in July 1997, researchers from the National Human Genome Research Institute in Bethesda found a gene in mice that appears to cause symptoms similar to those in Niemann-Pick type C disease. The scientists said that studies of these mice might shed additional light on the human disease. Both groups of Bethesda scientists expressed hope that their discoveries would eventually lead to a cure for Niemann-Pick type C disease and other cholesterol disorders.

Macular degeneration. A genetic factor behind some cases of macular degeneration, the major cause of vision loss in elderly people, was described by a team of U.S. researchers in September 1997. The investigators were led by molecular geneticist Michael Dean of the National Cancer Institute in Frederick, Maryland, and *ophthalmologist* (eye doctor) Richard Lewis of the Baylor College of Medicine in Houston.

In macular degeneration, cells of the *macula* (the region at the back of the eye that sees fine details) are destroyed. The disease is thought to have many causes, including smoking, a high-cholesterol diet, and genetic mutations.

The researchers found that of 167 elderly people with the disorder, 26 had a mutated copy of a gene that, when inherited in two copies, was known to be involved in a form of macular degeneration in younger people. The scientists said they were not sure what exact role the gene plays in the eye disorder.

The investigators hoped that further research into how the gene works would help them better understand how macular degeneration develops. This understanding, in turn, might lead to new drugs to halt the disease.

[David Haymer]

A report on conditions inside Old Faithful, the famous geyser in Wyoming's Yellowstone National Park, was published in October 1997, by geologists Roderick Hutchinson of the National Park Service, James Westphal of the California Institute of Technology in Pasadena, and Canadian geologist Susan Kieffer. They found that the geyser's inner workings were not as complex as had been thought. By studying geysers, scientists hope to better understand their relationship to other geologic events, such as earthquakes and volcanoes.

Anatomy of Old Faithful. Like all geysers, Old Faithful is a *conduit* (duct) in the ground that periodically spouts hot water and steam into the air when pools of underground water become superheated by Earth's internal heat. Old Faithful is famous because its eruptions are so regular, occurring about 18 to 22 times a day, and have never stopped in its recorded history.

Between eruptions, the scientists lowered a small video camera, along with temperature and pressure sensors, in insulated canisters into the nearly vertical conduit of Old Faithful to a depth of 22 meters (72 feet). Images from the camera showed that the narrowest part of the conduit is 7 meters (23 feet) below the surface. From there to a depth of 9 meters (29.5 feet), the conduit is only 15 centimeters (6 inches) wide. This constriction apparently greatly accelerates the flow of water in an eruption.

Above the constriction, the researchers saw water cascading down the sides of the conduit. These underground waterfalls come partly from water that had been forced into cavities off the main conduit during previous eruptions and partly from ground water seeping in from the Firehole River, which lies only 150 meters (500 feet) away. Below the constriction, the conduit widens to form a chamber more than 2 meters (6.6 feet) in diameter, but then narrows again to become a slot just 30 centimeters (12 inches) wide.

The information enabled the team to illustrate in detail how an Old Faithful eruption occurs. The elevation of Old Faithful is 2,233 meters (7,326 feet) above sea level, where the boiling point

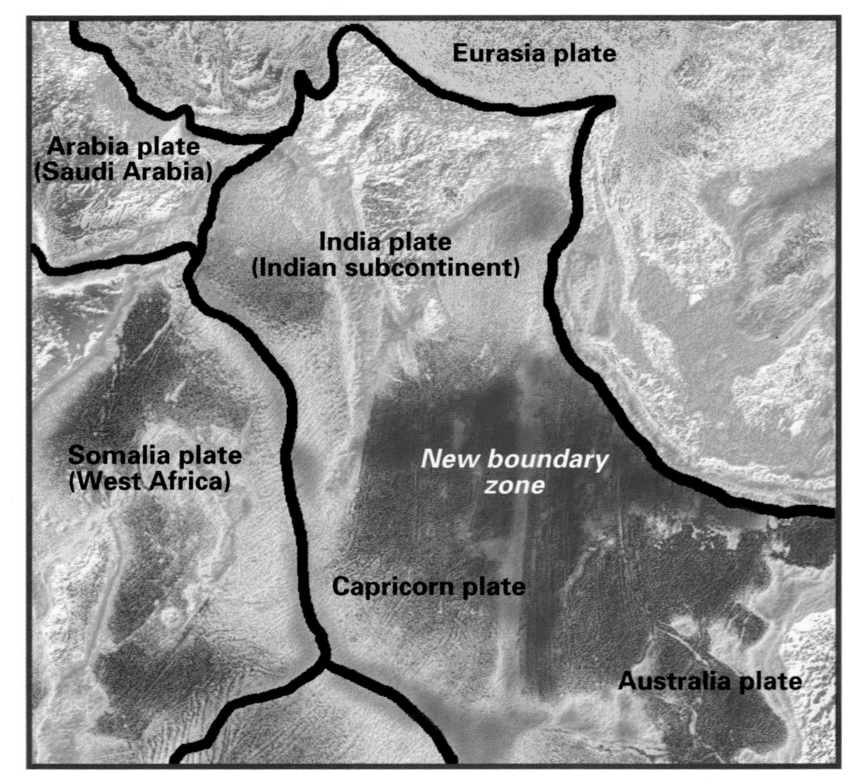

A study published in August 1997 suggested that the boundaries between the individual plates that form the Earth's crust are not always as clear-cut as geologists had thought. Using earthquake data and other clues, geologists Jean-Yves Royer of Geosciences Azur in France and Richard G. Gordon of Rice University in Houston found that the Indo-Australia plate is actually made up of three components—the India and Australia plates plus the newly recognized Capricorn plate. The scientists observed indications of stress in the middle of the Indo-Australia plate, far from any known plate boundary (black lines). This area of stress, they deduced, was a spot at which the shifting of the three plates has resulted in a diffuse boundary zone of molten crust thousands of kilometers wide (dark blue area).

A preliminary version of the first detailed map of Antarctica was created in late 1997 from data obtained by RADARSAT, a satellite operated by the Canadian Space Agency. RADARSAT took more than 8,000 high-resolution radar images of the continent. Previously, the best maps of Antarctica had been pieced together from data gathered by different satellites between 1980 and 1987. The new images will provide scientists with better information for the study of Antarctica. Cartographers at Ohio State University's Byrd Polar Research Center were combining the RADARSAT images to generate the final map, a project that was expected to be completed in late 1998 or early 1999.

of water is only 92 °C (197.6 °F). After an eruption, the conduit begins filling up with water, which comes in from the sides rather than from below. Geothermal heat from deep below quickly heats the water, which boils wildly and flashes into steam. The boiling water expands and rises up the conduit.

By the time the water level reaches 12 meters (39 feet) below the surface, the uppermost water has been cooled to only 86 °C (186.8 °F) by the underground waterfalls in the conduit. The water below continues to boil and expand, however, rising and increasing in pressure. Bubbles of steam and what may be carbon dioxide rise from below. As the water rises, the bubbles grow larger, pushing more and more water upward. An eruption occurs when the bubbles push a large quantity of water out of the conduit with such explosive force that it shoots high into the air.

The researchers found that the hottest water in the conduit, 118 °C (244.4 °F), was at the deepest point their instruments could reach. During an eruption, however, the water reached 130 °C (266 °F), indicating that suction from the eruption was drawing up and expelling water from great depths.

New asteroid impact sites. A number of places on Earth have been shown to be the sites of ancient asteroid impacts in recent years, and additional sites of major impacts are being discovered nearly every year. One newly found site, reported in November 1997, is in the Bellingshausen Sea, southwest of South America. Another, reported in August 1997, is called the Morokweng Structure and is located in the Northwest Province of South Africa.

The Bellingshausen Sea impact site was discovered in 1995 by an international group of scientists led by geologist Rainer Gersonde of the Alfred Wegener Institute for Polar Research in Bremerhaven, Germany. The first evidence for an impact in this area was the discovery in 1981 of an unusually high level of the element iridium in a sediment sample recovered by a U.S. Navy research vessel in the 1960's. The presence of iridium in the ocean floor signaled a possible asteroid strike, because the element is scarce in most parts of Earth but relatively abundant in rocks of extraterrestrial origin.

Gersonde and his team visited the site, about 1,500 kilometers (930 miles) southwest of Chile, to collect more sediment samples. They also surveyed the deep-sea floor with instruments that produced detailed images of the seafloor sediments. They found a region where sediments deposited over a period of 50 million years were highly disturbed, indicating a catastrophic event at some time in the past.

Analysis of the data and sediment samples pinpointed the time of the impact as 2.15 million years ago. The data also indicated that the asteroid was probably 1 to 4 kilometers (0.6 to 2.5 miles) in diameter. If it had struck on land, the impact would have left a crater 15 to 40 kilometers (9 to 25 miles) across, but it had less effect on the denser ocean crust. However, it may have affected the global climate by blasting huge quantities of water vapor into the atmosphere.

The other site, the Morokweng Structure, was studied by a team led by geologist Christian Koeberl of the University of Vienna and a group of researchers from South Africa, working independently. The two teams jointly reported discovering the large crater, near Morokweng, in South Africa.

Magnetic and gravitational data indicated that the crater, which is now buried beneath the sands of the Kalahari Desert, may be up to 340 kilometers (210 miles) in diameter, which would make it the largest known crater on Earth. Radioactive dating of rock samples indicated that an asteroid struck the Earth in this region about 145 million years ago, a time that marks the boundary between the Jurassic and Cretaceous periods of geologic history. Thus, the impact may explain a moderate wave of extinctions among reptiles and marine life that is known to have occurred at about that time.

Erosion of the Himalaya. Over the past 40 million years, Earth's climate has become cooler, and many geologists suspect that this cooling has resulted from a large drop in the amount of carbon dioxide in the atmosphere. (This con-

A computer model of Iceland, reported in December 1997 by Yang Shen of the Woods Hole Oceanographic Institution in Massachusetts and his colleagues, indicates that the island's *mantle plume* (a column of molten rock rising from the Earth's mantle, the layer below the crust) extends into the lower mantle, which reaches a depth of about 2,900 kilometers (1,800 miles). Iceland is one of many islands on Earth that rest atop a mantle plume, and geologists have debated the depth of these plumes. By measuring changes in earthquake shock waves traveling through the mantle, scientists have made models of mantle plumes before. However, these models are unreliable at the depth of the lower mantle. The new model was made by measuring not the shock wave itself but its effects on the upper/lower mantle boundary.

trasts with events of the past 100 years, during which an increase of atmospheric carbon dioxide has apparently contributed to a slight global warming.) A possible explanation for the long-term cooling was reported in November 1997 by geologists Christian France-Lanord of the Petrographic and Geochemical Research Center in Nancy, France, and Louis Derry of Cornell University in Ithaca, New York.

Atmospheric carbon dioxide warms the planet by preventing the reradiation of solar energy from the Earth's surface into space, so a drop in atmospheric carbon dioxide leads to global cooling. Geologists had long thought that the loss of atmospheric carbon dioxide over the last 40 million years was due to the weathering of rocks during the formation of the Himalaya, the highest mountain system in the world, which lies in southern Asia. In weathering, carbon dioxide reacts with water and minerals to produce carbonates. Over millions of years, this chemical reaction uses up atmospheric carbon dioxide.

However, France-Lanord and Derry realized that there was another explanation for the loss of the carbon dioxide—the burial of organic carbon in plants. Erosion caused by the uplift of the Himalaya would have caused the burial of billions of tons of plants and algae. These plants had absorbed carbon dioxide as they grew, so their burial would have prevented the carbon in their cells from ever being released back into the atmosphere as carbon dioxide when the plants died and decayed.

The researchers calculated both the current rate at which minerals were being weathered in the Himalaya and the rate at which organic carbon was being washed down into the Bay of Bengal. Then they estimated the total amount of organic carbon already buried beneath the bay. They found that the burial of organic carbon was two to three times more effective at removing carbon dioxide from the atmosphere than the weathering of minerals. Thus, they concluded that the burial of plant life during erosion of the Himalaya may have been the major factor in the loss of atmospheric carbon dioxide. [William W. Hay]

Medical Research

Several major medical news stories captured widespread public attention in 1997 and 1998. The developments attracting the most attention were a 1997 report that a popular diet-drug combination could cause heart damage and cancer-research findings announced in 1998 that raised hopes that a cure for cancer was on the horizon.

Diet drug scare. Millions of people who had used the weight-loss drugs fenfluramine and phentermine simultaneously were stunned in August 1997 by news that the popular drug combination may cause damage to the valves of the heart. A study published by researchers at the Mayo Clinic in Rochester, Minnesota, linked heart-valve disorders with the use of the two-drug combination popularly known as "fen-phen." In 1996, more than 18 million fen-phen prescriptions were written in the United States.

Damaged heart valves may not open properly or close completely, thus allowing blood to leak backward. The study reported 24 cases of women who took fen-phen and developed a heart murmur, an indication of irregular blood flow within the heart. None of the women had a history of heart problems.

In September 1997, in the aftermath of the Mayo Clinic study, the U.S. Food and Drug Administration (FDA) asked manufacturers of fenfluramine and a popular related drug, dexfenfluramine, to voluntarily withdraw the drugs from the market. In October 1997, the American Heart Association (AHA) recommended that patients who had taken either drug consult their physician. But the AHA's advisory report also noted that the Mayo Clinic study was "an initial description of a clinical association" and not a controlled clinical trial.

The link between weight-loss drugs and heart-valve damage was questioned in March 1998, when investigators at the Georgetown University Medical Center in Washington, D.C., presented contrary data. Their findings indicated that patients taking dexfenfluramine were not significantly more likely to develop heart-valve problems than people taking a *placebo* (inactive substance).

The Georgetown study was a large clinical trial involving more than 1,000 patients, half of whom were given dexfenfluramine. After two to three months of use, the participants showed that the incidence of heart-valve problems in the dexfenfluramine-treated group was not significantly different from that in the placebo group. The researchers said their study showed that people who had taken dexfenfluramine for a relatively short period appeared to run no risk of valve problems. However, they pointed out that their study did not evaluate related drugs, such as fenfluramine, and thus no conclusions could be reached about the safety of those drugs.

Cancer breakthrough? In May 1998, front-page stories in many major newspapers raised hopes that a cure for cancer might soon be available. The articles described research by M. Judah Folkman, a cancer researcher at Harvard Medical School and Children's Hospital, both in Boston. Folkman had demonstrated that two experimental drugs, angiostatin and endostatin, eliminated tumors in mice without causing any side effects. The drugs worked by cutting off the blood supply that tumors need to survive.

Folkman had published some of his findings in the journal *Nature* in November 1997, and by mid-1998 his work had sparked international media interest. Many researchers, who are normally guarded in their comments about any new advance in cancer research, were also enthusiastic. For example, Richard Klausner, director of the National Cancer Institute, stated that the new drugs were "the single most exciting thing on the horizon" for cancer therapy.

A number of other cancer experts, however, were quick to caution the public—and especially cancer patients—against too much optimism. It was possible, they warned, that the drugs would not work as well in humans as they had in mice. In his first public statement after his research made national headlines, Folkman also sought to temper public enthusiasm. Even if angiostatin and endostatin proved to be effective in humans, he said, they would be used in conjunction with existing treatments. A controlled clinical evaluation of the drugs was not expected until 1999.

Preventing breast cancer. A study by the National Cancer Institute published in April 1998 found that a synthetic hormone, tamoxifen, cut the risk of breast cancer by 45 percent in 13,000 women participating in a study of the drug. Half of the women received tamoxifen and the others received a placebo in daily doses for five years. While 154 of the women taking the placebo eventually developed breast cancer, only 85 women in the tamoxifen group did. Because of the positive results, the study was halted 14 months early, and women in the placebo group were given the drug. However, experts said it was too early to tell whether tamoxifen prevented breast cancer or simply delayed its onset.

There were also risks associated with tamoxifen. Thirty-three women in the treated group developed cancer of the *endometrium* (lining of the uterus), compared to 14 in the placebo group, and the risk of blood clots was also higher.

More good news was reported in May 1998 at the annual meeting of the American Society of Clinical Oncology. Researchers from Northwestern University in Evanston, Illinois, and the University of California at San Francisco announced findings about raloxifene (sold as Evista), a drug previously approved by the FDA to treat *osteoporosis,* a disease in which bones become weak and brittle. The researchers reexamined the data from Evista's preapproval trials and found that raloxifene significantly reduced the risk of breast cancer without raising the risk of endometrial cancer.

The Northwestern University study included 10,553 postmenopausal women with osteoporosis who took either the drug or a placebo for an average of 28 months. The Evista group had a 54 percent lower incidence of breast cancer than the placebo group. In the University of California at San Francisco study, 7,705 postmenopausal women with osteoporosis were followed for 33 months. The Evista group showed a 70 percent reduction in breast cancer onset compared with the placebo group. Neither study showed a rise in the risk of endometrial cancer.

New uses for gene therapy. In November 1997, cardiologist Jeffrey Isner and colleagues at St. Elizabeth's Medical Center in Boston reported on an innovative genetic technique to bypass severely blocked blood vessels in the legs by growing new vessels around them. This therapy, called therapeutic angiogenesis, may be useful in treating critical limb ischemia, a condition in which fatty *plaques* (deposits) build up in the blood vessels in the leg and interfere with blood circulation.

The most common means of restoring blood flow in the legs is surgery to graft a blood vessel from another part of the body onto the blocked leg vessel to channel blood around the blockage. In many patients, however, this procedure must be repeated when blockages recur in the newly attached vessels.

The new procedure employs a gene responsible for the production of a protein called vascular endothelial growth factor (VEGF). VEGF is responsible for promoting the development of blood vessel cells in the body. The researchers injected copies of the gene coding for VEGF into the calf or thigh muscle of the affected leg, near the blockage. The genes became incorporated into the muscle cells, which then began to produce VEGF. Within two to three weeks, the doctors confirmed new blood-vessel growth and improved circulation in most patients.

In a March 1998 update of their progress, Isner and his colleagues reported that treatment with the VEGF gene was successful in about two-thirds of the 29 patients treated. The researchers considered this an encouraging result, since these people might otherwise have faced amputation of the affected leg. Isner was also testing a similar therapy on patients with coronary heart disease, and early reports were encouraging but not yet conclusive.

Boosting heart function. The results of a landmark five-year study of patients with congestive heart failure (CHF) were presented at the American College of Cardiology meeting in March 1998 by Milton Packer, director of the Center for Heart Failure Research at Columbia-Presbyterian Medical Center in New York City. Although drugs called ACE inhibitors have been shown to reduce hospitalization and mortality rates in CHF patients, Packer found that many patients were receiving inadequate doses of these drugs. In CHF, the heart works inefficiently because of faulty heart valves or damage from such condi-

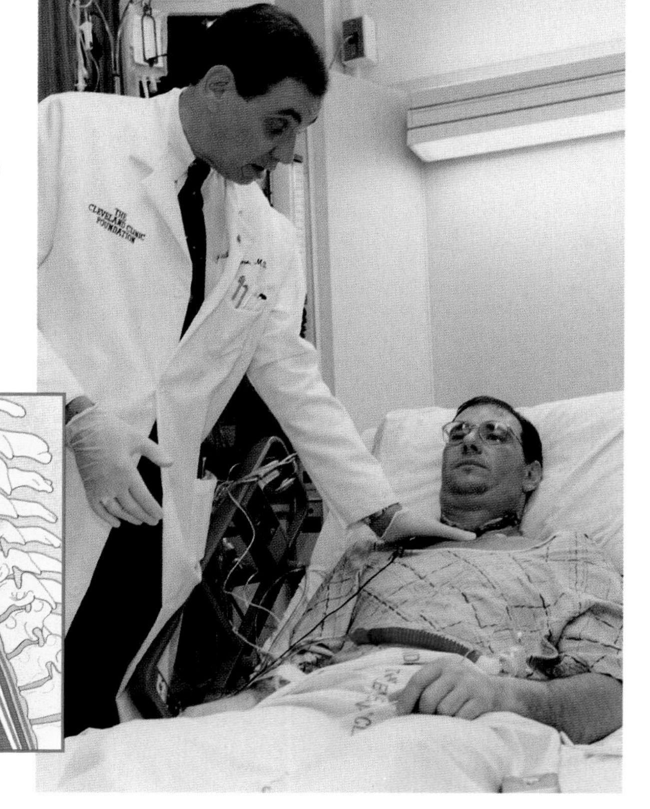

Speaking of transplants
A surgeon at the Cleveland Clinic in Ohio checks the progress of a transplant patient who received a *larynx* (voice box) and part of the *trachea* (windpipe) from an unidentified donor in January 1998. The larynx and upper trachea are located in the throat just below the jaw, *below*. Three days after the procedure, the recipient, a 40-year-old man whose throat was crushed in an accident in 1979, uttered the words "Hello" and "Hi Mom." A month later, he was speaking full sentences. Transplants of such "nonvital" organs as the larynx are rare because doctors have felt that the risks of such procedures outweigh the potential benefits.

tions as heart attack or high blood pressure. CHF affects 5 million Americans, often impairing their quality of life and shortening their life expectancy.

The study compared the effects of low doses (2.5 to 5 milligrams per day) and high doses (32.5 to 35 milligrams per day) of the ACE inhibitor lisinopril on more than 3,000 patients with moderate to severe CHF. The high-dose treatment was shown to reduce the risk of death and hospitalization by 12 percent compared to low-dose therapy. In addition, patients given high doses experienced only a slight increase in side effects.

According to Packer, the main reason that ACE inhibitors are often prescribed in low doses is that many doctors believe low doses are as effective as higher ones. By raising the dosage, he said, doctors could prevent at least 60,000 CHF deaths in the United States each year.

Calming tremors. In July 1997, the FDA approved Activa, a brain implant device similar to a heart pacemaker, for use in controlling muscular tremors in patients with essential tremor or Parkinson disease. Essential tremor is a poorly understood nerve disorder characterized by the involuntary trembling of the hands and arms. Parkinson disease, a degenerative disorder of the brain, typically causes similar trembling.

Involuntary trembling occurs when the thalamus, a portion of the brain that helps control movement, emits abnormal signals. The new device, which includes a thin electrode implanted into the brain and a battery-powered pulse generator implanted below the skin of the chest, transmits mild electrical impulses to the thalamus, interfering with signals that trigger the trembling. Thus, it permits patients to lead a more normal life. However, doctors stressed that Activa was not a cure for either Parkinson disease or essential tremor.

After four years of study, data presented to the FDA showed that 67 percent of severely impaired Parkinson patients and 58 percent of severely impaired essential-tremor patients had a significant reduction in their tremors after receiving the Activa implant. However, other Parkinson symptoms, such as muscle rigidity, did not improve. Side effects from the device included tingling in the hands, though fine-tuning the device could reduce that problem.

Weighing the benefits of HRT. The debate over whether postmenopausal women should take hormone replacement therapy (HRT) received some clarification from a study published in June 1997. Researchers at Brigham and Women's Hospital in Boston found that women who received HRT had a significantly lower risk of death than those who did not. The study essentially confirmed previous research indicating that HRT minimized the risk of heart disease but raised the risk of breast cancer.

To clarify the overall risk, the researchers studied surveys of 38,000 postmenopausal women in what was the most definitive assessment of the pros and cons of HRT to date. Results of the 18-year-long study indicated that the risk of death was 37 percent lower in women who received HRT compared with those who had never taken it. The risk of death from heart disease was 53 percent lower in patients who received HRT. However, the investigators found that this benefit was offset by an increased risk of breast cancer in women who took HRT for 10 years or more. But that group still had an overall death rate 20 percent lower than the group that did not receive HRT. "On average," the researchers concluded, "the survival benefits appear to outweigh the risks."

In a separate study published in October 1997, British researchers analyzed the effect of HRT on breast cancer risk using data from 51 previous studies of more than 160,000 women. This study found that women taking HRT for five years or more had a 35 percent greater risk of developing breast cancer, compared with women who did not receive HRT. The risk tended to be higher in thinner women. Once women stopped taking HRT, however, the increased cancer risk declined or disappeared completely after about five years.

Overprescribing antibiotics. Doctors have known for many years that antibiotics are not effective against *viral* (virus-caused) illnesses, including such ailments as the common cold. However, a study in the September 1997 *Journal of the American Medical Association* found that many American physicians inappropriately prescribe antibiotics for viral infections.

Researchers at the University of Colorado Health Sciences Center and the

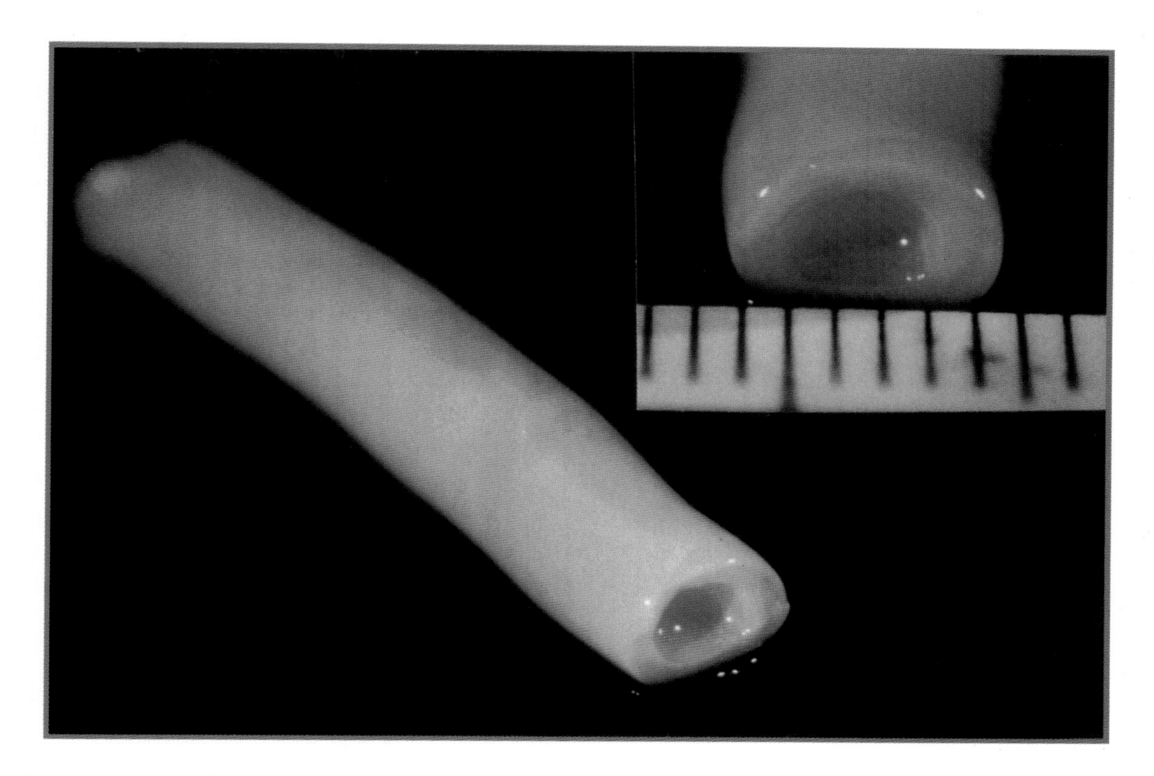

Researchers at Laval University in Laval, Quebec, reported in January 1998 that they had successfully manufactured blood vessels in the laboratory. The lab-grown vessels are made entirely from human cells but are about 20 times stronger than natural blood vessels. Replacement blood vessels grown from a patient's own cells may one day represent an alternative to vein grafts for bypass surgery, in which doctors channel blood around a blockage using veins taken from elsewhere in the patient's body. In some cases they may also be superior to synthetic blood vessels, which other researchers were developing in 1998.

University of Utah studied the prescribing habits of 1,500 doctors and concluded that the overuse of antibiotics was "widespread." About 12 million antibiotic prescriptions were written for respiratory infections in 1992, even though the cause of more than 90 percent of such infections tends to be viral.

This finding was particularly worrisome because the overuse of antibiotics was contributing to the growing problem of antibiotic-resistant bacteria. In an editorial accompanying the study, physicians at the Centers for Disease Control and Prevention in Atlanta, Georgia, wrote that doctors overprescribe antibiotics because of "unrealistic patient expectations, coupled with insufficient time to discuss with patients why an antibiotic is not needed." They emphasized that the spread of antibiotic-resistant bacteria required an "immediate and aggressive response" by doctors in changing their prescribing habits.

New hope for dementia. According to a study published in October 1997, EGb 761, an extract of the ginkgo tree, can improve the mental functioning of people with Alzheimer's disease and other forms of *dementia* (a progressive deterioration of a person's mental capacities). Ginkgo extracts have been popular medicines in China for almost 4,000 years, most often used to improve blood circulation. The trial was led by investigators at the New York Institute for Medical Research in Tarrytown.

The study involved 309 people with dementia. About half were given EGb 761, while the others received a placebo. After six months to a year of treatment, 27 percent of evaluated patients in the EGb 761-treated group showed modest improvement in tests to measure factors such as memory and reasoning ability, compared to 14 percent in the placebo group. In the EGb 761 group, 37 percent also showed improvement in their social behavior and daily living ability, such as holding conversations or managing money, compared to 23 percent of those taking the placebo. Other studies were underway in mid-1998, but researchers cautioned that they had found no evidence that EGb 761 could prevent dementia. [Richard Trubo]

The 1997 Nobel Prizes in science were awarded for advances in understanding the basic chemistry of life, the manipulation and study of atoms, and the discovery of a previously unknown infectious agent.

The Nobel Prize for chemistry was shared by Jens Skou of Aarhus University in Denmark, John Walker of the Medical Research Council in Cambridge, England, and Paul Boyer of the University of California at Los Angeles. The three chemists were honored for pioneering work on adenosine triphosphate (ATP), a compound that stores energy in all living organisms and supplies the energy needed for a heart to beat, a brain to function, and muscles to contract.

Skou was awarded half of the $1-million prize for his discovery of an enzyme that works with ATP to regulate the concentrations of sodium and potassium in cells.

Boyer and Walker split the other half of the prize for their studies on how an enzyme, ATP synthase, manufactures ATP. Boyer spent three decades studying how the enzyme forms a chemical bond that links three phosphate groups and how that bond, when broken, releases energy to cells. Using X-ray *crystallography* (beaming X rays through crystals to analyze their structure), Walker unraveled the structure of ATP synthase by creating an atomic-scale map of certain portions of the enzyme. The three-dimensional map allowed researchers to see exactly how the enzyme functions, confirming Boyer's theory of the production of ATP.

The prize for physics was awarded to Steven Chu of Stanford University in Stanford, California, William D. Phillips of the National Institute of Standards and Technology in Gaithersburg, Maryland, and Claude Cohen-Tannoudji of the College de France and Ecole Normale Superieure in Paris. The physicists were honored for developing a method of cooling atoms by greatly restricting their motion in a laser-light "trap." Through the use of highly concentrated laser light, Chu and Phillips cooled a gas to within a millionth of a degree of absolute zero (−273.15 °C [−459.67 °F]), the temperature at which the atoms and molecules would possess the least possible energy. The laser beams, with Phillips's introduction of magnetic traps, held the atoms in a state sufficiently slow to allow for the first close examination of atomic properties.

Cohen-Tannoudji proposed a theory that explained Chu and Phillips's experiments and then performed experiments of his own that slowed atomic movement further. The techniques developed and refined by Chu, Phillips, and Cohen-Tannoudji led to the 1995 creation of a Bose-Einstein condensate, a form of matter whose existence was predicted by Albert Einstein but had remained unproved. The work also laid the foundation for a more precise atomic clock, increasing the accuracy of global and space navigation.

The prize for physiology or medicine was awarded to Stanley B. Prusiner of the University of California at San Francisco. Prusiner was honored for his hypothesis that certain neurological disorders—scrapie in sheep, bovine spongiform encephalopathy ("mad cow disease") in cattle, and kuru and Creutzfeldt-Jakob disease in humans—are caused by a previously unsuspected infectious particle, which Prusiner named the prion *(PREE ahn)*. Prusiner theorized that prions stem from normal proteins (prion proteins) in the brains of all animals, including humans. They trigger neurological disorders only after changing to take on abnormal shapes. He documented that prion proteins in contact with infectious prions take on the abnormal shape. The defective proteins gradually multiply and riddle the brain with small holes, creating a spongelike pattern of destruction and destroying mental functioning.

Prusiner theorized that prions, unlike other infectious agents—bacteria, fungi, protozoa, and viruses—reproduce without *nucleic acids* (genetic material). When Prusiner published his theory in 1982, it met with great resistance, because it was universally accepted among biologists that only nucleic acids are capable of self-duplication. However, since 1982, a growing body of research has supported Prusiner's theory, and in October 1996, English scientists confirmed that prions were involved in "mad cow disease." Some researchers, however, remained skeptical that prions could infect and reproduce without nucleic acids. [Scott Thomas]

Nobel Prize winners in physics
Steven Chu of Stanford University in Stanford, California, *above,* Claude Cohen-Tannoudji of the College de France and Ecole Normale Superieure in Paris, *right,* and William D. Phillips of the National Institute of Standards and Technology in Gaithersburg, Maryland, *below,* were awarded the 1997 Nobel Prize in physics. The three were cited for separate, complementary efforts at slowing the movement of atoms sufficiently to cool them to within a millionth of a degree of absolute zero (−273.15 °C [−459.67 °F]), allowing for a close examination of atomic properties.

Two researchers at at the Christchurch School of Medicine in Christchurch, New Zealand, reported in January 1998 that children who were breast-fed as infants appear to do better in school for longer periods than children who were bottle-fed. Doctors had long theorized that breast-feeding may lead to small yet detectable improvements in the learning ability and educational achievement of children. Most studies, however, focused on preschool children. The New Zealand study concluded that such improvements are not confined to early childhood but rather extend into adolescence and young adulthood.

Researchers L. John Horwood and David M. Fergusson tracked a group of more than 1,000 New Zealand children as part of an 18-year study. Horwood and Fergusson collected information on whether the children had been bottle-fed or breast-fed from birth to age 1. If the children were breast-fed, the researchers then charted for how long. The researchers evaluated the children from the ages of 8 to 18 on their intelligence and a variety of academic skills. They discovered that those children who had been breast-fed for eight months or longer had test scores higher than those children who had been fed formulas.

The researchers reported that they measured consistent and statistically significant increases in the childrens' intelligence quotient (IQ) assessed at ages 8 and 9; reading comprehension, mathematical ability, and scholastic ability assessed between the ages of 10 to 13; teacher-ratings of reading and mathematics assessed at ages 8 and 12; and higher achievement levels at the end of high school.

Throughout their study, the researchers adjusted their data to allow for certain factors, including the age, education, income-level, and marital status of the mother. After making such adjustments, they still concluded that breast-feeding remained a significant indicator of higher childhood and adolescent school achievement.

Dietary fat and heart disease. The relationship between different types of fat in the diet and the risk of heart disease has been the subject of several studies. One recent study by researchers at the Harvard School of Public Health, Brigham and Women's Hospital, and Harvard Medical School—all in Boston—published in November 1997 concluded that replacing saturated fats with unsaturated fats may be more effective in preventing heart disease in women than just reducing overall fat intake.

A fat is composed of three chainlike molecules, called fatty acids, connected to a backbone glycerol molecule. A fatty acid consists of a string of carbon atoms to which hydrogen atoms are attached. In a saturated fat, each fatty acid has as many hydrogen atoms as possible attached to its carbon chain, and the carbon atoms are linked together by single bonds. In an unsaturated fat, at least one pair of carbon atoms in each of the fatty acids is joined by a double bond. For each such bond, the carbon chain is missing a pair of hydrogen atoms. One type of unsaturated fat—trans unsaturated fat—contains fatty acids in which the hydrogen atoms surrounding a double bond are on opposite sides of the carbon chain. In ordinary unsaturated fatty acids, the hydrogen atoms are on the same side of the chain.

During the 14-year study, the team of investigators followed more than 80,000 women between the ages 34 and 59. The women had no known heart disease, stroke, cancer, or diabetes when the study began. Information on the women's dietary habits was obtained at the study's onset and updated during the following years by questionnaires. The researchers also documented heart attacks and deaths from coronary disease.

At the conclusion of the study, the investigators found that total fat intake was not significantly related to the risk of coronary heart disease. Women who had consumed the most total fat—46 percent of all calories—had no greater risk of heart attack than those with the lowest consumption of total fat—29 percent of all calories.

However, the researchers discovered that each 5 percent of calories from saturated fat, as compared with an equivalent amount of calories from carbohydrates, resulted in a 17-percent increase in the risk of coronary heart disease. A 2-percent increase in calorie intake from trans unsaturated fat, compared to the equivalent calories from carbohydrates, increased the risk for coronary heart disease by 93 percent.

New milk names

Old name	New name	Total fat per 240 milliliters (1 cup)
Whole milk	Milk	8 grams
Low-fat 2 percent milk	Reduced fat or less-fat milk	4.7 grams
Not applicable	Light milk	4 grams or fewer
Low-fat 1 percent milk	Low-fat milk	2.6 grams
Skim milk	Fat-free, skim, zero-fat, no-fat, or nonfat milk	Less than 0.5 grams

Source: U.S.Food and Drug Administration.

New nutrition labels were placed on milk containers nationwide after new Food and Drug Administration (FDA) regulations took effect in January 1998. Unlike other foods, milk had been exempt from declaring its fat content. The regulations require nutrition labeling on milk to conform with the Nutrition Labeling Act of 1990.

From these results the researchers estimated that replacing 5 percent of calories from saturated fat with calories from unsaturated fat would reduce the risk of coronary heart disease by 42 percent. Replacing 2 percent of calories from trans unsaturated fat with calories from unsaturated fat would reduce the risk by 53 percent.

Folic acid deficiencies. Researchers at the Oregon Regional Primate Research Center in Beaverton announced in April 1998 that people who eat breakfast cereal enriched with 400 micrograms of folic acid a day can lower blood levels of homocysteine, an amino acid that has been linked to heart disease. The researchers assessed the effects of breakfast cereals fortified with folic acid—a B-vitamin found in dried beans and peas, fruits, and green, leafy vegetables—on 75 men and women diagnosed with coronary artery disease. In January, new guidelines set by the U.S. Food and Drug Administration took effect that required cereal, bread, and pasta to be enriched so the average adult would receive 140 micrograms of folic acid daily.

A separate study conducted by researchers at the University of California at Berkeley, published in 1997, showed that a diet low in folic acid can also cause nicks in genes. Genes are made of a molecule called DNA (deoxyribonucleic acid). They are strung together to form long chains in cellular structures called chromosomes. Nicks in genes can result in the breaking of chromosomes, which could contribute to cancer and other diseases.

Folic acid is needed to synthesize thymine, a DNA *base* (subunit). If an insufficient amount of folic acid is present, thymine is not made, and uracil—a precursor of thymine—is inserted in its place along the two-stranded DNA molecule. The cell's repair mechanisms remove the molecules of uracil, resulting in nicks in the DNA, which can cause the chromosome to fall apart. The University of California researchers concluded that increasing the amount of folic acid in a person's diet may prevent this genetic damage—and the diseases that result from it.

[Phylis B. Moser-Veillon]

Scientists in 1997 and 1998 observed what appeared to be the strongest El Nino ever recorded. El Nino is a periodic warming of surface waters of the eastern tropical Pacific Ocean that occurs an average of every four years and disrupts weather patterns worldwide. Above-average water temperatures and unusual wind patterns alter global storm patterns, resulting in droughts in some regions of the world and severe flooding in other parts.

Scientists investigating the El Nino of 1997-1998 made great strides in predicting the impact of the phenomenon and understanding how interactions between the ocean and the atmosphere affect global weather patterns. With the aid of computer *models* (simulations), climate researchers were able to predict the latest El Nino several months before the surface waters of the Pacific peaked in temperature.

Medicine from under the sea. Scientists at the Scripps Institute of Oceanography in San Diego reported in September 1997 that they had isolated a potentially valuable chemical, called eleutherobin, from a rare species of coral. The researchers said eleutherobin might someday be useful for treating certain forms of cancer.

Marine chemist William Fenical, director of the Center for Marine Biotechnology and Biomedicine at Scripps, discovered the coral on an undersea boulder in 1993 while diving in the shallow tropical waters of northwestern Australia. Scientists often collect soft-bodied sea creatures such as corals and sponges, because animals without shells or spines defend themselves using chemicals. Researchers have found that some of these chemicals are useful as medicines for the treatment of some human diseases.

Using potent solvents to dissolve unwanted chemicals, the Scripps researchers extracted eleutherobin from the coral samples and then tested its effectiveness against human cancer cells. The chemical was so extraordinarily potent that it was dangerous to handle, Fenical reported.

However, Fenical said that the toxicity of eleutherobin appeared to be selective

An image of the Earth created with data obtained from a new satellite called the Sea-Viewing Wide Field-of-View Sensor (SeaWIFS) gives a detailed view of the world's oceans. Red and yellow in the image represent stirred-up sediments, blue corresponds with clear water, and green shows where microscopic plant-like organisms called phytoplankton are abundant. SeaWIFS, operated by the Orbital Sciences Corporation in Dulles, Virginia, was launched in August 1997 to monitor the status of the oceans.

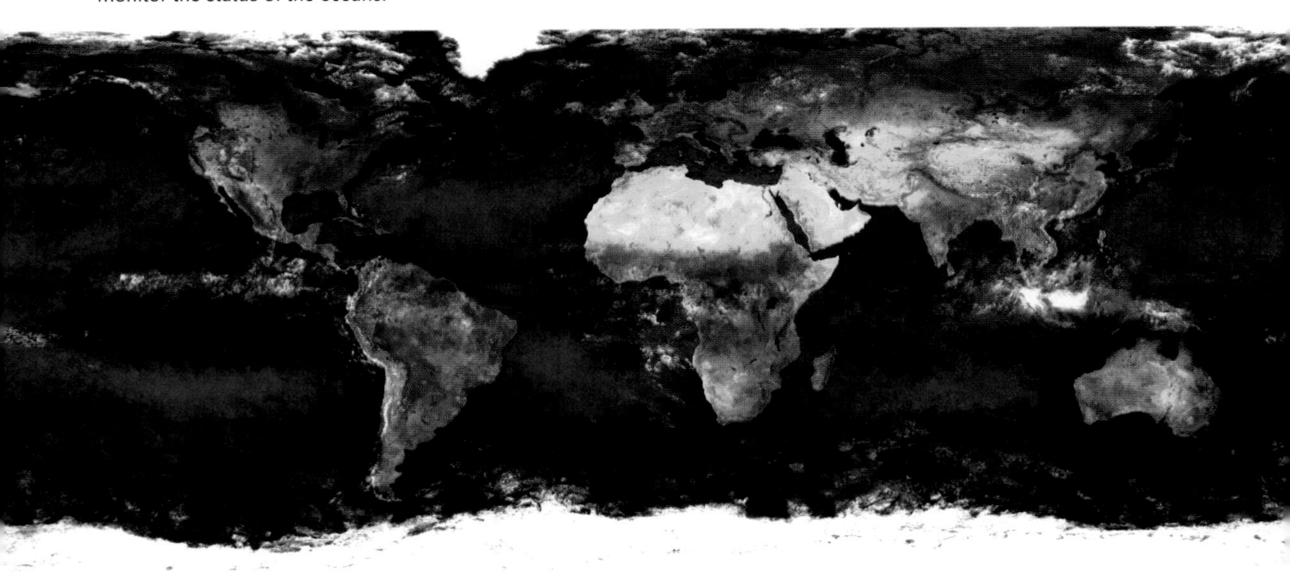

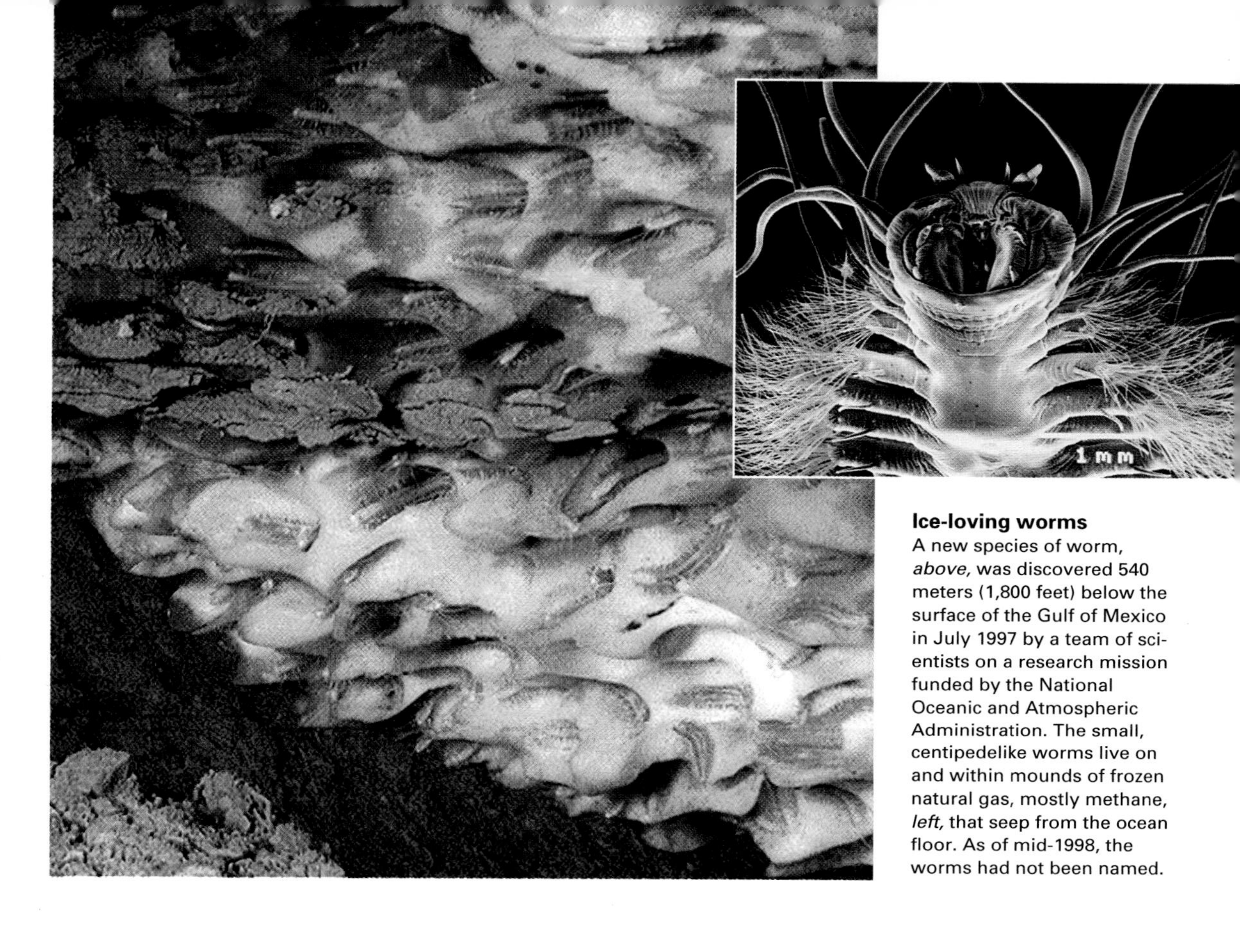

Ice-loving worms
A new species of worm, *above,* was discovered 540 meters (1,800 feet) below the surface of the Gulf of Mexico in July 1997 by a team of scientists on a research mission funded by the National Oceanic and Atmospheric Administration. The small, centipedelike worms live on and within mounds of frozen natural gas, mostly methane, *left,* that seep from the ocean floor. As of mid-1998, the worms had not been named.

toward breast and ovarian cancer cells. As a result, the scientists theorized that any potential toxic effects toward normal cells might be reduced—a necessary characteristic for an anticancer drug. The researchers said eleutherobin functions in much the same way as Taxol (paclitaxel), a cancer-fighting drug found in the bark of certain trees that also stops cancer cells from dividing. Scientists at Bristol-Myers Squibb, a New York City-based drug manufacturer, in 1998 were conducting clinical trials on eleutherobin to determine the safety and effectiveness of the drug as a treatment for various types of cancer.

Monitoring the oceans. The National Oceanographic Partnership Program (NOPP), a consortium of 12 federal agencies with links to oceanographic research, in 1998 purchased more than 200 floating monitoring devices to map the ocean's surface currents. The project was undertaken to celebrate the United Nations' designation of 1998 as the International Year of the Ocean.

The first batch of the devices, which are called drifters, was released in the Caribbean Sea in March. Additional drifters were to be released in the Caribbean Sea and the tropical Atlantic Ocean in late 1998. The National Oceanic and Atmospheric Administration's (NOAA) Global Drifter Program in Miami, Florida, which monitors the more than 500 drifters similar to those deployed by NOPP in 1998, helped coordinate operations for the NOPP drifter program.

A drifter is a buoyant instrument that floats at a specified depth. The drifters being used in the NOPP program were designed to float near the surface in order to trace the circulation of surface currents.

Drifters are made of a fiberglass ball, housing electronic monitoring devices, connected by a short cable to a submerged, socklike sail 15 meters (50 feet) long. The sail, called a drogue, acts much like a kite carried by the wind and allows the drifter to travel at the speed and in the direction of a current. Without the underwater sail, the drifter would be pushed by winds and waves and give scientists inaccurate readings.

Understanding El Nino

In the spring of 1997, weather experts at the United States National Oceanic and Atmospheric Administration (NOAA) began to detect the first stirrings of a well-known but still mysterious shift in the climatic conditions of the central Pacific Ocean. In the weeks that followed, the analysts watched as these changes began to affect weather patterns around the globe. It was clear that "El Nino," a warm ocean current that can trigger far-reaching changes in weather patterns, had returned. By the time the world's weather situation began to normalize in spring 1998, El Nino had dominated international news reports for months. Its effects had triggered droughts in Indonesia and Australia; floods in Chile, Peru, and Ecuador; a crippling ice storm in New England and eastern Canada; and many weeks of rain in California. By mid-1998, many climatologists were calling it the most severe El Nino of the century.

Although scientists lacked a complete understanding of an El Nino's causes, they had learned much from the El Nino of 1982-1983, which was so intense that even subtle effects on the weather became easy to see. The devastation caused by this El Nino convinced world leaders that more needed to be learned about the El Nino phenomenon. Thus, the 1997-1998 event also became the most closely studied ever.

Since at least the mid-1800's, Ecuadorian and Peruvian fishermen have noticed a slight annual warming of the coastal ocean waters in which they work. The warming is most noticeable around the end of December—hence the name El Nino, which in Spanish means *the child* and refers to the Christ child. Normal sea-surface temperatures in this region of South America are cool for the tropics, about 21 to 23 °C (70 to 73 °F). This warming, which usually lasts for one or two months, raises the temperature by about 1 °C (1.8 °F). But every few years, an El Nino develops that is more extreme and widespread and lasts much longer than usual. On average, these severe El Ninos occur every four years, but the interval has ranged from two to seven years. They typically begin in March or April and last 12 to 18 months, with maximum warming in the eastern and central equatorial Pacific tending to occur in December.

Under normal conditions, warm Pacific trade winds blow westward, dragging along the surface of the ocean and inducing a westward-flowing surface current. In the western Pacific, the current creates a vast pool of warm water around Indonesia and New Guinea. Temperatures in this pool can exceed 30 °C (86 °F), making these the warmest surface ocean waters in the world. Strong thunderstorms tend to form above this pool of warm water.

The first sign of El Nino is a weakening of the normally steady Pacific trade winds, which leads to a weakening of the ocean current. In an intense El Nino, the trade winds weaken dramatically and can even reverse direction for a time, blowing from west to east. The ocean current can disappear almost completely. Without the ocean current to contain it, the warm water in the western Pacific spreads eastward in a narrow band along the equator and extends to the west coast of South America. Sea-surface temperatures rise well above normal in the eastern and central Pacific. The eastward shift of warm water also shifts the focal point of the strong tropical thunderstorms to the central Pacific, altering the location and intensity of the subtropical jet streams that influence weather patterns all over the world.

An El Nino probably occurs when conditions in the atmosphere over the western Pacific become unstable. At this point, even a relatively small event in the atmosphere or in the ocean at just the right time appears to release this instability, leading to a weakening of the trade winds that signals the onset of El Nino. It is also possible that a combination of causes triggers an onset. Although scientists have developed several theories of what causes El Nino, none of their explanations has been shown to apply in all cases, and so research continued in 1998.

To learn about El Nino, scientists call upon data from sensors placed all across the Pacific and

A satellite image of Earth, taken in the summer of 1997, clearly shows a band of unusually warm water in the Pacific Ocean, spreading eastward along the equator. During a severe El Nino, this band spreads across the entire eastern Pacific, turning north and south when it reaches the coast of South America.

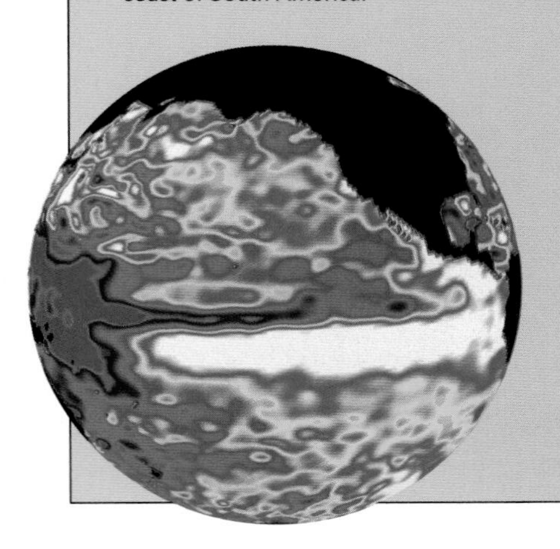

How an El Nino develops

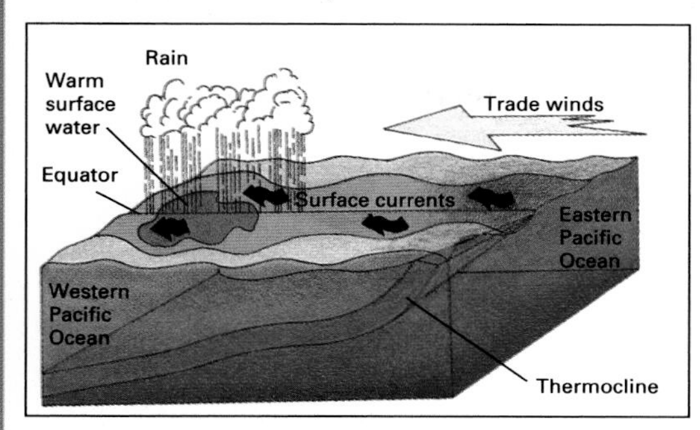

For most of each year, the Pacific trade winds blow steadily toward the west, maintaining normal weather conditions. The winds drag surface waters across the ocean, creating a pool of very warm water in the western Pacific. The warm pool pushes the ocean *thermocline* (the boundary between surface waters and the cold waters of the deep ocean) downward. In the eastern Pacific, the thermocline is much shallower, which allows cold, nutrient-rich water from the deep ocean to well up and sustain huge populations of fish and other sea life.

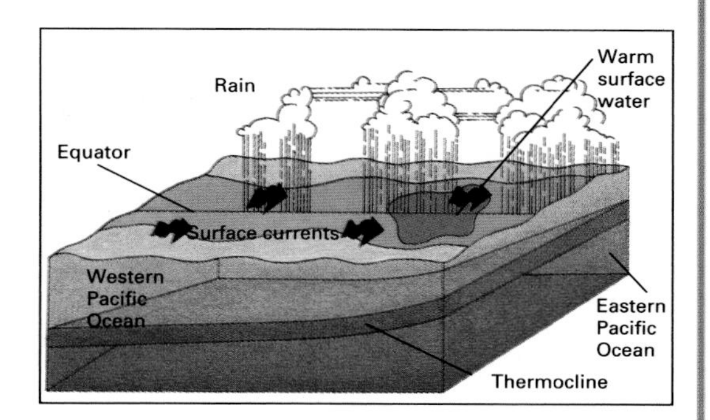

The first sign of a severe El Nino is a dramatic weakening of the Pacific trade winds. The ocean surface current slows, allowing the pool of warm water in the western Pacific to spread far to the east. The waters of the eastern Pacific become significantly warmer. The thermocline flattens, reducing the amount of deep-sea nutrients being brought to the surface. The focal point of tropical storms also shifts eastward, following the warm pool. This imbalance begins a ripple effect that leads to weather changes thousands of miles away.

from satellites in Earth orbit. During the 1990's, information from these instruments helped scientists fill in many pieces of the El Nino puzzle, paving the way for more accurate forecasts of the phenomenon. By 1997, meteorologists had learned enough and had access to enough data to enable them to predict the onset of El Nino several months in advance.

Forecasters monitoring the 1997 El Nino soon realized that an intense event was possible. On May 15, NOAA announced that conditions in the tropical Pacific were consistent with the start of El Nino. A month later, on June 17, the agency confirmed that strong El Nino conditions were developing. From November 1997 through March 1998, NOAA said, cooler weather would affect the Southern United States, with warmer than normal temperatures in the northern part of the country from the western Great Lakes to Washington state. This was the first forecast of its kind.

In 1997, the trade winds reversed their flow three times—in June, October, and November—indicating an exceptionally strong event. This El Nino was also notable for the speed of its development. As early as October 1997, sea-surface temperatures along and just south of the equator were 4 to 5 °C (7 to 9 °F) above normal from about Easter Island to the coast of South America. These were the highest October water temperature readings in this region of the world since 1950, when such records first began to be kept.

From mid-January through mid-March 1998, California was battered by more than a dozen heavy rain systems that caused widespread flooding across the state. Many homes in southern California were destroyed as water-soaked hillsides gave way in mudslides, and there was extensive damage to farm crops in central California.

El Nino's effects were felt in many other parts of the world as well. Droughts caused by El Nino's influence on global weather patterns increased the severity of fires that consumed large swaths of grasslands in Australia and of rain forests in Indonesia, Brazil, and southern Mexico. Heavy rains soaked the normally dry equatorial west coast regions of North and South America, bringing mudslides and flooding to Chile, Peru, and Ecuador.

By early April 1998, sea-surface temperatures in the Pacific were dropping, and it appeared that conditions were beginning to return to normal. The U.S. Federal Emergency Management Agency (FEMA) reported that, as of the end of March 1998, it had committed more than $289 million to 1997-1998 winter disasters, compared to $280 million the previous winter and $294 million in 1996-1997. Given the intensity of this latest El Nino, many people expected the total to be much higher. However, the early predictions were instrumental in helping government officials prevent or minimize potential losses. [John T. Snow]

Every drifter is equipped with a transmitter that communicates its position via satellite to land-based receivers. In addition to mapping the ocean's surface currents, the NOPP drifters were designed to measure water temperature. Some of the drifters were also equipped to estimate wind speed and pressure, and a few carried instruments for measuring water clarity, readings that scientists planned to compare with data from satellites. The drifters were designed to collect and transmit data into 1999.

Ellen Prager, a geologist with the United States Geological Survey in Reston, Virginia, and the NOPP project coordinator, said the drifters were being deployed in some of the least understood and most important areas of the world's oceans. Those areas included the tropical Atlantic Ocean, which is the breeding ground for hurricanes that strike the East Coast of the United States, and the Caribbean Sea, an area important for fisheries and climate studies. By monitoring the movements of the drifters, scientists hoped to improve their hurricane forecasts and develop better computer models of the Earth's climate.

In addition, drifter data from the Gulf of Mexico were expected to help predict where oil would flow in the event of a tanker accident. A few drifters were also due to be placed in the waters of the North Atlantic to help scientists track the movement of icebergs. In conjunction with the project, students and teachers were to be able to download the drifter data from the Internet.

Retreating Antarctic sea ice. Evidence indicates that the amount of sea ice in the oceans around Antarctica decreased by as much as 25 percent between the mid-1950's and the early 1970's due to increased melting, William K. de la Mare, a fisheries biologist with the Australian Antarctic Division, reported in September 1997. Oceanographers in 1998 were examining possible reasons for the melting. Although they were having trouble coming up with a comprehensive explanation for the phenomenon, the findings challenged two common beliefs of experts—that sea ice has remained largely unaltered throughout the 1900's and that climate change is a gradual process.

Scientists have estimated that sea levels are rising about 2 millimeters (0.08 inch) a year, but they did not know that sea ice had also diminished in some areas. The amount of sea ice changes naturally with the seasons. Large ice sheets cover vast areas of the South Atlantic Ocean, South Pacific Ocean, and Indian Ocean in June, July, and August—winter months in the Southern Hemisphere. During the Southern summer, much of the ice melts as air temperatures rise.

During the same months, many whale species migrate to Antarctica to feed on *krill,* a shrimplike crustacean. Whalers used to track the whales as they traveled south in search of krill and other food sources and recorded their ships' positions. Because the whalers typically followed the edge of the ice sheets while searching for whales, de la Mare theorized that their logs would reveal the extent of the summertime sea ice over a period of years. He studied data collected by British and Norwegian whalers between 1931 and 1987. The latter year is when the world's nations, obeying a 1982 moratorium adopted by the International Whaling Commission 1982, ceased all commercial whaling.

To be certain that the whalers' logs were accurate, de la Mare compared the whalers' data with satellite images of sea-ice cover available after 1973. He discovered that the two sources of information were largely in agreement. After reviewing all of the logs, he reported that although from 1931 to 1954 the edge of the sea ice had remained in the same position, by 1973 it had receded 2.8 degrees—about 310 kilometers (190 miles)—to the south. The change represented a 5.65-million-square-kilometer (2.18-million-square-mile) reduction in the amount of sea ice covering the area. By 1987, de la Mare reported, the southern boundary of the sea ice had again stabilized.

Oceanographers agreed that the cause of the melting sea ice will require further examination. Although some experts theorized that the decrease may have been linked to atmospheric warming, it was unclear why the melting did not continue through the 1980's. Because the ice has stabilized, some oceanographers believed that the melting was simply part of a natural cycle.

[Christina S. Johnson]

See also OCEANOGRAPHY (Close-Up).

On Dec. 8, 1997, the United States Department of Energy (DOE) announced final approval of U.S. participation in an international effort to build the Large Hadron Collider (LHC), the world's most powerful particle accelerator. The accelerator, a huge machine designed to explore the structure of matter at its smallest levels, will be constructed at the European Laboratory for Particle Physics, known as CERN, near Geneva, Switzerland. The DOE agreed to provide $450 million toward construction of the LHC, and the National Science Foundation (NSF) pledged $81 million. A total of 24 nations planned to share the approximately $6 billion cost, which will make the LHC one of the world's most international research facilities.

The announcement ended a long period of uncertainty about United States involvement in particle physics research. That question had been up in the air since 1993, when Congress halted construction of an even larger accelerator in central Texas. About 25 percent of U.S. high-energy researchers are expected to use the LHC facility. The other 75 percent planned to continue working at U.S. facilities, which were being upgraded but would not match the capabilities of the LHC.

Particle accelerators boost electrically charged atoms and subatomic particles to extremely high speeds and cause them to collide with one another or with stationary targets. The collisions allow physicists to study the behavior and properties of elementary particles and to create new kinds of particles from pure energy. The LHC is to be built in an existing circular tunnel 27 kilometers (17 miles) in circumference. In 1998, the tunnel still housed another, less powerful accelerator, the Large Electron-Positron ring. The new accelerator was expected to be completed by 2005.

The LHC will produce two beams of protons—each beam with an energy of 7 trillion *electronvolts* (TeV)—circulating in opposite directions. (Electronvolts are units of measurement for the amount of energy a particle gains when it is accelerated through an electric field. One electronvolt is approximately the energy acquired by a single electron

A microscopic guitar about the size of a human blood cell demonstrates the possibilities of nanotechnology, an emerging field of study that aims at constructing objects from individual atoms and molecules. Physicists Dustin Carr and Harold Craighead of Cornell University in Ithaca, New York, carved the nanoguitar from crystalline silicon with a beam of electrons. Each guitar string is about 100 atoms wide, and the guitar would be playable by a "nanomusician."

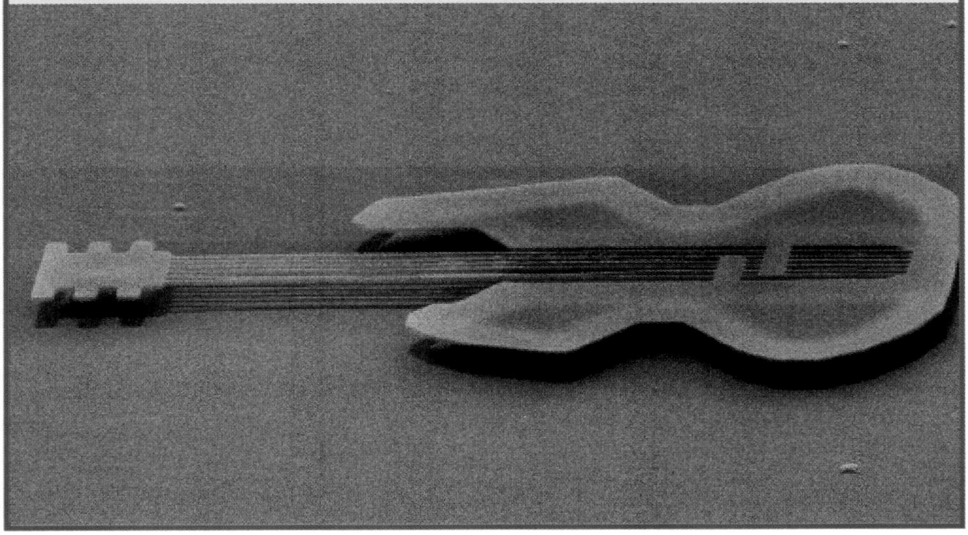

as it moves through a flashlight battery.) At four different points in the LHC, the beams will cross, producing violent near-head-on collisions. The LHC will also be capable of accelerating heavier nuclei than other accelerators have been able to do.

The main scientific goal for the LHC will be the discovery of the *Higgs boson,* a particle that is thought to create mass for other particles that carry it. Although the LHC should have enough energy to produce the Higgs boson, physicists were uncertain whether the machine would be able to generate the bosons in sufficient-enough numbers to have them stand out from other particle reactions. Another goal was to search for new and exotic states of matter.

Quantum teleportation. In the television series "Star Trek," characters were routinely broken down into subatomic particles and then reassembled elsewhere just moments later. Although such a phenomenon exists only in the realm of science fiction, European scientists in December 1997 reported on experiments that brought to mind the familiar phrase "Beam me up, Scotty."

In two independent studies—one led by physicist Anton Zeilinger at the University of Innsbruck in Austria and the other by physicist Francesco De Martini at the University of Rome—the researchers demonstrated the ability to communicate a "quantum state." The experiments, however, were not even a tiny step toward the mode of transportation featured on "Star Trek." Instead, they demonstrated what the German-American physicist Albert Einstein called the "spookiness" of *quantum theory,* the branch of physics dealing with the behavior of atoms and subatomic particles. Modern physicists call this spookiness "quantum entanglement."

The quantum world does not operate the same way as the everyday world. An object that we can observe has certain properties, such as its position, to which we can assign a definite value. Quantum theory asserts, however, that a subatomic particle does not have such properties until it interacts with something that allows a property to be measured. For example, a *photon* (particle of light) does not have a definite value of *angular momentum,* a property that is most simply defined as a state that is either parallel or opposite to the motion of the particle. (These two possible states are often referred to as spin-up and spin-down, respectively.) But a photon is not in either one of those states until it is measured. Also, the process of measuring a photon's angular momentum changes other information about the particle in an unpredictable way.

Quantum entanglement, an even more unusual feature of the quantum world, is a sort of "invisible link" between two particles that share a common source. Because their properties are entangled, any action on one particle creates a certain state in the other. A measurement performed on one particle would result in the other particle being in a complementary, and predictable, state. The experiments reported in 1997 were using this entanglement as a means of "transportation."

The researchers began by "splitting" a single photon in a special type of crystal, thereby creating a pair of entangled photons. One of these photons, which could be referred to as X, was given to a part of the experimental apparatus referred to as the "sender." The other photon, Y, was given to the "receiver." A third photon, not a part of the entangled pair, was the "passenger" in the teleportation setup.

True to the unusual nature of quantum physics, the passenger photon was not actually transported from X to Y. Instead, the teleportation system allowed the *polarization* (orientation of a light wave determined by the angular momentum of its photons) of the passenger photon to be manifested in Y.

In the teleportation system, X and the passenger photon were passed through a device called a beam-splitter, which mixed the two together. This process entangled them even though they were not created from a single source. Since X and Y were already an entangled pair, the passenger photon and Y were simultaneously entangled even though they had never interacted—a sort of "long-distance" entanglement.

After the passenger photon and X passed through the beam splitter, the sender jointly measured the polarization of the two photons. (It is important to remember that whatever occurred with X and the passenger photon because of the measurement also occurred with Y.)

Harnessing the power of sound

The vibration of a specially shaped cavity creates sound waves that are strong enough to create an "acoustic compressor." The technology, called resonant macrosonic synthesis, was developed by researchers at the MacroSonix Corporation in Richmond, Virginia.

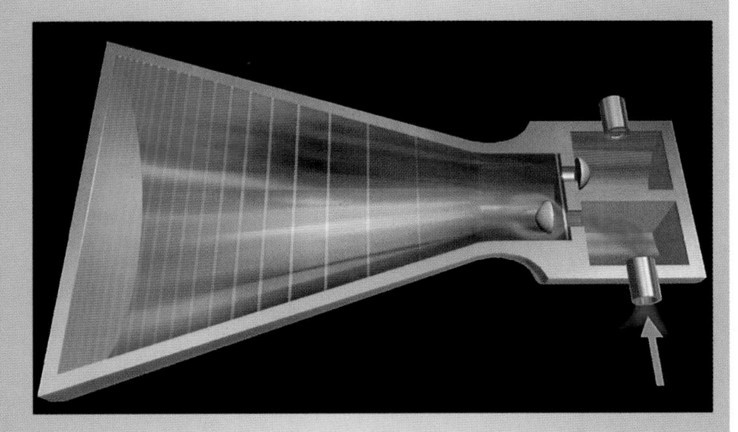

The intense sound waves are trapped inside the bell-shaped cavity of the compressor, *above.* The areas of highest and lowest pressure of the sound waves—represented by red and blue, respectively—shift rapidly back and forth. When the pressure is low at the small end, a valve opens, allowing a gas to flow into the cavity.

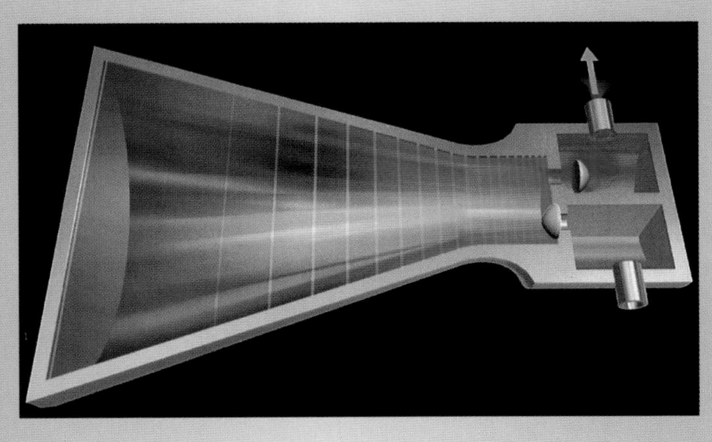

When the high pressure shifts to the small end, *above,* another valve opens, forcing the gas out under extremely high pressure. MacroSonix engineers said acoustic compressors could have many uses, including powering refrigerators and air conditioners.

The sender then had to complete the teleportation through more conventional means of communication—electronic devices. The sender relayed results of the joint measurement to the receiver. Those results were a message for the receiver to perform a certain operation on Y, such as changing its angular momentum. This last step assured that the polarization of Y matched the original polarization of the passenger.

The incredibly elaborate experiment all took place in a single laboratory, but in theory it could be re-created over great distances. In 1998, however, the process was still experimental, and the practical application of the process in fields such as quantum computing was a long way off. Creating entangled states, and maintaining that entanglement over a long period of time, can only be done with very simple objects such as photons or perhaps electrons. Even a single atom would not remain entangled for long. So, creating a pair of entangled humans—as would be needed for the transporter on "Star Trek"—remains an unimaginable process.

Putting sound to work. Not all discoveries in physics deal with exotic theories. Every so often, research in an area that was regarded as well understood turns up surprises with great practical potential. Such a surprise occurred in December 1997 when researchers at the MacroSonix Corporation, a small research laboratory in Richmond, Virginia, revealed a few tricks that make it possible to put sound to work. The head of MacroSonix, physicist Tim Lucas, called the technique resonant macrosonic synthesis, or RMS.

The average air pressure at sea level is measured as one atmosphere—about 1.03 kilograms per square centimeter (14.7 pounds per square inch). Our ears are sensitive to tiny variations in air pressure, which we perceive as sound. For example, the ear-splitting roar of a nearby jet engine represents an increase from normal atmospheric pressure of only about 0.0002 percent. Consequently, sound does not seem like a likely candidate for work since machines such as pumps and compressors must generate pressures of several atmospheres.

In free air, it is impossible to generate sound waves of such high pressure because of a process called acoustic saturation. This process can be considered a sort of breaking point at which layers of strong sound waves combine to form shock waves. The sonic boom generated by a supersonic aircraft is one example of a shock wave. The MacroSonix team hoped that intense sound waves could be confined to a suitably shaped box called a "resonant cavity" that would prevent shock-wave formation.

One thing the MacroSonix team knew was that their resonant cavity could not be a simple shape, such as a cylinder. All musical wind instruments are cylindrical resonant cavities. Such a cavity generates a fundamental note, the basic note a musician is playing, and many *overtones.* The overtones are higher-frequency sound waves that stay in step with the wavelength of the fundamental note. This property of sound is called "consonance." Although consonance contributes to the mellow sound of many wind instruments, it is this same layering of sound that, at a high enough intensity, leads to the formation of shock waves. The trick, therefore, was to devise a cavity in which the fundamental note and its overtones would all remain out of step—a property called "dissonance." In other words, the cavity had to create very loud sounds with no shock waves.

The researchers developed computer programs that enabled them to simulate the generation of sound waves in variously shaped cavities. As they had expected, they found that small changes in the shape of a cavity made a big difference in the relationship between the fundamental tone and the overtones. The cavity that proved to maintain dissonance at high frequencies of sound was another familiar musical shape—the flared bell of a trumpet, transformed into a cavity by closing it at the end.

The researchers then built a working model of the RMS cavity. They generated the sound waves by vibrating the entire cavity at high frequencies. By experimenting with a variety of gases inside the cavity, including the regular air we breathe, the MacroSonix group generated sound waves that created pressures as high as 10 atmospheres. This is within the practical operating range of many useful devices, such as pumps and compressors. Of course, those high pressures meant that the sound trapped inside the cavity was inconceivably loud. On the outside of the resonant cavity, however, all that was heard was a gentle hum because the "outside" sound waves were operating in a different acoustic environment—the free air in the room.

The researchers expected the RMS cavity to have several practical applications and to provide better alternatives than some conventional machines. One possible application for the technology would be for compressors in home refrigerators and air conditioners. The compressors in these devices circulate various gases. The compressing and decompressing of a gas converts the material back and forth from liquid to gas—a process that creates the cooling effect in refrigeration. The RMS technology was expected to create a compressor that works more efficiently than those currently used in cooling systems.

Another advantage of an acoustically driven compressor over conventional ones is that RMS devices would have fewer moving parts that require lubrication. MacroSonix licensed an RMS compressor design in 1997 to an unnamed major appliance manufacturer, and an RMS refrigerator was expected to be on the market in a few years.

Advances in fusion power. The quest to harness nuclear fusion, the power source of the sun and other stars, took a significant step in September 1997. An experimental fusion device called the Joint European Torus (JET) in Abingdon, England, set a new world record of 12.9 megawatts of fusion power. This record eclipsed the previous one of 10.7 megawatts set in 1994 by the Tokamak Fusion Test Reactor (TFTR) in Princeton, New Jersey.

A fusion reaction occurs when two lightweight nuclei combine to form the nucleus of a heavier element. The resulting nucleus has less mass than that of the two original nuclei, and the "lost" mass is converted to energy. The most common fuel mix for fusion experiments contains two different *isotopes* (forms) of hydrogen. One isotope, called hydrogen-2 or deuterium, has a nucleus made up of one *proton* (positively charged subatomic particle) and one *neutron* (subatomic particle with no electric charge). About 1 of every 6,700 naturally occurring hydrogen atoms is deuterium. Hydrogen-3 or tritium, which has one proton and two neutrons, is produced in nuclear fission reactors.

In an artist's depiction of a black hole, powerful X rays of various wavelengths (twisted white lines) are emitted from above and below the hole. In November 1997, physicists at the Massachusetts Institute of Technology in Cambridge reported the detection of such X rays from a source in the Milky Way Galaxy believed to be a black hole. The scientists argued that the findings were evidence of a phenomenon called *frame dragging*, the twisting of space and time (blue-green grid) by the gravitational pull of a massive rotating object. The pattern of emission is consistent with predictions of a wobbling disk of matter in the twisted space around the center of a black hole.

The potential payoff for mastering fusion power is significant because the enormous supply of hydrogen atoms in the Earth's oceans would provide a nearly inexhaustible source of deuterium. But several obstacles remained in 1998 that scientists had yet to overcome.

Normally, hydrogen nuclei repel one another so strongly that they cannot come close enough to react. But at high temperatures, such as those found in the center of a star, the nuclei are moving fast enough to occasionally overcome this repulsion and stick together.

Any gas at a high temperature, however, generates pressures that cause the nuclei to fly apart—thus ending the fusion reaction—unless there is a force strong enough to overcome the pressure. In a star, that force is gravity. On Earth, some other force must hold the fusion reaction together. Devices like the JET and the TFTR use magnetism to confine the nuclei. Each machine consists of a large doughnut-shaped vacuum tank wound with electric coils to generate a powerful magnetic field.

In a fusion reactor, a combination of powerful radio waves and particle beams heats the hydrogen fuel, stripping away the electrons from the nuclei. The resulting combination of protons and free electrons is called a plasma, which is kept confined by the magnetic field. Consequently, the hydrogen nuclei can collide and produce the fusion energy.

Although experiments have created fusion energy for many years, the process continued to consume more energy than it produced. In 1994, the TFTR in Princeton produced only 28 percent of the amount of energy it consumed. The 1997 JET experiment narrowed that gap by producing 50 percent of what it consumed. In December 1997, JET moved even closer to "scientific break-even" by producing 16 megawatts of fusion power—65 percent of what it consumed.

Nonetheless, this scientific milestone did not yet make fusion power a practical scheme. Many engineering problems remained to be solved. The JET facility was nearing the end of its useful life in 1998 and was expected to continue only as a means to study designs for future fusion devices. [Robert H. March]

During the late 1700's, the Austrian physician Franz Anton Mesmer used iron and glass rods to "mesmerize" his patients—an attempt to harness the body's magnetic properties to relieve symptoms of illnesses, including mental conditions. In 1997, magnetism was once again evaluated as a possible treatment. The use of a procedure called repetitive transcranial magnetic stimulation (rTMS) for alleviating symptoms of major depression was reported in December by neuroscientist Mark S. George of the Medical University of South Carolina in Charleston and researchers at the National Institute of Mental Health in Bethesda, Maryland.

Major depression is a mental disorder in which individuals experience both psychological symptoms—such as sadness, hopelessness, and a diminished interest in and enjoyment of normal activities—and neurological symptoms, including changes in sleep patterns, appetite, and energy level. Common treatments for depression include medications and *psychotherapy* (talking therapy).

The new rTMS treatment was based on the theory that repetitive pulses from a small but powerful electromagnet placed on the scalp would cause *neurons* (nerve cells) in the brain to discharge. That is, neurons in the area below the magnet would temporarily malfunction and lose their ability to send and receive chemical messages.

The researchers used rTMS for 10 days on 12 individuals diagnosed with major depression. The electromagnet was focused above the left eye to affect a cluster of neurons called the dorsolateral prefrontal cortex, the brain circuitry thought to be involved in the regulation of mood, sleep, and appetite. The same 12 people also received an additional 10 days of a placebo, or "sham" magnetic stimulation. In a placebo-controlled trial, the subjects of the experiment do not know when they are receiving the actual treatment, so the scientists can be sure that any improvements are actually caused by the procedure.

The researchers found that the rTMS treatment produced an overall decrease of about 30 percent in the intensity of depressive symptoms in the group. Some individuals' responses to the treatment were even more pronounced. The side effects of rTMS were minimal, with four subjects reporting mild headaches. There were no seizures, and patients reported no loss of memory or ability to concentrate. George's team said further research would be required to fully evaluate rTMS, but the results of the study suggested that it is a safe method for the treatment of major depression.

An ancient remedy. An extract from the gingko tree was shown in a 1997 study to be a safe and effective treatment for patients suffering from dementia. The study was reported in October by researchers led by neuroscientist Pierre Le Bars at the New York Institute for Medical Research in Tarrytown. The extract, called EGb 761, stabilized or improved the mental skills and social functioning of patients for up to one year.

Dementia is a general term to describe a progressive deterioration of mental abilities. It disrupts short-term memory, language, and the performance of previously learned skills. Alzheimer's disease, which affects about 4 million elderly people in the United States, is the most common form of dementia. Although the cause of Alzheimer's disease is unknown, scientists do know that collections of abnormal proteins are linked to the death of neurons in the brain. Another form of dementia, vascular dementia, is the result of a chronic lack of blood to the brain.

Extracts of the gingko tree have been used as medicines in China for almost 4,000 years. Since the 1970's, many European doctors have prescribed gingko extracts to treat problems of impaired blood flow and dementia.

The active ingredient of EGb 761 is a molecule called flavonoid glycoside, which is believed to protect the neurons from *free radicals,* destructive molecules that are produced by the body. EGb 761 may also enhance the chemicals that the brain uses to encode memory.

Le Bars's group studied more than 300 patients diagnosed with either Alzheimer's disease or vascular dementia for one year. The patients received either EGb 761 or a placebo. Using rating scales to measure each patient's mental performance, as well as the caregiver's impressions of the patient's functioning, the investigators found a significant improvement in the group taking EGb 761. The study also indicated that the extract was a safe treatment alternative.

Brain activity of bilingual speakers
A technique called functional magnetic resonance imaging (fMRI) revealed that language areas in the brains of fluent bilingual speakers function differently depending on whether an individual learned the second language during early childhood or as an adult. Neuroscientists Karl Kim, Joy Hirsch, and their colleagues at the Memorial Sloan-Kettering Cancer Center in New York City reported these findings in July 1997.

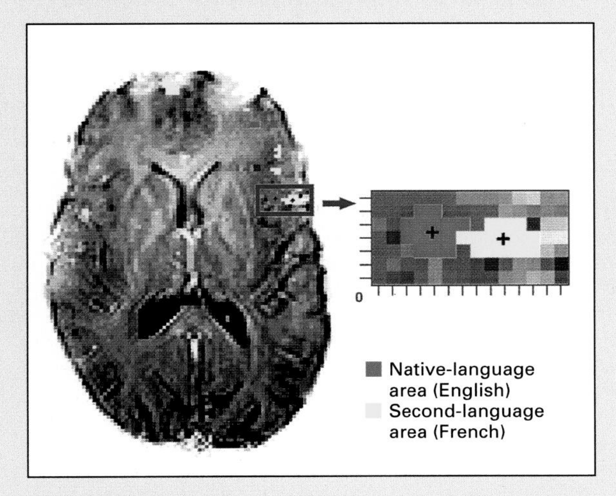

The brain scan, *above,* of a native English speaker who learned French as an adult reveals that two distinct parts of Broca's area, a language center in the brain, were used when the person was silently reciting in one language or the other. The plus mark indicates the center of brain-cell activity.

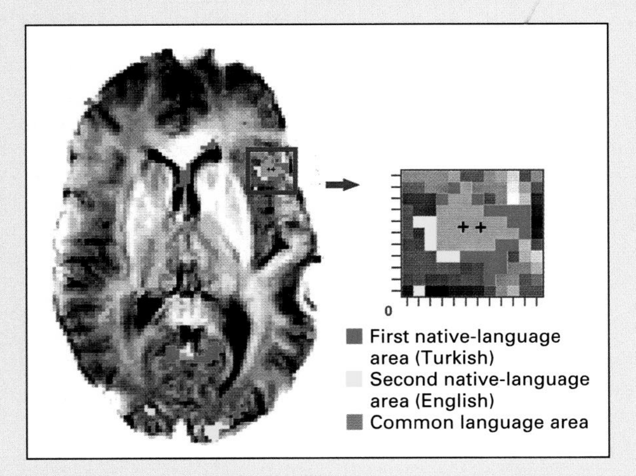

The brain scan of an individual who learned Turkish and English during early childhood, *above,* reveals that a common region of Broca's area is used for both languages. The scientists reported that the images offered clues to the significance of Broca's region in early language acquisition and the difference between how adults and children learn languages.

Antidepressant drug for PMS. A medication for treating depression can also help women who experience the most severe form of premenstrual syndrome (PMS). Researchers, led by psychiatrist Kimberly A. Yonkers at the University of Texas Southwestern Medical Center in Dallas, reported in September 1997 that the drug sertraline alleviated symptoms of PMS and prevented disruptions in women's work and personal life.

PMS is a common disorder associated with the menstrual cycle, which prepares a woman's body to carry a child. A part of this cycle is the menses, or menstrual period, during which a woman's uterus sheds blood and tissue if she does not become pregnant. PMS symptoms usually appear about 3 to 10 days before the period begins. The physical symptoms of PMS include breast pain and a feeling of abdominal swelling. The psychological symptoms include mild depression, anxiety, and irritability.

In some cases, symptoms are severe enough to interfere with a woman's daily functioning and quality of life, a condition known as premenstrual dysphoric disorder (PMDD). Treatments are available for PMDD, but many offer only partial relief, cause *sedation* (extreme drowsiness), or can be addictive.

In an effort to find a better and safer treatment for PMDD, Yonkers's research group conducted tests with sertraline, one of a class of antidepressant medications called serotonin uptake inhibitors. Serotonin is a chemical messenger in the brain and is involved in the regulation of mood, anxiety, appetite, and sleep. Proteins in the brain, called serotonin transporters, control serotonin levels by removing excess amounts of the chemical between neurons. Serotonin uptake inhibitors block the transporters and allow for higher levels of serotonin. This action is effective in diminishing the symptoms of depression.

The study followed 243 women between the ages of 24 and 45 who had suffered PMDD symptoms for at least two years. The women took either sertraline or a placebo daily during three menstrual cycles, beginning with the first day of menses. Many of the women receiving the drug noted an improvement during the first cycle, and they experienced increasingly greater relief during their second and third cycles.

About two-thirds of the women were considered greatly or considerably improved by the end of the third cycle. Only 8 percent of subjects withdrew from the study because of side effects.

Children of depressed parents. Individuals whose parents suffered from major depression are at a greater risk for developing severe and impairing forms of depression than are people whose parents were not depressed. These conclusions were reported in October 1997 by a team of researchers led by epidemiologist Myrna M. Weissman at Columbia University in New York City.

Major depression is a common ailment that often initially affects people while they are in their teens or early 20's. Specialists do not agree on whether depression stems more from hereditary factors or from a person's life experiences. But regardless of the cause, children of depressed parents have a high risk of developing major depression. Previous studies of such high-risk individuals had followed subjects for relatively short periods, 1½ to 3 years. Weissman's group was the first to follow such children into adulthood. This sort of study was considered particularly important because major depression can often coexist with other mental illnesses such as anxiety disorders, as well as with alcohol and drug dependence.

The researchers began their study in 1982 with an evaluation of 220 offspring between the ages of 6 and 23. This group included 153 individuals from 65 families with 1 or more depressed parents and 67 individuals from 26 families with no depressed parents. The families were reassessed 10 years later.

Weissman's group found that the incidence of serious depression was much greater among the children of depressed parents. As these people matured, they had a three-times-higher rate of major depression and five-times-higher rate of alcohol dependence. They also functioned less well in work, family, and marriage and were less likely to seek treatment. The researchers said the findings demonstrated the need for the early detection and treatment of people at risk for developing major depression. [Robert A. Lasser]

Public Health

From mid-1997 through early 1998, a new strain of influenza challenged previous ideas about the flu. The new strain, designated A(H5N1) but commonly called the bird flu, was found in chickens in Hong Kong between March and May 1997. The first human victim of the virus, a 3-year-old boy, died in May. In all, 18 people in Hong Kong became ill from the bird flu, and 8 of them died.

Investigators from Hong Kong and the U.S. Centers for Disease Control and Prevention in Atlanta, Georgia, determined that the new influenza strain was transmitted directly from live chickens to humans. That was a new phenomenon. All previously known animal viruses transmitted to people had first incubated and changed form in another animal host. In December 1997, to prevent a *pandemic* (spread of a disease throughout the world), Hong Kong officials ordered the slaughter of about 1.5 million chickens.

No further cases of bird flu were reported after Dec. 28, 1997. Nonetheless, laboratory scientists in 1998 were working to develop a vaccine, in case it was found that A(H5N1) could be spread by human-to-human transmission as well.

Beef recall. In August 1997, the U.S. Department of Agriculture forced meat processor Hudson Foods, Incorporated, of Rogers, Arkansas, to recall an estimated 11 million kilograms (25 million pounds) of ground beef. This was the largest U.S. beef recall ever conducted.

The meat, which had been processed at Hudson's Columbus, Nebraska, plant, was feared to contain a deadly strain of bacteria, *Escherichia coli* 0157:H7. At least 15 people had become ill in June and July 1997 after eating hamburger meat traced to the Nebraska plant. Although food-safety officials speculated that the bacteria may have been introduced into the beef at the slaughterhouses that supplied the plant, Hudson closed the factory in late August.

The hidden hunger. The United Nations Children's Fund (UNICEF) reported in December 1997 that between 6 million and 7 million children under age 5 die of conditions related to malnutrition each year. According to *The State of the World's Children,* UNICEF's an-

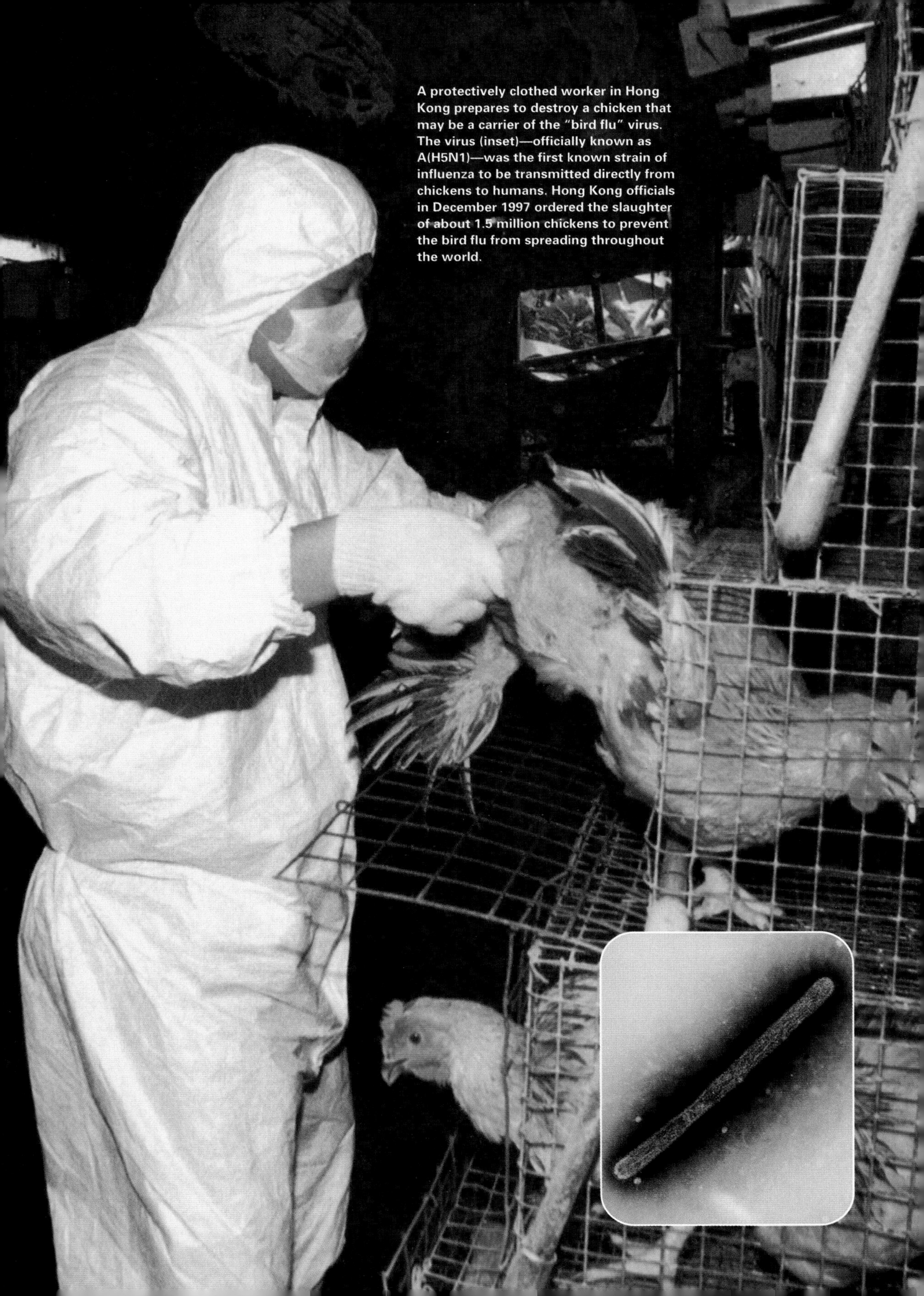

A protectively clothed worker in Hong Kong prepares to destroy a chicken that may be a carrier of the "bird flu" virus. The virus (inset)—officially known as A(H5N1)—was the first known strain of influenza to be transmitted directly from chickens to humans. Hong Kong officials in December 1997 ordered the slaughter of about 1.5 million chickens to prevent the bird flu from spreading throughout the world.

nual publication, more than half of the nearly 12 million deaths of children under age 5 result from poor nutrition.

Malnutrition is often called the "hidden hunger" because three-fourths of those who die from the condition are only mildly or moderately malnourished and show no outward signs of ill health. Malnutrition develops when a person's diet lacks essential nutrients such as vitamin A, iodine, iron, and zinc.

UNICEF reported that although the number of malnourished children decreased in Latin America and East Asia during the mid-1990's, it increased worldwide. The agency said the problem was worst in areas racked by war or poverty, such as central Africa, North Korea, and Iraq. In the United States, the report stated, more than 13 million children under the age of 12 were inadequately nourished and at risk of adverse health consequences.

Birth weight and race. A study published in October 1997 cast doubt on the theory that birth weight is determined by genetic predisposition, particularly by race. The study showed that while the babies of American-born black women are three times more likely than white infants to be premature and of low birth weight, they also suffer those problems at a higher rate than the babies of black women from sub-Saharan Africa. Low birth weight and prematurity are leading causes of poor health and death in infants.

Pediatricians James W. Collins, Jr., of Northwestern University Medical School in Chicago and Richard J. David of the University of Illinois at Chicago School of Medicine compiled the report. The physicians surveyed the birth weights of all children born in Illinois from 1980 to 1995. They then compared the risk of delivering a low-birth-weight infant among women of European, African American, and purely African descent. The researchers assumed that, on average, women of African American descent have a greater percentage of European genes than women from Africa. Thus, if birth weight were determined by race, the babies of women from Africa should be at greatest risk of being born prematurely with low birth weight.

Instead, the study found that the percent of infants born underweight was 4.3 among European American women, 13.2 among U.S.-born black women, and 7.1 among African-born black women. The researchers speculated that in the United States, a woman's childhood exposure to poverty or the effects of racial discrimination may have an impact on later health.

Obesity and health. In January 1998, researchers reported the results of the largest study ever conducted on the relationship of obesity to premature death. The study found a low increased risk of death from obesity in older people but a significant risk in those aged 30 to 54.

Nutritionist June Stevens of the University of North Carolina at Chapel Hill and her colleagues reviewed follow-up data on almost 325,000 nonsmoking men and women over the age of 30 who participated in an American Cancer Society study in 1959 and 1960. Stevens and her group compared each participant's *body mass index* (BMI—the weight in kilograms divided by the square of the height in meters) to the individual's health status or cause of death over 12 years. Many physicians believe that BMI is the most accurate measure of obesity. The ideal BMI is about 21.

The researchers found that people aged 30 to 54 with a BMI of 32 had double the risk of premature death that leaner individuals in the same age group had. There was no excess risk of death associated with a BMI below 27. By age 74, persons with a BMI of 32 had only a slight risk of premature death from obesity.

Fewer unplanned pregnancies. The Alan Guttmacher Institute, a reproduction-research organization in New York City, reported in January 1998 that the number of unintended pregnancies in the United States had fallen. The group based its findings on data from the 1995 National Survey of Family Growth.

According to the survey, 49 percent of the 5.4 million pregnancies in 1994 (the most recent year analyzed) were either mistimed or unwanted. In 1987, 57 percent of pregnancies were unplanned. The institute attributed the decline to the increased, and more effective, use of condoms and other contraceptive devices and methods. Despite the decline, the number of unplanned pregnancies in the United States remained higher than that in many other developed countries. [Deborah Kowal]

Bioethics issues remained in the headlines in 1997 and 1998. The controversy that began in early 1997, when Scottish scientists announced that they had developed a new cloning technique—a way of creating exact genetic duplicates of animals—intensified later in the year. In December, a Chicago physicist, Richard Seed, announced that he was planning to clone human beings. Although the extensive media coverage that Seed's announcement received seemed to give it credibility, Seed lacked credentials in genetics and did not have adequate financing for his idea.

Nevertheless, after Seed's announcement, the U.S. Food and Drug Administration (FDA) asserted that it had the authority to regulate human cloning and warned that anyone trying the procedure without FDA approval would be violating federal law. Throughout the first half of 1998, leaders in the U.S. Congress considered legislation to outlaw human cloning.

Animal-organ transplants raise fears. In October 1997, the FDA temporarily halted an effort by an Ohio-based company to develop genetically engineered pigs. The company planned to offer the pigs' hearts, livers, and kidneys for use in *xenotransplants* (the transplantation of organs from one species to another). The FDA was concerned that a pig virus could infect organ recipients, possibly setting off an epidemic.

Researchers had been working to improve xenotransplantation since the early 1960's, when the first cross-species transplant was attempted. In 1984, a California surgeon tried unsuccessfully to use a baboon heart to save the life of a dying baby. By the 1990's, developments in genetics and immunology had improved the odds of success by allowing researchers to modify the genes of other organisms, such as pigs, to keep their tissues from being rejected by the human immune system. Experimenters had implanted cells from a pig's pancreas into diabetic patients, and brain cells transplanted from fetal pigs into people had shown some success in treating patients with Parkinson disease.

As xenotransplants became more feasible, researchers began to explore potential dangers and ethical quandaries. Research published in early 1997 suggested that a hereditary pig virus—a virus that has inserted its genes into the genetic material of pigs—was capable of infecting human cells in a test tube. In January 1998, researchers and policymakers met at the National Institutes of Health (NIH) in Bethesda, Maryland, to discuss the pros and cons of xenotransplantation. At the meeting, government officials rejected calls for a total halt to xenotransplantation research and announced plans to allow limited clinical trials of xenotransplants under carefully controlled conditions. The U.S. Department of Health and Human Services announced a proposal to establish an advisory committee to monitor cross-species transplants.

The government hoped that xenotransplant research would help the thousands of people who die each year because of a shortage of human organs for transplantation. Whether the benefits of xenotransplantation outweigh the risks is an issue that is likely to be a source of controversy for years to come.

Manipulating human genes. Researchers and bioethics specialists debated the societal issues raised by *human germline intervention* in meetings held in late 1997 and early 1998. Human germline intervention involves altering human genes and then inserting them into sperm or egg cells, so that the traits those genes control are passed on to future generations.

In the genetic therapies in use in the mid-1990's, scientists inserted genes into specific tissues so that the genes improved the functioning of those tissues. For example, scientists were working to correct the genetic defect that makes it difficult for people with cystic fibrosis to breathe. However, creating genetic changes that could be passed on to future generations was beyond geneticists' capability. By 1998, however, a growing number of scientists believed that the prospect was in sight.

The American Association for the Advancement of Science, a federation of scientific organizations located in Washington, D.C., held a public forum on human germline intervention in September 1997. Ethicists and religious leaders raised questions about "playing God," the connection of the soul to the organism, and the potential misuses of genetic engineering techniques for *eugenics* (the manipulation of human reproduction to produce a supposedly superior population).

In March 1998, a meeting on germline intervention at the University of California at Los Angeles brought together leading scientists in the field. Many of the scientists spoke glowingly about the possibility of inserting genes into sperm and egg cells that could provide immunity from colds, enhance intelligence, or extend life. Most of the scientists seemed to think that the new techniques offered great prospects without raising serious ethical issues.

Kennewick Man. A U.S. district court in July 1997 ordered that a case involving the remains of "Kennewick Man"—the oldest and most complete remains of a prehistoric human ever found in the Pacific Northwest—be reopened. The U.S. Army Corps of Engineers, which had custody of the 9,300-year-old bones discovered in July 1996, had planned to turn them over to the Confederated Tribes of the Umatilla Indian Reservation for reburial. However, scientists, who were particularly interested in the skeleton because it was reported to have Caucasian features rather than those more characteristic of Native Americans, contested the move. The scientists wanted to analyze Kennewick Man's DNA (deoxyribonucleic acid—the molecule that genes are made of) in order to determine his genetic make-up.

Controls on cryptography. In December 1997, arguments began in a federal appeals court on the question of U.S. government control of *cryptography* (the encoding or scrambling of messages to keep them private) in electronic mail (e-mail) and other forms of electronic communication. The case against the government was brought by a California graduate student who had sought a license to publish a cryptography program he developed.

The student had been denied the license because the U.S. government considers cryptographic software to be related to national security. Federal laws forbid the export of all but the most easily broken codes, and U.S. law enforcement agencies say they need a key to any encryption software so that they can intercept communications between sus-

An auctioneer at Sotheby's in New York City accepts a bid for the largest and most complete skeleton of a *Tyrannosaurus rex* dinosaur ever found. The fossil, which sold for a record $8.4 million, was purchased by the Field Museum of Natural History in Chicago. The auction raised concerns among scientists, who feared that a valuable fossil would be lost to science if an educational institution was unable to raise the funds that would enable it to make the top bid at the auction.

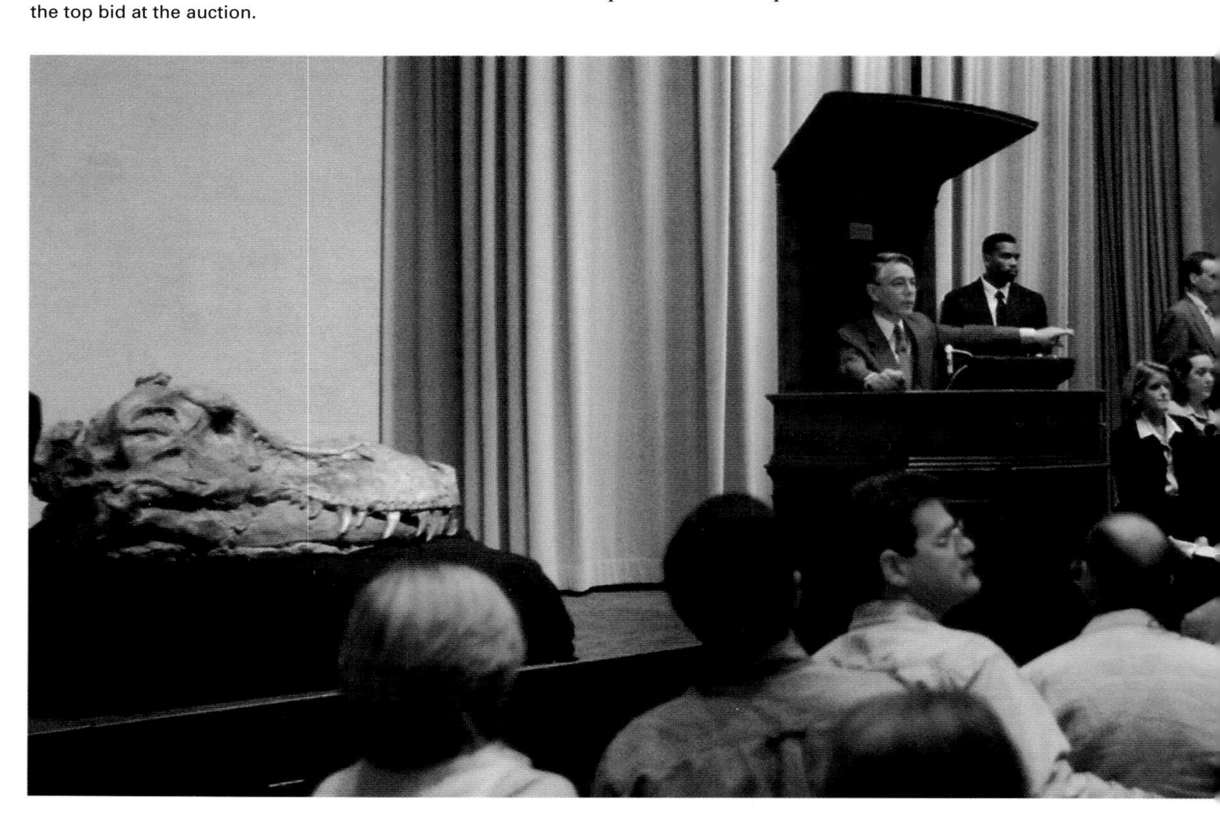

pected terrorists or other criminals. In December 1996, a federal court had ruled in the student's favor, stating that the laws forbidding the export of cryptographic software were unconstitutional. The government then appealed the decision.

Under federal law, it has been a crime in the United States for researchers to publish their work on cryptography on the World Wide Web (the Web) or to present it at meetings where non-U.S. citizens are present without an export license. The U.S. Federal Bureau of Investigation (FBI) has not only opposed lifting export controls on cryptographic software but has suggested domestic controls as well.

However, companies that want to sell products over the Internet regard strong cryptography as essential for protecting credit card information on-line. Privacy advocates regard cryptography as necessary to assure confidentiality in personal communications. Human-rights groups claim that that they need unbreakable codes in order to communicate by e-mail with political dissidents in other nations.

The international nature of modern commerce and communications technology may decide the case before the courts do. The same cryptographic software that U.S. firms are forbidden to export is available worldwide from Web sites in other countries. And in March 1998, Network Associates, the largest independent maker of computer security software in the United States, announced that it would circumvent U.S. laws by permitting its Dutch subsidiary to sell a strong encryption program called Pretty Good Privacy (PGP).

Public funding of research. The outlook for U.S. government funding for research and development (R & D) changed dramatically during the second half of 1997 and the beginning of 1998. In the early 1990's, the need to cut spending in order to balance the federal budget had led to reductions in many research programs and expectations of deeper cuts in future years. A stronger-than-expected economy in the mid-1990's, however, produced tax revenues that eliminated the deficit and even generated a surplus in 1998, much sooner than most government accountants had expected. The surplus, together with growing political support for research—particularly health research—turned the downbeat projections into calls for major increases in research spending.

The new optimism began in the summer of 1997 with an agreement between U.S. President Bill Clinton and the U.S. Congress that allowed for increases in discretionary spending, the part of the budget in which R & D programs fall. As a result, most R & D agencies received budget increases for fiscal year 1998 instead of the cuts they were expecting.

With the encouragement of several science lobby groups, Congress in late 1997 began to consider legislation that would authorize doubling much of the government's nonmilitary R & D over the next decade. Although the legislation, even if passed, would not guarantee such funding increases, most observers in the scientific community saw it as a very encouraging sign.

The budget that President Clinton proposed in February 1998 fell short of the doubling track, but it included large increases for a number of research programs. In a move viewed by some as a clever political strategy and by others as an attempt to avoid difficult choices, the president's budget tied future increases in research funding to a proposed settlement with the tobacco industry. That settlement stemmed from a dispute over the industry's responsibility for health problems caused by smoking.

Under the plan, the tobacco companies would pay several hundred billion dollars in compensation to the states and the federal government. Steep increases in cigarette taxes would produce additional billions while hopefully deterring smoking, especially by teens. Some of these revenues would be used to pay for research, including health research, allowing government spending to exceed the limits on future expenditures contained in the 1997 budget agreement. Whether Congress would succeed in passing tobacco legislation and, if so, whether it would allow the money to be spent on research rather than on such programs as Social Security or Medicare was not clear in mid-1998.

[Albert H. Teich]

See also ARCHAEOLOGY. In the Special Reports section, see CLONING ISN'T SCIENCE FICTION ANYMORE.

Beginning in mid-1997, the U.S. National Aeronautics and Space Administration (NASA) began to test a new approach to the design and construction of robotic spacecraft. The agency had decided to launch many inexpensive probes and satellites to gather data on a larger number of objects in the solar system, rather than to continue flying only a few highly capable but much more costly missions. The results of this decision through 1998 were mostly positive. Successful missions using lower-cost spacecraft flew to Mars, the moon, and an asteroid. In human spaceflight, however, there were further troubles on Russia's space station, Mir, and Russia faced difficulties in building hardware for a new international space station.

More woes for Mir. After a fire on the Russian space station earlier in the year, Mir had another major accident in June 1997. The mishap occurred while cosmonauts Vasili Tsibliyev and Alexander Lazutkin and U.S. astronaut C. Michael Foale were preparing for the arrival of an unmanned Progress cargo spacecraft. Such supply ships had docked at Mir many times, relying on their own automatic guidance system. However, Russian mission controllers wanted Tsibliyev to test a new technique for docking the Progress, using radio controls and a device similar to a computer "joy stick."

Unfortunately, the cargo ship approached at too high a speed, and Tsibliyev was unable to prevent it from crashing into the station. The Progress seriously damaged an array of solar power cells and punctured one of Mir's modules, Spektr. Air inside the station began to escape, forcing the crew to seal off Spektr from the rest of Mir.

During the months that followed, cosmonauts repaired much of the damage. However, they were unable to find the hole in Spektr. Mir's main computer failed several times as well. The accidents forced NASA to reconsider whether it was safe to keep astronauts aboard Mir. The agency finally decided that it was safe, replacing Foale with astronauts David A. Wolf in October and Andrew S. W. Thomas in January 1998.

International Space Station. Problems with a key Russian component for the new International Space Station—a project involving 16 nations—also cast doubt on the wisdom of NASA's partnership with its former rival. Work on the module fell behind schedule in 1997 and 1998. By mid-1998, it was apparent that NASA would have to postpone the launch of the first components of the station to late 1998 or 1999.

Meanwhile, an independent review of the station program estimated that NASA's costs would ultimately reach $24.7 billion, not $17.4 billion as the agency had promised. The review also said station assembly would not be completed until 2005 or 2006, not by 2003 as NASA had originally planned.

Shuttle missions. On July 1, 1997, the orbiter Columbia lifted off to repeat a Spacelab mission that had failed earlier in the year. The earlier Columbia flight was forced to land after just four days when a sensor indicated that one of its electricity-producing fuel cells had malfunctioned. In July, the same crew of seven astronauts flew for 16 days, performing the experiments slated for the previous Columbia mission. Many of the experiments involved setting small fires to study combustion in space.

The next shuttle mission was a 12-day flight begun on August 7 by the orbiter Discovery. Six astronauts, including Canadian Bjarni V. Tryggvason, tested a Japanese robotic arm and released and retrieved a German satellite that collected data about Earth's atmosphere.

On September 25, the orbiter Atlantis was launched on a mission to Mir. The shuttle's crew included Russia's Vladimir G. Titov and France's Jean-Loup Chretien. Atlantis picked up U.S. astronaut Foale, who had been on Mir for 145 days, and dropped off his replacement, David Wolf. The crew also included two astronauts who had trained to live and work aboard Mir but were unable to do so because of their height: Wendy Lawrence, who was originally scheduled to replace Foale, was too short for the Russian space suits that needed to be worn to repair the damaged Spektr module. Scott Parazynski was too tall to fit into the Soyuz escape capsule if an emergency occurred.

Columbia was launched again on November 19, for a 16-day mission devoted mainly to materials science. The crew of six included Leonid Kadenyuk, the first Ukrainian cosmonaut to fly on a U.S. shuttle, and Takao Doi, an aerospace engineer from Tokyo. Doi became

Solar arrays on the Russian space station Mir are battered after being struck in June 1997 by a Progress supply ship as it docked with the station. The accident also damaged one of Mir's modules, called Spektr, puncturing it and denting its radiator, at center.

the first Japanese astronaut to perform a spacewalk. He and American Winston E. Scott used their hands to grab a slowly spinning satellite that had malfunctioned when the crew deployed it.

On Jan. 22, 1998, Endeavour began a nine-day mission to Mir—its first such flight. Endeavour was the last U.S. orbiter to be fitted with a mechanism enabling it to dock at Mir. The orbiter's crew of seven included Wolf's replacement, Andrew Thomas, who was to be the last U.S. astronaut on Mir. Russia planned to allow Mir to gradually lose altitude and fall to Earth after assembly of the International Space Station begins in orbit. Endeavour brought Wolf back to Earth after 128 days in space.

On April 17, Columbia began a 16-day Spacelab mission to study the effects of very low gravity on the brain and nervous system. The crew of seven, which included Canadian physician Dafydd Rhys Williams, ran tests of their own sensory and thinking skills in space and performed experiments on more than 2,000 animals, including fish, snails, crickets, mice, and rats.

Return trip. NASA announced in January 1998 that it would add U.S. Senator John Glenn to a space shuttle crew in late 1998. The agency said it wanted to gather physiological data on the aging process in space. But most Americans saw it simply as a way to honor the first American to orbit Earth. Glenn had not flown in space again after 1962, when his Mercury capsule made three orbits. At the age of 77, Glenn would be by far the oldest person ever to fly in space.

Close encounter. In June 1997, a small spacecraft called NEAR, for Near Earth Asteroid Rendezvous, produced detailed images and measured the mass of 253 Mathilde, an asteroid that orbits the sun in an area between Mars and Jupiter. NEAR, launched in 1996, came as close as 1,200 kilometers (750 miles) to the extremely dark object.

Mathilde, which is about 60 kilometers (37 miles) wide, was only the third asteroid ever imaged in a fly-by. (The Galileo spacecraft photographed the other two.) It was also the largest and the first of the C-type asteroids—those presumed to be made of *hydrocarbons* (materials similar to those in tar)—to be photographed. The feat was remarkable for a simple $122-million spacecraft that weighs less than a ton. After photographing Mathilde, NEAR continued on to its main target, asteroid 433 Eros, which it was expected to reach in 1999.

Mars missions. The star in NASA's cast of new low-cost missions was a spacecraft called Mars Pathfinder. It reached the red planet on July 4, 1997. Because it was so light, Pathfinder was able to land using a parachute and inflatable airbags rather than the rockets used by larger spacecraft. After landing, Pathfinder released a wheeled rover about the size of a child's wagon called Sojourner. Though Pathfinder was designed to work for only a month, and the rover for just 10 days, both worked for months, sending scientists on Earth detailed images of the planet's surface.

Mars Global Surveyor fared less well. It began orbiting Mars on September 11. However, braking problems forced mission controllers to push back the start of Surveyor's mapping mission from March 1998 to March 1999. As a result, the duration of the mission was cut by more than half.

Lewis and Clark, two planned Earth-observing satellites, were NASA's biggest disappointment in 1997 and 1998. Lewis was launched in August 1997 from Vandenberg Air Force Base in California. But the $71-million spacecraft malfunctioned and began spinning, draining its battery. Meanwhile, the construction of Clark fell way behind schedule. NASA halted the project in early 1998, though $55 million had already been spent.

Off to Saturn. On Oct. 15, 1997, a Titan 4B rocket lifted off from Cape Canaveral, Florida, to launch Cassini/Huygens, the most complex and expensive planetary science effort ever undertaken. After being released by the Titan, Cassini/Huygens, a U.S.-European spacecraft which cost $3.4 billion, began a circuitous 3.5-billion-kilometer (2.2-billion-mile) voyage to Saturn. Over seven years, the spacecraft was to loop close to Venus twice and back by Earth once to use those planets' gravity to gain speed before proceeding to Saturn. It was scheduled to reach Saturn in July

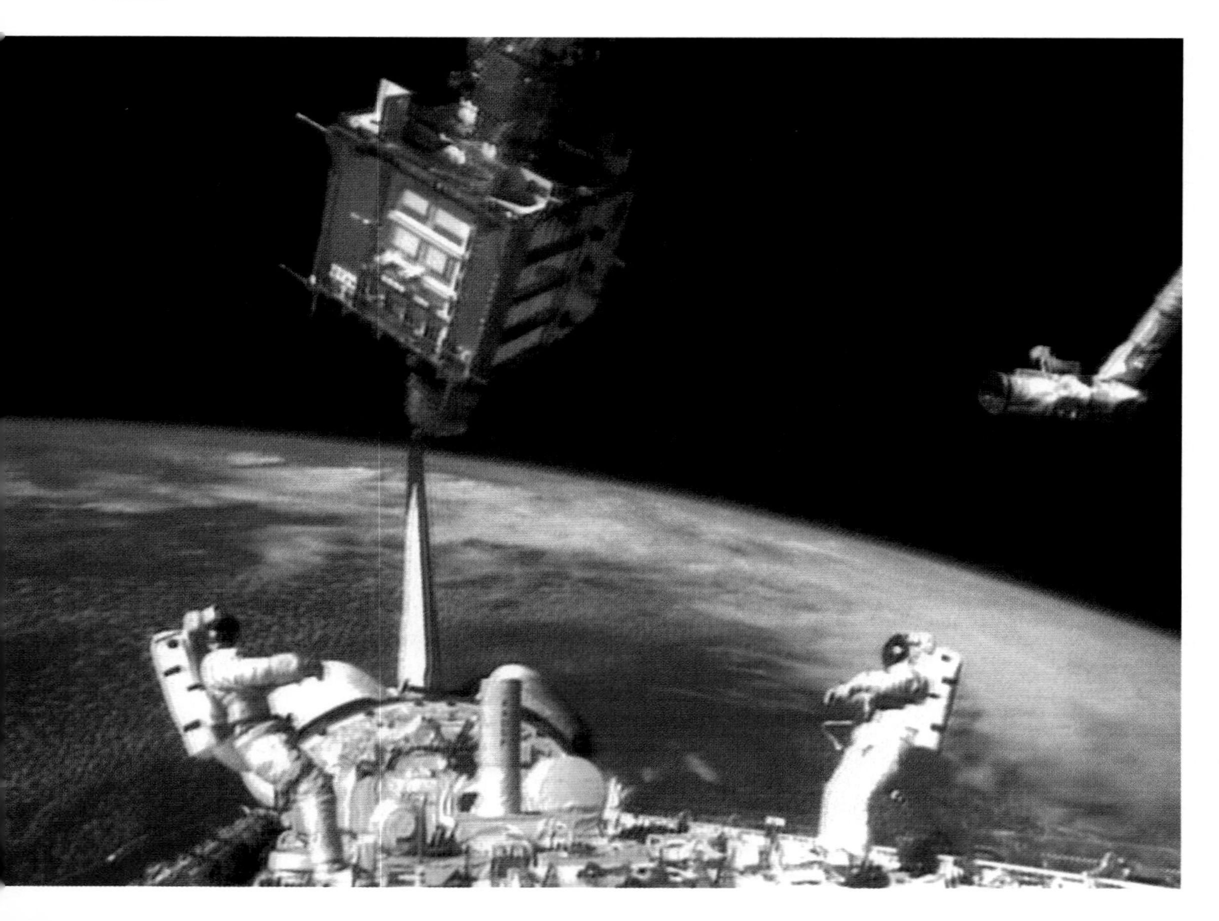

Standing in the cargo bay of the U.S. space shuttle orbiter Columbia, U.S. astronaut Winston Scott, left, and Japanese astronaut Takao Doi prepare to retrieve a malfunctioning satellite in November 1997. The satellite had been launched from Columbia earlier in the mission.

Mission to Saturn

The sky above Cape Canaveral, Florida, is lit up by the trail of a Titan rocket on Oct. 15, 1997, as a two-part spacecraft called Cassini/Huygens, *inset,* is launched toward Saturn. Cassini/Huygens was expected to reach Saturn in 2004. Before going into orbit around Saturn, Cassini—the parent spacecraft—will release the Huygens probe. Huygens will make a landing on Titan, one of Saturn's moons.

2004. The 5.6-metric-ton (6.2-ton) spacecraft included a U.S.-built orbiter, Cassini, and a European-built probe, Huygens. Huygens was to land on one of Saturn's moons, Titan.

Back to the moon. On Jan. 7, 1998, NASA launched the Lunar Prospector spacecraft to the moon on an Athena 2 rocket. At a cost of $63 million, Lunar Prospector was dirt cheap for a planetary science mission. The 1.4-meter (4.5-foot), drum-shaped spacecraft was extremely simple, with only a few parts that had to be *deployed* (moved into place) once it was in space. Lunar Prospector produced strong evidence that there is water in the form of ice at the Moon's poles.

Galileo, NASA's mission to Jupiter and its moons, continued to send back data and pictures in 1998. Images of the moon Europa suggested that a slushy ocean may lie below its icy surface. To explore that possiblility further, NASA extended the spacecraft's mission beyond its original 1997 termination date to 1999. Galileo, which began orbiting the planet in 1995, was to study Europa before going on to another moon, Io.

Ariane tries again. In October 1997, Europe's large Ariane 5 rocket lifted off from Kourou, French Guiana, placing two test *payloads* (cargoes) into orbit. The flight of the 740-metric-ton (816-ton) booster came a year and a half after another Ariane 5 exploded on its first flight. The 1997 flight placed a dummy communications satellite into orbit around the Earth, though 9,000 kilometers (5,600 miles) lower than planned. The rocket's main engine was cut off early when Ariane 5 rolled more strongly during its ascent than the onboard flight-control computer expected. Though the mission was declared a success, engineers began work to prevent the roll problem on future flights.

The Japanese space agency in November 1997 successfully launched a satellite called the Tropical Rainfall Measuring Mission from its Tanegashima complex off the coast of the island of Kyushu. The mission was a joint program with the United States, conducted to gain a better understanding of global weather patterns.

But Japan's space program also encountered difficulties. In June 1997, its Advanced Earth Observation Satellite

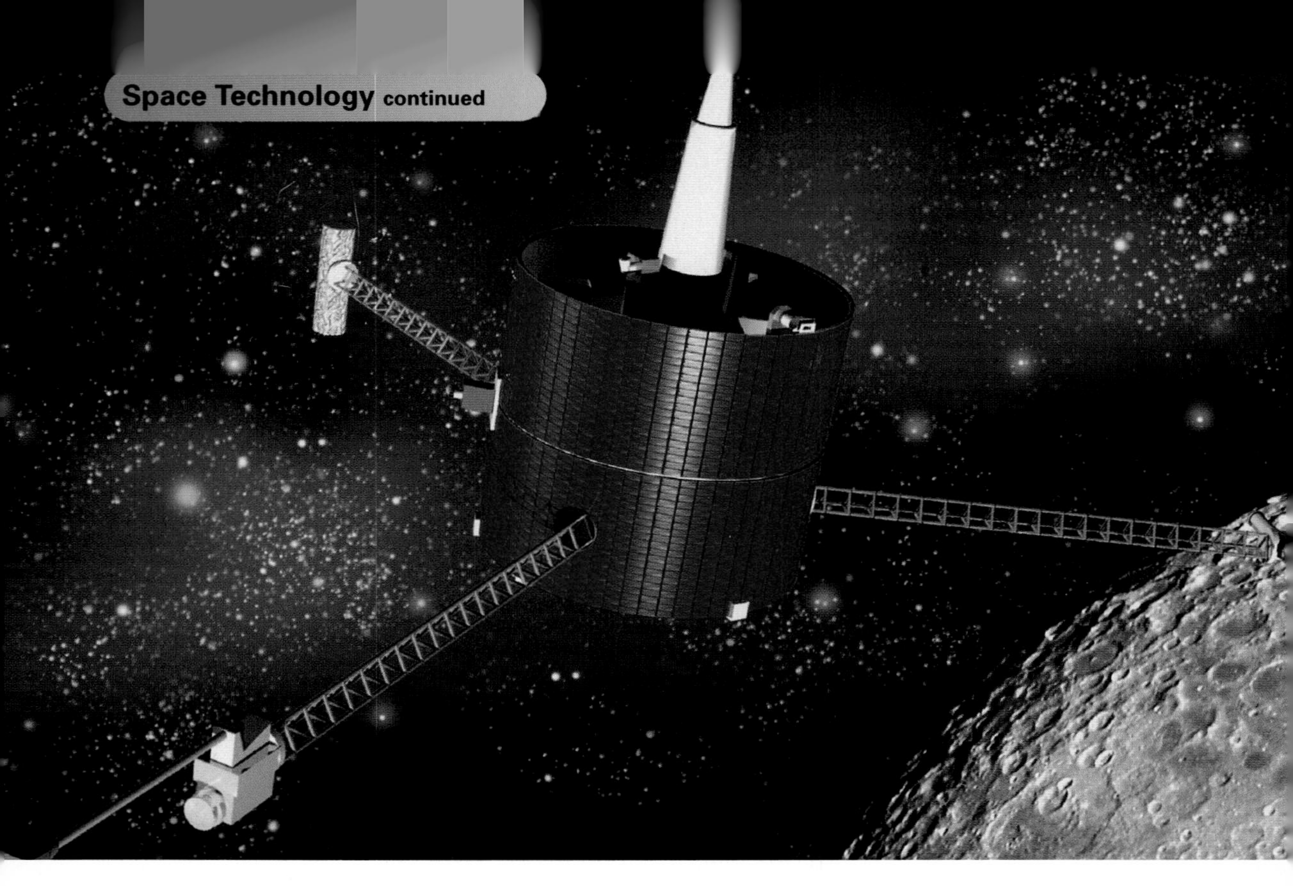

The Lunar Prospector spacecraft, launched in January 1998 by the National Aeronautics and Space Administration, approaches the moon in this artist's rendering of the flight. The spacecraft carried instruments to study the lunar surface, particularly to search for signs of water—which it detected in large quantities at both poles of the moon.

abruptly stopped relaying data. The $1-billion spacecraft, part of an environmental science program in which the United States and France had also taken part, was considered lost. In February 1998, a flawed flight of Japan's H-2 launch vehicle placed a $375-million communications test satellite into a largely useless orbit.

Curious route. In April 1998, a commercial satellite was sent on an unusual path to place it into orbit—a flight by way of the Moon. The spacecraft, called AsiaSat 3, was built by Hughes Electronics Corporation of Los Angeles for a Hong Kong telecommunications company. It was launched from Kazakhstan on a Russian Proton booster rocket on Dec. 25, 1997. But the rocket malfunctioned, placing the satellite in a useless orbit. The satellite's insurers declared it a loss and paid the owners $200 million.

Then Hughes engineers tried an experiment: They used the satellite's onboard rocket motor to send the satellite around the moon. The moon's gravity acted like a slingshot, accelerating the satellite and flinging it back into a usable orbit around the Earth, though not the one originally planned. A second swing by the moon was planned to refine the orbit. While NASA had often used such gravity assists in its interplanetary missions, the experiment marked the first time that a private company had done so—and the first time that a private company had ever sent a spacecraft to the moon.

Silent pagers. Hughes engineers faced another challenge in May 1998 when a Galaxy 4 communication satellite they built for PanAmSat Corporation of Greenwich, Connecticut, failed. The malfunction shut down about 90 percent of the almost 50 million electronic pagers in the United States, as well as some radio and television broadcasts and business networks. PanAmSat quickly shifted its customers to another satellite and moved a spare satellite to a new orbit to replace Galaxy 4. However, service for some customers was disrupted for up to six days. [James R. Asker]

See also ASTRONOMY. In the Special Reports section, see RETURN TO MARS and THE INTERNATIONAL SPACE STATION.

SCIENCE YOU CAN USE

Topics selected for their current interest provide information that the reader as a consumer can use in understanding every-day technology or in making decisions—from buying products to caring for personal health and well-being.

The Global Positioning System Charts New Territory

What do the following scenarios have in common?

- A band of hikers wanders off a poorly marked mountain path as night approaches and the temperature is falling. But the readout from a small, portable receiver—about the same size as a cellular telephone—allows the hikers to make a safe return.
- Scientists trying to bring snow leopards back from the brink of extinction are able, for the first time, to gather accurate information about where these large predators roam on the remote Mongolian plains.
- Workers operating earth-digging machines at a large construction site receive signals that enable them to follow a blueprint to an accuracy of a few centimeters.
- The driver of a rental car, lost in an unfamiliar city, finds his destination using a dashboard-mounted audio and visual display panel showing the vehicle's location on a map of the city.

All these solutions are made possible by a worldwide navigation system called the Global Positioning System (GPS). The GPS was developed in the 1970's and grew out of the United States Department of Defense's need during the Cold War—a period of intense rivalry between the United States and the former Soviet Union—to be able to guide missiles with extreme accuracy.

The GPS brought a wide range of benefits to consumers when it was made available for civilian use in the mid-1990's. The technology became so popular and widespread that by 1998, some models of handheld, computerized radio receivers were available in stores for as little as $100.

The main premise behind the GPS—using objects in the sky to determine one's exact position on Earth—dates back centuries. Travelers once used the sun and stars to determine their location and plot a course. The GPS relies on a constellation of time-synchronized satellites for this purpose.

The GPS uses 24 satellites—21 operational and 3 spares—in six orbits about 17,500 kilometers (10,900 miles) above

The Global Positioning System (GPS) allows motorists to find their way on a computerized map by using an on-board receiver. The system pinpoints the motorist's exact location.

the Earth. Each satellite orbits the Earth every 12 hours. Five Defense Department monitoring stations in the United States and elsewhere measure the altitude, position, time, and speed of each satellite to detect any minor course errors that may be caused by solar radiation or the gravitational pull of the Earth or the moon. If a satellite is found to be off course, the monitoring stations get it back into the correct orbit by activating an on-board propulsion system.

GPS satellites broadcast a continuous radio signal containing two pieces of information necessary for people on the ground to calculate their location: the satellite's position in space and the exact time, which is determined by an ultraprecise atomic clock accurate to within billionths of a second. Radio signals from the satellites travel at the speed of light, or 299,792 kilometers (186,282 miles) per second.

On the ground, a GPS receiver "listens" for these signals. Each receiver picks up signals from at least four of the satellites and compares the time transmitted from each satellite with a reading from its own internal clock. The time comparisons reveal how long it took for the individual signals to reach the receiver—tiny fractions of a second. Multiplying the time intervals by the speed of light gives the distance of each satellite from the receiver.

The receiver uses the combined data from at least three satellites to calculate the user's *latitude* (distance north or south of the equator) and *longitude* (position east or west of the prime meridian, the line of longitude passing through Greenwich, England). The signal from a fourth satellite provides users with additional data for error correction and also enables pilots of aircraft to determine their altitude.

Navigation is by far the most common civilian use of GPS. Boaters use GPS units to chart their course in the water. Private pilots use the system to check their course. And drivers boldly go where they have never been before, secure in the knowledge that their dashboard GPS display will keep them from getting lost.

More expensive GPS receivers for cars give motorists turn-by-turn directions for reaching a destination. Because these devices cost several thousand dol-

Navigating with the GPS

The Global Positioning System (GPS) uses 24 orbiting satellites that broadcast data to receivers on Earth, enabling people to navigate with precision.

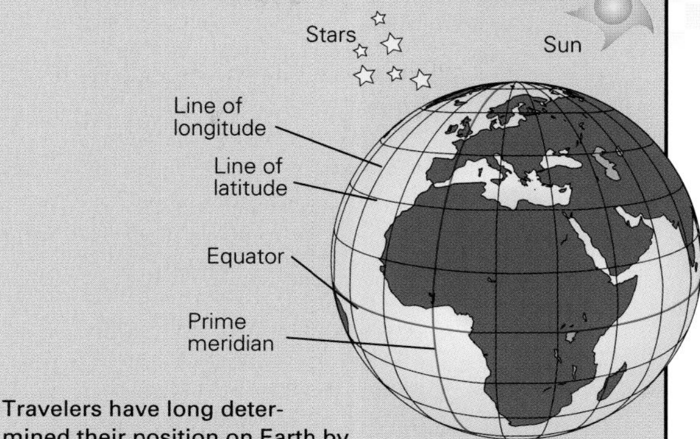

Travelers have long determined their position on Earth by taking fixes on objects in the sky—traditionally the sun and stars. One's position is figured in terms of imaginary lines of latitude, or distance north or south of the equator, and longitude, or distance east or west of the *prime meridian,* the line of longitude passing through Greenwich, England.

The GPS makes navigation fast and simple. Each satellite broadcasts its precise orbital position and the time of day, accurate to within billionths of a second. The time reading, compared with a reading from a clock within the receiver, tells how long it took the signal to reach the receiver and thus how far the receiver is from the satellite. A receiver correlates signals from four satellites. The combined data from three satellites gives the receiver's latitude and longitude. The signal from the fourth satellite provides additional data for error correction and also enables airplanes to determine their altitude.

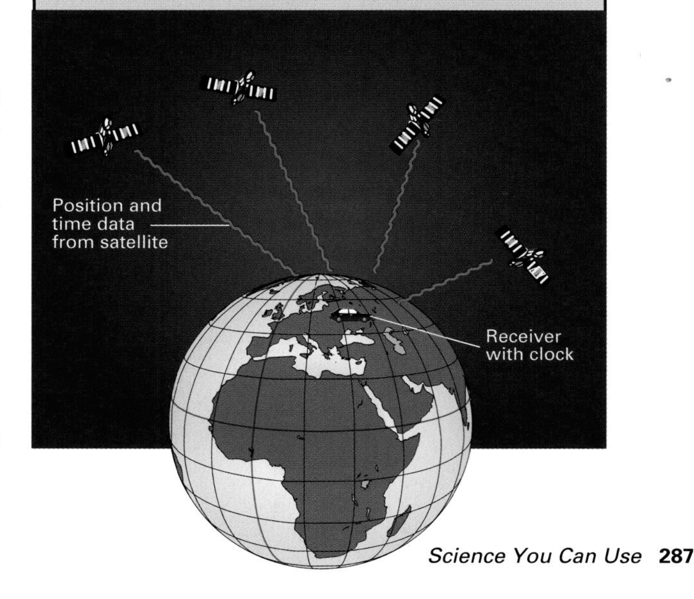

Uses of the Global Positioning System

- **Locating:** Users can determine the exact location of any stationary object on Earth, such as a particular power pole, tree, or structure.
- **Navigating:** People can get from one point to another with confidence. Handheld units allow hikers to pinpoint their location and avoid getting lost; boaters, drivers, and pilots can navigate safely and efficiently in any light or weather condition.
- **Cataloging:** The precise location of an object and other information about it can be recorded in a GPS database for later use, such as in mapmaking.
- **Tracking:** When a receiver is coupled with a wireless communications link, users can monitor the movement of any object worldwide, which enables businesses to track packages or company vehicles.
- **Coordinating:** A GPS time-data system enables users to "time stamp" various types of information—film images shot from different angles by different cameras, for example—so the information can later be synchronized.

Source: Trimble Navigation.

lars they were offered as an option mainly on luxury cars. Many rental car companies in 1998, however, were offering cars equipped with GPS receivers. Drivers unfamiliar with an area were often willing to pay a few extra dollars a day for a rental car that could pinpoint their location on a dashboard-mounted electronic map.

Automobile GPS systems do have limitations, though. For example, their 100-meter accuracy limit will not bring a driver to the exact building he or she is looking for. Also, GPS satellite signals are blocked by tall buildings, making the systems less reliable in the concrete canyons of most large cities. GPS receivers compensate for this problem by estimating a position until satellite-signal reception is reestablished. In addition, vehicle GPS devices supplement satellite data with calculations from "dead reckoning systems." These onboard computer programs monitor a vehicle's speedometer and steering wheel movements to determine how far the vehicle has traveled and in what directions.

But the GPS has other uses besides navigation. GPS units that broadcast their position have been installed in many taxicabs, ambulances, and public buses to enable dispatchers to know the locations of all the vehicles in their fleet from minute to minute. This information can greatly enhance service. With an ambulance company, for instance, a dispatcher can respond to an emergency call by sending the vehicle that is closest to the scene.

In many emergency situations, with the aid of the GPS, summoning help may become automatic. By 1998, the General Motors Corporation had already developed an automotive emergency GPS system that signals a car's position whenever an air bag deploys, alerting emergency crews that an accident has occurred at that location.

Cellular phones could be equipped in the same way so that emergency 911 callers automatically transmit location information to police and fire dispatchers. Such enhanced 911, or E911, service is already required in some states for emergency calls made from wired telephones.

GPS satellites also serve as timekeepers. Because each satellite carries an atomic clock, the most accurate timepiece that has ever been devised, the signals beamed down to Earth can serve as master clocks for the entire world. GPS receivers could be built into wristwatches and clocks, which would synchronize themselves with the timing signals emanating from the satellites.

As GPS electronics get smaller and cheaper, high-precision location information may become a part of many appliances. A camera that "knows" where it is at all times, for example, could record that information on each exposure, much the way that some cameras now record the time and date.

The GPS also has important military uses. Each GPS satellite transmits two signals, called L1 and L2, at different frequencies. The L1 signal, which is broadcast at 1572.42 *megahertz* (million cycles per second), can be picked up by any GPS receiver.

However, to prevent the GPS from being used against the United States by hostile forces, the L1 signal includes an operational code implanted by the U.S. military that limits the signal's data content. This security measure prevents civilians from tapping into the full accuracy of the GPS. As a result, most civilian GPS users can pinpoint their location only to within 100 meters (330 feet) and their altitude to within 150 meters (495 feet).

The system reserves its greatest precision for the L2 signal, which is available primarily for military use. The L2 signal, broadcast at a frequency of 1227.60 megahertz, enables receivers on the ground to correct for time errors created by the disruption of satellite signals by their passage through the *ionosphere,* an electrically charged layer of the upper atmosphere.

However, the L2 signal is *encrypted* (coded) so that only receivers with a decoding key can use the data. According to the U.S. military, authorized civilian receivers using an L2 signal can pinpoint locations to within 20 meters (66 feet)—approximately five times better accuracy than with the L1 signal. Authorized military personnel, using a security module, can improve the accuracy to 7 meters (23 feet).

There was pressure from civilian users in 1998 to make this higher-precision GPS more widely available. President Bill Clinton had already announced a goal in 1996 of allowing full public access to all GPS data by the year 2006.

In the spring of 1998, Vice President Al Gore announced plans for the addition of two new civilian signals to the GPS. The signals were to be located in the existing L2 band. Experts said that the additional signals, which were scheduled to be activated by the year 2005, would significantly improve navigation, positioning, and timing services for users.

Even without such new signals or access to more advanced GPS data, however, many civilian GPS users by the late 1990's could calculate their whereabouts with an accuracy superior to that provided by an L2 reading. This was accomplished by using a technique called differential GPS (DGPS).

DGPS uses a second, stationary GPS receiver whose geographical position has been determined with great precision. That receiver calculates its supposed location from the GPS L1 transmissions and compares the result to its known position. The difference between these two positions is then broadcast to portable GPS receivers in the vicinity so that they may apply the same correction to their own readings. The system can be used to determine locations to an accuracy of 2 to 5 meters (6.5 to 16 feet). The U.S. military allows civilians to access this highly accurate system, because the corrected signals can be electronically jammed, and thereby rendered useless, in the event of a military emergency.

The U.S. Coast Guard maintains a network of GPS receivers along the coastlines and transmits DGPS corrections over radio beacons to boats and ships at sea. Inland, private DGPS services use leased FM radio frequencies, satellite links, or private radio beacons to broadcast the signals. However, the advanced system can be useless if the receiver is not within range of a DGPS transmitter.

DGPS can also be more expensive than GPS because users need two receivers—the conventional GPS unit that listens for signals from the satellites and a second unit that tunes into a DGPS transmitter. The additional DGPS receiver costs between $500 and $1,000.

To further enhance DGPS accuracy, the Federal Aviation Administration planned to offer, by 1999 or 2000, a "wide-area augmentation system" that would provide DGPS service to the entire continent. A network of ground stations across the United States would receive, analyze, and refine the GPS signals and transmit the information to receivers.

With the refined information that this system will provide, ordinary civilian receivers throughout the country should be able to figure their location to within 1 meter (3.3 feet).

The military also makes use of DGPS when it needs extreme accuracy. Authorized military users in 1998 were able to use additional codes that allowed them to figure their precise location to within just a few centimeters.

The GPS promises to make the everyday knowledge of location as commonplace as knowledge of the time of day. Just as the invention of the clock brought precision to everyone's daily schedules, so will the GPS continue to enable travelers to easily make their way through unfamiliar territory.

[Herb Brody]

Wet Cleaners: the Latest Wave in a Green Revolution

Dropping clothes off at any of the more than 34,000 dry cleaners in the United States is one of those routine tasks we do without thinking very much about it. We leave our dresses, jackets, or pants with a clerk, take the claim check, and we're on to other things.

Yet behind that simple ritual is a cleaning process that has drawn the fire of many environmentalists, who claim it relies on chemicals that pose potential hazards to human health and the environment. Such concerns have given rise to an alternative cleaning method: wet cleaning. This system makes use of a familiar formula—soap and water—to remove dirt and stains from fabrics. Since wet cleaning poses no known hazards to the environment, establishments offering the process are often known as "green cleaners."

Professional cleaning has always been a necessary service because regular laundering can shrink, stretch, or fade some kinds of clothes, such as garments made of wool or silk. Draperies, bedspreads, and other household items also may be ruined if simply tossed into a washing machine.

Professional wet cleaners, often called "green cleaners" because the process they use is considered to be environmentally safe, are growing in popularity as consumers search for alternatives to dry cleaning.

Dry cleaning has long been the preferred alternative to washing for cleaning easily damaged garments and fabrics. However, the term "dry cleaning" is misleading, because the process is not really dry. It involves the use of a liquid chemical *solvent* (a liquid in which substances will dissolve) to remove soil and stains from fabrics. The process is called dry cleaning because little or no water is used.

Clothes to be dry cleaned are first separated according to color and type of fabric. If a garment contains spots or stains, workers brush or spray those areas with special chemicals that loosen or remove the discolorations.

Garments are then put into a dry-cleaning machine, which resembles an oversized front-loading washing machine. The dry-cleaning machine has a rotating drum that fills with the solvent. As the drum rotates, the solvent penetrates the clothes and dissolves dirt and stains. A dry-cleaning detergent is often added to the solvent to help remove water-soluble spots.

After the cleaning cycle, the solvent drains from the machine to be filtered and reused in other dry cleaning loads. The drum then spins rapidly to remove most of the remaining solvent from the clothing.

Next, the clothes are tumble-dried. This is usually done in the same machine, though some dry cleaners use older equipment with a separate machine for drying. Finally, the garments are pressed to remove wrinkles and restore their shape and texture.

The solvent used by almost 90 percent of all dry cleaners is called perchloroethylene, or "perc." The remaining 10 percent of cleaners use petroleum-based solvents. According to the International Fabricare Institute, a professional association representing the cleaning and laundering industry, dry cleaners in the United States in 1996 used almost 5 million kilograms (11 million pounds) of perc. In 1989, however, dry cleaners used 108 million kilo-

Dry cleaning

In dry cleaning, garments are placed in a machine with a rotating drum filled with a synthetic chlorine-containing solvent called perchloroethylene, or "perc." As the drum rotates, the compound penetrates the clothes and dissolves dirt and stains. After the cleaning cycle, the perc drains from the machine (to be filtered and reused), and the drum spins rapidly to remove most of the remaining solvent from the fabrics. The garments are then tumble-dried in the same machine.

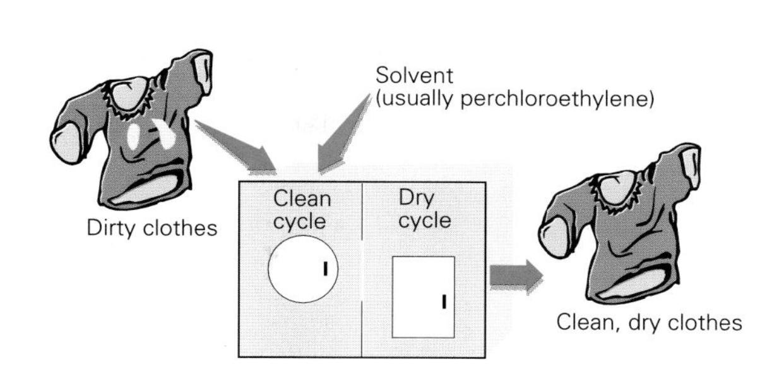

Wet cleaning

In wet-cleaning, special washing machines are used to clean garments with water and a nontoxic soap or detergent. A computer regulates such variables as how fast the machine's drum rotates, the temperature of the water, and the amount of soap or detergent used. After the wash cycle, clothes are either dried on hangers or placed in a dryer that can be programmed to precisely regulate temperature, humidity, and drying time in order to limit shrinkage.

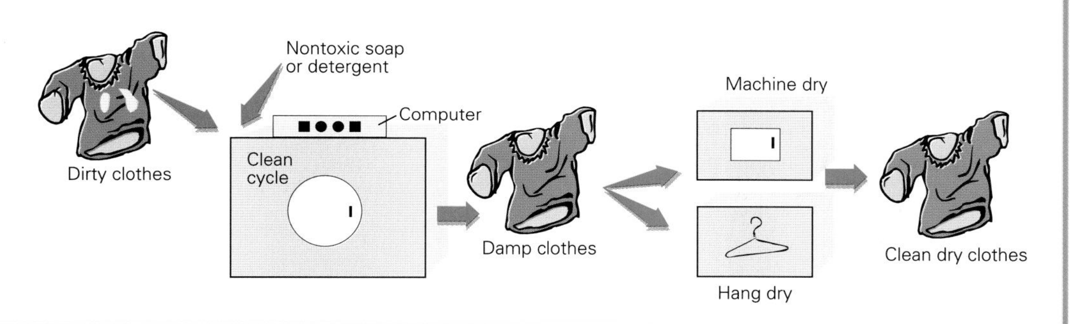

grams (240 million pounds) of the compound. Experts explained that the reduction was due to improved dry-cleaning machines, which more efficiently used and recycled perc.

Many scientists and environmentalists have questioned the safety of this widely used chemical. Because perc often accumulates in the air inside dry-cleaning establishments, it is regulated as an air pollutant under the Clean Air Act.

Perc also poses problems outside of dry cleaners' premises. The chemical has been discovered in underground water systems as well as in the air of apartments and restaurants adjacent to dry-cleaning establishments.

Some controversial studies have also linked continued exposure to perc with cancer. Other studies have concluded that perc affects the *central nervous system* (brain and spinal cord), causing such problems as dizziness, nausea, headaches, and memory problems, especially in workers who are constantly exposed to the chemical. Scientists, however, have agreed that it is highly unlikely that perc poses a health problem for people who wear clothing cleaned with the chemical.

The EPA in the early 1990's began working with the dry-cleaning industry to reduce the amount of perc being used. Concern over potential health risks also led governmental and professional organizations to explore new, environmentally friendly ways of cleaning garments.

Wet cleaning has been the most popular of the new cleaning methods. Al-

The pros and cons of wet cleaning

	Uses and advantages	Limitations
Effects on clothes	■ No chemical odor ■ Whiter whites ■ Better than dry cleaning for removing water-based stains	■ Can shrink some garments ■ Can cause color change ■ Not as good as dry cleaning at removing oil-based stains
Environmental and health effects	■ No residue of toxic chemicals in clothes ■ No hazardous chemicals released into air, water, or soil	■ Increased water use
Cost	■ Comparable for most items to dry cleaning	■ Can cost more for some items because of increased labor in pressing and finishing
Types of clothing	Best on: ■ Wool ■ Cotton ■ Leather/suede ■ Wedding gowns ■ Garments with beads or sequins	Not good for: ■ Acetate linings ■ Antique satin ■ Gabardine ■ Highly tailored garments
Availability	■ All cleaners have the capacity to wet-clean some items with existing equipment. Shops devoted exclusively to wet cleaning, with specialized equipment and trained personnel, are growing in number.	■ The spread of wet cleaning is limited by the need for specialized equipment and for more highly trained employees.

though wet cleaning relies on soap (or detergent) and water, it is the way those cleaning agents are used that sets the process apart from regular machine laundering.

A washing machine ordinarily damages delicate fabrics and garments because it uses excess agitation and does not properly control water temperature to the same degree as professional wet cleaning. Also, laundering is done with detergents that are inappropriate for woolens or delicate fabrics. Wet cleaners use specially formulated, gentler soaps or detergents, and they also take steps to prevent fabrics from shrinking, stretching, or losing color.

In one respect, wet cleaning is not really new. Water-based cleaning methods were used on natural-fiber garments by about one-quarter of all professional cleaners during the 1930's and 1940's. But the introduction in the 1950's of perc and other new chemical solvents used in dry cleaning caused most cleaning establishments to abandon the wet process until the 1990's. Modern wet cleaning, however, is to a large extent new because it uses computer-controlled machines.

Wet cleaning allows workers to customize treatment for each type of garment. The process can be used on many fabrics, including silk, wool, and leather, that have traditionally been dry cleaned.

In wet cleaning, workers inspect each garment for dirt and stains and determine how to treat it. The computer controls can be programmed for many variables, including how fast the drum inside the washing machine rotates, the temperature and amount of water used, and the amount of detergent added to the wash cycle.

The machines can have up to 100 different programs for various kinds of garments. For example, the washer can be programmed to make as few as six revolutions per minute to prevent especially delicate items from being damaged by excessive agitation.

In addition to using gentle detergents, wet cleaners may add an agent to the wash cycle that prevents dye from washing out of garments. They sometimes also use mild bleaching detergents, which remove tough stains without causing colors to fade, and fabric softeners, which restore garments' texture and body. To safely clean fabrics that can shrink when washed in water and then dried, cleaners can increase the amount of water spun out of wet garments after the final rinsing cycle so that minimal drying is needed.

Once garments complete their wash cycle they are moved to a separate, computer-controlled dryer. Dryers can be programmed to regulate the air temperature and drying. The computers monitor the moisture content of the garments continuously during the drying cycle in order to limit shrinkage. Especially delicate clothing is often simply placed on a hanger and allowed to dry.

After drying, garments are pressed and finished with the same equipment used by dry-cleaning professionals. Some wet cleaners also use specially designed pressing equipment, which also stretches clothing back into shape after it has dried.

Although wet cleaning yields garments with no chemical smell and is environmentally friendly, it has some drawbacks. For example, the process requires workers to have greater knowledge of fibers and fabrics, and it requires specialized equipment that many cleaners did not have in 1998. Also, it is not considered appropriate for garments with elaborate tailoring, such as a man's suit coat or sports jacket that may have been custom-tailored in order to fit properly.

The limitations of wet cleaning, however, are fairly minor, and the process seemed to be gaining popularity in 1998. The EPA encouraged the use of wet cleaning by conducting studies of the process and holding conferences where researchers could share information about it. The results indicated a growing acceptance of wet cleaning.

In a 1997 University of California, Los Angeles, study on wet cleaning, consumers had 34,950 garments cleaned at a professional wet-cleaning establishment. More than 90 percent of the customers surveyed rated the quality of the cleaning job as "good" or "excellent." Customers also rated the wet-cleaning method as equal to or better than dry cleaning in areas such as smell.

A similar study conducted by Environment Canada, the Canadian government's environmental agency, showed that 97 percent of customers whose clothes had been wet-cleaned were satisfied with the way their clothing looked and smelled.

While wet cleaning was the only nontoxic alternative to perc commercially available in 1998, chemists were exploring the use of liquid carbon dioxide (CO_2) as a dry-cleaning agent, and the first commercial liquid CO_2 cleaning machines were expected to be available by the end of 1998.

Many cleaning professionals hailed the use of CO_2 as a potential breakthrough because the compound is nontoxic, inexpensive, and abundant. Although CO_2 is normally either a solid (dry ice) or a gas, the gas can be liquefied by putting it under high pressure. The liquid would be delivered to cleaners in heavy metal canisters.

To be used as a cleaning agent, liquid CO_2 must be mixed with a special detergent. The detergent organizes into clusters called micelles that capture dirt and grease from garments.

In the cleaning process, the CO_2-detergent mixture is fed into newly designed machines in which clothes tumble in a rotating drum, just as they do in a conventional washing or dry-cleaning machine. The units, however, operate at a pressure of about 405 kilograms (900 pounds) per square inch to keep the carbon dioxide liquefied. Cleaning can be done at room temperature, which eliminates the possibility of heat damage to garments.

Following the wash and rinse cycles, the carbon dioxide is evaporated and collected for reuse. The process leaves a residue of detergent, dirt, and grime in a container inside the machine. The container can be disposed of without any special precautions.

The cleaning industry in 1998 was making important strides toward reducing its impact on the environment and health. Although processes such as wet cleaning were still a small part of the industry, it appeared that both cleaners and consumers were moving toward greener pastures. [Harvey Black]

Golf Ball Technology Slices into a Mystery of the Game

The once remote game of golf found a new legion of fans in the mid-1990's, thanks in part to the young sensation Tiger Woods, whose prowess on the course made the game look almost too easy. Watching Woods, it appeared that all you have to do is whack the ball a few times and then putt it into the cup.

Everyone knows that's not true, of course. And people spend lots of money on golfing lessons, golf clubs, and various kinds of accessories in an effort to improve their game. But the one thing that most people don't think that much about is arguably the biggest element of the game, and it's also undeniably the smallest: the golf ball.

In the late 1990's, some 25 million people in the United States were golfers, and they were buying about 850 million golf balls annually. That's roughly 34 balls per player. Yet, the particulars of golf balls remain a mystery to most people.

Mastering the game of golf involves more than just choosing the right clubs and shoes—it also involves a solid understanding of what golf ball is the best one for you.

There are significant differences between golf balls, and the design of the ball can make a difference in your golf score. The smallest details of a golf ball's interior and exterior construction, its size, and the number of those small indentations—dimples—on the cover play a part in how it performs.

There are three types of golf balls: wound balls, which have been in existence the longest; two-layer balls, the most commonly used ball; and multilayer balls, the most recent innovation in golf-ball technology. Each type of ball has its proponents.

The wound ball, first developed in the early 1900's, is constructed much like a baseball. Wound golf balls contain a maze of thin rubber thread wound tightly around a core. The core may consist of hard rubber or a sac filled with water or another liquid. The white outer cover of the ball is made of *gutta-percha* or *balata*—the dried and hardened gumlike extracts of certain kinds of tropical trees.

When a golfer hits a wound ball, the rubber-wrapped core and the soft outer covering change shape slightly from the impact with the golf club head. Although all golf balls change shape to some extent when hit with a golf club, the flexible core and soft cover of a wound ball make the effect more pronounced. As a result, these balls are heavily influenced by the angle, speed, and direction of contact with the club.

When struck with a full swing, a wound ball travels with minimal spinning movement, allowing it to fly and roll farther. When struck with a softer or a half swing, wound balls produce more spin. This enables a skillful golfer to exercise considerable control over the path of the ball, both in the air and when it lands. Because the performance of wound balls can be so precisely controlled, many professional golfers prefer them over the newer types of balls.

But the soft construction of a wound ball also makes the ball more apt to exaggerate the effects of a bad hit, making it spin faster in the wrong direction—something many less-skillful golfers would consider a disadvantage. Also, because a wound ball's cover is soft, it cuts or tears easily from glancing blows and deteriorates in wet weather.

To counteract some of these problems, golf-ball manufacturers in the late 1960's developed the two-piece ball. This type of ball consists of a large, solid rubber core, much like a toy "superball," with no rubber thread. A cover made of a hard, strong plastic—which still allows the ball to bounce—surrounds the core.

The hardness of the cover makes a two-piece ball much more durable than a wound ball. It also makes the ball more rigid, so its shape changes very little when it is struck. A two-piece ball is thus less influenced by the club, and so it is harder to control. The effect, compared with hitting a wound ball, is akin to the difference, when using a tennis racquet, between hitting a tennis ball and hitting a hard plastic ball. Even if both balls were the same size and weight, each would perform differently and feel different against the racquet.

Because they are harder to control, two-piece balls are not popular with professional golfers. Less-skillful players prefer them, however, mainly because of their durability. A two-piece ball is also less prone to flying off at an odd angle when hit with an improper swing.

In the mid-1990's, golf manufacturers introduced the multilayer golf ball, a ball that combines the strengths of the wound and two-piece balls. Like the wound ball and the two-piece ball, a multilayer golf ball has a solid rubber core. This core is surrounded by a hard plastic inner cover, which is encased in an outer cover of softer plastic.

Many golfers, both experts and duffers, find multilayer balls to their liking. The combination of the two covers gives these balls the best characteristics of wound balls and two-piece balls. The hard inner cover provides the distance and durability of a two-piece ball, while the soft outer cover provides the feel and control of a wound ball.

Since the introduction of multilayer balls, their popularity has grown with both professionals and average golfers. Some golf-ball makers predicted in 1998 that within 10 years every golfer would be using multilayer balls.

But the success of the multilayer ball doesn't mean that golf-ball designers have nothing left to work on—there are other aspects of golf balls that are being continually refined. One of the more prominent is dimple design.

The cover of the typical golf ball contains anywhere from 300 to more than 500 dimples of various sizes and shapes laid out in a geometric pattern. Although there are no specific regulations covering golf-ball dimples, most manufacturers produce balls with dimples that are 3.75 millimeters (0.15 inch) in diameter.

Dimples help increase a ball's dis-

The three types of golf balls

There are three kinds of golf balls: wound balls, two-piece balls, and multilayer balls. Each type of ball has its proponents.

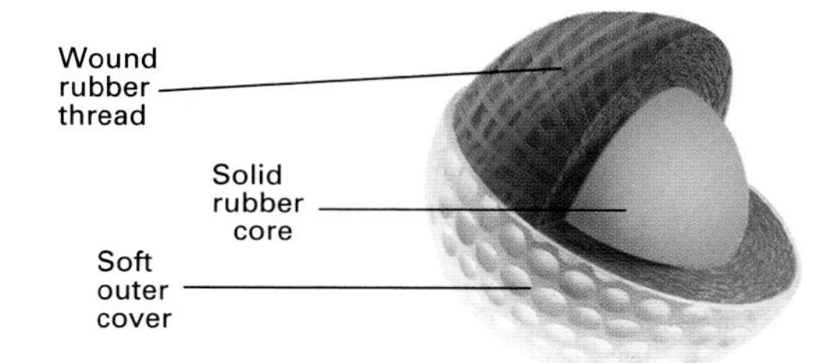

A wound ball consists of a core of solid rubber—or in some cases liquid in an enclosed sac—surrounded by a maze of thin rubber thread and a soft outer cover made from solidified tree gums. Many golfers prefer the control of a wound ball, but the cover can easily cut or tear.

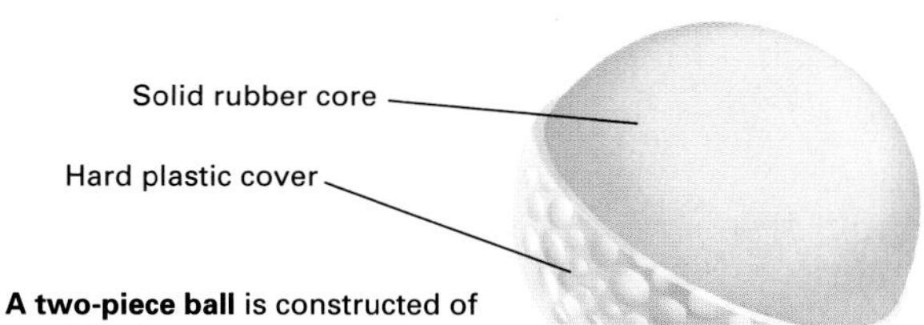

A two-piece ball is constructed of a solid rubber core surrounded by a hard plastic cover. Although the cover makes the ball more durable, some golfers find the two-piece ball more difficult to control than the wound ball.

A multilayer ball contains a solid rubber core encased in a hard plastic cover, which is in turn surrounded by a soft plastic cover. Many players think this ball offers the best combination of spin control, distance, and durability.

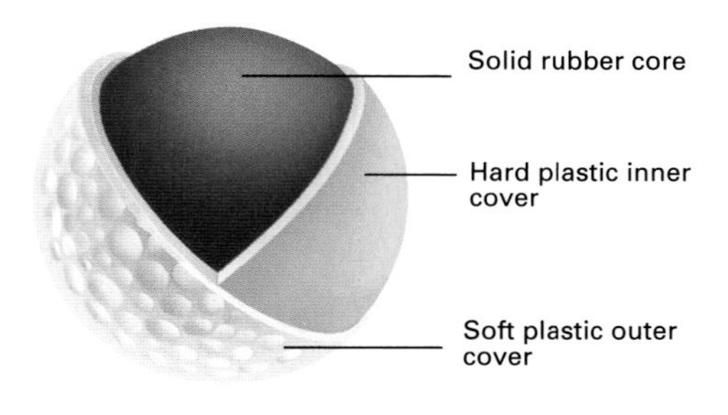

Effects of Dimpling

Dimples are an essential element of a golf ball. A smooth ball, *top,* is slowed in flight when air passing over the ball breaks up into *eddies* (circular flows) near the midpoint of the ball. With a dimpled ball, *bottom,* air clings to the indentations and breaks up closer to the back of the ball, allowing the ball to fly farther. Dimples also cause the flow of air to give the ball a backspin (curved red arrows), rotating air to the bottom of the ball. This motion produces a lifting force (top red arrow) that causes the ball to rise.

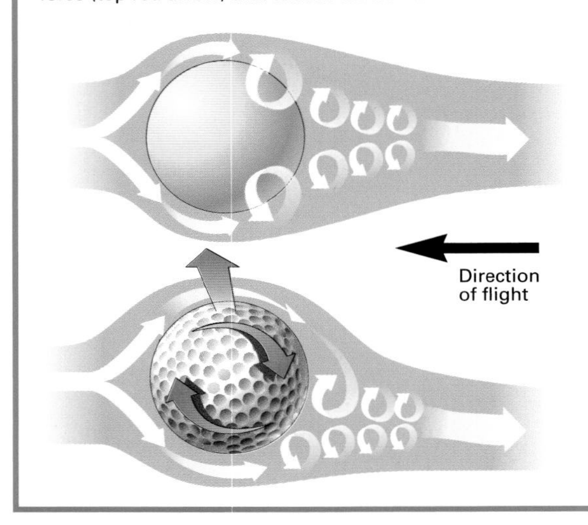

tance and accuracy. When an undimpled ball travels at high speed, air rushes over the front of the ball in a smooth, steady flow. Once passing the front of the ball, however, the air flow begins to whirl in small *eddies* (circular flows) over the top and bottom of the ball, creating back pressure that slows the flight of the ball. As a result, even a skillful golfer can hit an undimpled golf ball no more than about 120 meters (130 yards).

A properly dimpled ball, however, can travel about 250 meters (275 yards). The dimples cause the air flow to cling to the ball, keeping the air from breaking up into eddies until it curves over the back side of the ball, rather than at the midpoint. This creates a smaller area of back pressure, allowing the ball to travel faster and farther.

Dimples also cause a ball to acquire a back-spin as it leaves the club face, something that does not occur with a smooth-surfaced ball. This spinning action rotates the area of back pressure toward the bottom of the ball and causes more air to flow underneath the ball than over it. The difference in air flow creates an aerodynamic lifting force, just as occurs with the air flow around an airplane wing. So dimples not only make a golf ball fly farther, they make it fly higher.

But the effect of dimpling is subtle. If the dimples are too big, they create too much lift, and the ball won't go very far at all. Smaller dimples don't give enough lift. In fact, changing the depth of the dimples by 0.00005 centimeter (0.00002 inch) can alter the ball's trajectory. Using this knowledge, ball makers can design balls that have more lift for people who hit too low and balls with less lift for those who hit too high.

Although golf-ball manufacturers never cease trying to improve their products, they must work with certain well-defined limitations. The United States Golf Association (USGA), a governing body that interprets the rules and standards of the game, enforces five standards for golf balls to ensure that golf remains a game of skill rather than of technology.

According to the USGA, a ball cannot fly more than 256 meters (280 yards), plus or minus 6 percent, in a standard ball-driving test. The velocity of the ball is limited to no more than 75 meters (250 feet) per second. Each ball and its dimple pattern must also have perfect symmetry.

The USGA also requires a golf ball to measure at least 4.3 centimeters (1.68 inches) in diameter. Most balls are exactly that size, but some manufacturers produce oversized golf balls. These balls vary in size, and some are as large as 4.5 centimeters (1.75 inches) in diameter. However, all golf balls, regardless of their diameter, must be the same weight: 46 grams (1.62 ounces).

Picking the right ball can be a difficult task, but most experts agree that less-skillful golfers should not use the same type of ball that professional or more experienced golfers favor. The weekend golfer, experts say, should choose a ball with less spin, since spin can shorten the distance the ball travels and make the ball harder to control. For that reason, a two-piece or multilayer ball, rather than a wound ball, may be the choice for players who just want to lower their score. [Jim Gorant]

Controlling the Causes of Underarm Odor

Anyone who has had to speak in public or go out on a blind date knows about perspiration. The anxiety generated by such experiences can turn one's underarms into something akin to a tropical rain forest. And that sweating is usually followed by a not-so-pleasant odor.

It's something most of us try to avoid. In fact, smelling good and staying dry has become an obsession in the industrialized world. But for most of the 1900's, people have not been content with simply masking their bodily odors with perfumes, as our ancestors did for centuries. Rather, we control underarm odor and wetness by using deodorants and antiperspirants.

Nowhere are people more preoccupied with underarm hygiene than in the United States, where in the 1990's people spent more than $1.6 billion a year—about half of the global total—on antiperspirants and deodorants. As public demands for pristine underarms have increased, so have manufacturers' efforts to develop more effective products to combat perspiration odor.

There are two types of sweat glands that contribute to perspiration odor: eccrine glands and apocrine glands. Most perspiration consists of water, salt, and trace chemicals and is secreted by the eccrine glands. There are about 2 million of these glands, located on all parts of the body. However, eccrine glands are found in highest concentrations under the arms and on the forehead, the palms of the hands, and the soles of the feet. The eccrine glands serve primarily to help people maintain a constant body temperature, and the evaporation of eccrine sweat creates a cooling sensation on the skin. However, this type of sweat is not a major source of odor.

It is the secretions produced by the other group of sweat glands, the apocrine glands, that are the primary raw material for body odor. The average person has about 2,000 of these glands, most of which are located in the armpits and the groin. Researchers have determined that apocrine secretions affect the female menstrual cycle, but the overall purpose of apocrine sweat glands is not known. Some scientists theorize that chemicals in apocrine secretions may once, in the far-distant past, have also influenced human social and sexual behavior.

The apocrine glands remain inactive until puberty and later decline with advancing age, which is why children and older people rarely have a problem with body odor. At puberty, the body produces hormones that stimulate the apocrine glands to become active. These glands secrete only microscopic amounts of sticky, milky sweat in response to emotional stress.

Body odor is produced when bacteria, which thrive on warm, moist skin surfaces, consume organic substances in apocrine secretions. The purpose of deodorants and antiperspirants is to keep that from happening, or at least to minimize it.

The daily use of an antiperspirant or deodorant, as part of one's regular hygienic routine, can help prevent body odor.

Why perspiration odor occurs

Body odor is produced when skin bacteria consume organic substances in perspiration secreted by the body's apocrine glands, one of two kinds of sweat glands. These glands are especially numerous in the armpits, which is why those areas are prone to developing an odor. Bacteria thrive there because they obtain warmth, abundant nutrients, and moisture—including moisture produced by the other source of sweat, eccrine glands. Deodorants combat perspiration odor by reducing the bacterial breakdown of apocrine sweat, chemically neutralizing odor, and masking with fragrance any odor that gets produced. Antiperspirants are designed to reduce perspiration by temporarily blocking both types of sweat ducts.

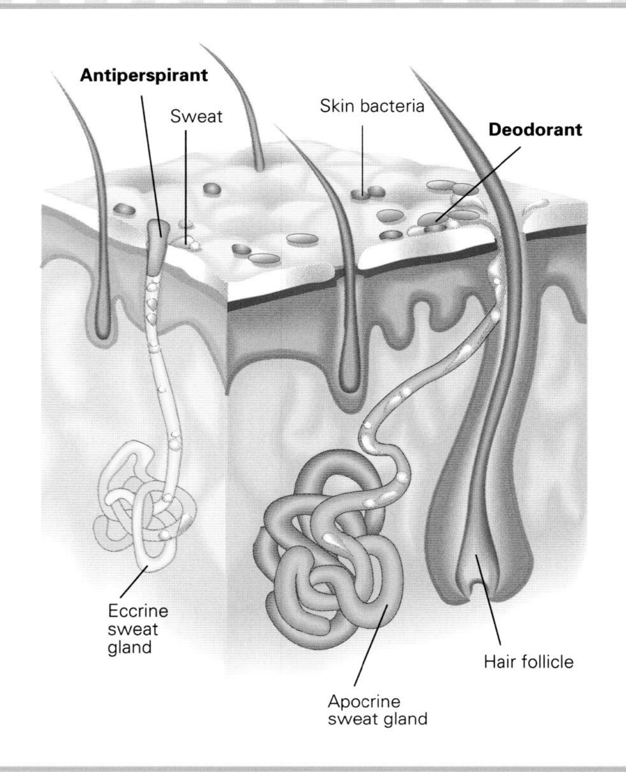

Deodorants are designed to suppress the bacterial breakdown of apocrine sweat and to mask with a pleasant fragrance any unpleasant odor that does occur. Antibacterial compounds in deodorants include various chemicals, such as triclosan and benzethonium chloride, that help reduce bacterial growth.

Antiperspirants, like deodorants, usually contain fragrances to mask unpleasant odors and compounds to slow bacterial growth. But, as the name suggests, antiperspirants are also designed to temporarily stop or reduce perspiration. For that reason, these products are regulated by the U.S. Food and Drug Administration (FDA) as over-the-counter drugs. (Deodorants, which do not alter any type of bodily function, are classified as cosmetics.)

The manufacturer of an antiperspirant must prove both that the product is safe and that it reduces perspiration by at least 20 percent. Many antiperspirants greatly exceed FDA requirements for effectiveness, some reducing perspiration by as much as 65 percent.

The active ingredients in antiperspirants create gelatinous plugs to temporarily block sweat ducts. These harmless seals prevent perspiration from reaching the skin's surface. Since the plugs form near the top of the sweat ducts, they are sloughed off with dead skin cells. Otherwise, pressure from the sweat glands will push the plugs out of the pores, usually after 24 hours.

Antiperspirants usually contain aluminum salts, sometimes in combination with zirconium salts. These salts dissolve in sweat or moisture from a bath or shower to create the plugs. Some antiperspirants contain an ingredient called aluminum chlorohydrate, or ACH. In some formulas, ACH has been able to decrease perspiration by as much as 45 percent.

The strongest antiperspirants also contain a compound called ziroconium chlorohydrate, along with an amino acid called glycine. This combination of ingredients is listed on labels as "aluminum zirconium tri- (or tetra-) chlorohydrex gly," but is referred to as ZAG by industrial chemists. ZAG has been able to reduce perspiration by as much as 65 percent.

Both ZAG and ACH are complex *polymers* (long-chain molecules) that can assume a wide range of structures, some of which are far more effective against perspiration than others. Manufacturers of antiperspirants have devoted significant effort to identifying and producing the most active polymers. Chemists refer to these more-effective configurations as "superactive" polymers.

Superactive polymers have been purified and separated from less-active forms in a process called spray-drying, which removes water from the molecules. In the presence of water, superactive polymers convert into less-active forms. Therefore, these ingredients cannot be used in a water-based product such as certain roll-ons and creams.

Although active ingredients largely determine an underarm product's effectiveness, whether the product is a solid, a liquid, or spray also can affect how well it works. For example, spray antiperspirants are the least effective form of these products, since they cannot, by law, contain ZAG.

Because of concerns on the part of the medical community that inhaling ZAG might be harmful, the FDA prohibits manufacturers from using this ingredient in aerosol antiperspirants. Therefore, antiperspirant sprays are made only with ACH, and so they are less likely to keep a person dry than a solid or a roll-on antiperspirant.

Superactive products also work most effectively in "dry" roll-ons—ones formulated without water. Stick antiperspirants tend to be slightly less effective than dry roll-ons, because the waxes in stick products impede the ability of the ingredients to block the sweat ducts. Researchers have found that both white and clear stick antiperspirants appear to be equally effective at reducing perspiration.

Not all clear underarm products are sticks, however; many are water-based gels. These products, in the case of antiperspirants, do not contain superactive polymers and thus are less effective than either sticks or dry roll-ons.

In addition, a person's body chemistry may influence the effectiveness of certain deodorants. For example, if a person's skin is very acidic, it can make the fragrance in a deodorant less effective in masking odor.

Patterns of antiperspirant and deodorant use have changed since the mid-1900's. In the United States in 1998, most consumers favored solid stick deodorants and antiperspirants over other forms of these products. The solid form was followed in popularity by gels, roll-ons, and sprays.

That breakdown was just the opposite of what it had been 20 years earlier. Aerosol sprays had dominated the underarm market from their introduction in the early-1960's until the mid-1970's, when concerns about the safety of inhaling antiperspirants arose.

Also during the mid-1970's, federal agencies concluded that chlorofluorocarbon propellants (CFC's) were depleting Earth's ozone layer and banned CFC's from use in aerosols. Manufacturers solved that problem by introducing sprays with ozone-safe propellants.

Despite the changes made to aerosols, however, they never regained their former popularity. Part of the reason for that decline, according to experts, is that consumers discovered that using a solid or a roll-on product was easier than using an aerosol.

Although there was some resistance to aerosol sprays during the 1960's, consumers eventually turned to the product because they were easier to apply than the stiff solids and sticky roll-ons of the era. But by 1998, consumers in the United States had accepted the improved versions of solids and roll-ons as far more pleasant to use. In other parts of the world, however, such as Australia and some European nations, deodorants and antiperspirants sold in aerosol forms were more popular.

A major improvement in both the feel and effectiveness of roll-on and stick antiperspirants came with the introduction of products with *anhydrous* (without water) formulas based on slick silicone molecules. As a result of this reformulation, most roll-ons and sticks now glide on easily and dry quickly, rather than leaving a person's underarm cold and sticky for several minutes following an application. In addition,

Combating perspiration odor

Perspiration level	Best option
Low	Daily use of a deodorant that suppresses the growth of bacteria or application of an antiperspirant every other day.
Average	Daily use of an antiperspirant after bathing or showering; application to damp skin allows aluminum compounds in the antiperspirant to form temporary plugs in sweat pores.
High	Use of an antiperspirant at least twice daily. Also, use of a product that does not contain water but does contain both aluminum and zirconium as part of the ingredients.

these new formulas allowed manufacturers to use the superactive forms of ACH and ZAG.

Despite the value of antiperspirants and deodorants as part of a daily hygiene routine, they do have drawbacks for some people. Dermatologists note that these products occasionally cause skin irritations. The most common irritant is the fragrance included in the ingredients. Fortunately, for most people, particular fragrances are easy to avoid. Those who develop rashes from antiperspirants or deodorants can usually find relief by simply switching brands or by using an unscented formula.

People can also experience allergic reactions, such as skin rashes and inflammation, to these products. Doctors usually advise patients allergic to the aluminum salts found in some brands of antiperspirants to use only deodorants.

Preservatives and other inactive ingredients in both deodorants and antiperspirants also cause allergic reactions in some people. Such problems—for those who know the source—can easily be avoided by checking the label and choosing a product with different ingredients. Another problem is that mist from aerosol sprays can sometimes irritate the eyes or, if inhaled, the respiratory tract.

There are some consumers who believe that temporarily stopping the natural flow of perspiration is harmful. However, medical experts contend that there is no danger, since people who use an antiperspirant continue to perspire through other skin areas. So long as a person perspires through other areas of the body, the decrease of perspiration in a relatively small area is not a major health factor.

Moreover, researchers maintain that concerns raised during the 1980's over the safety of aluminum compounds were unfounded. Those concerns centered on whether aluminum compounds, including the aluminum salts found in antiperspirants, might contribute to Alzheimer's disease, a brain disease resulting in the gradual destruction of brain cells.

In the 1990's, however, the supposed link between Alzheimer's disease and aluminum appeared very weak. Many scientists said that because antiperspirants remain in the upper layers of skin and do not usually enter the bloodstream, they do not represent a significant source of aluminum in the body.

Thus, deodorants and antiperspirants are considered by experts to be both effective and safe. Despite the rows of odor-reducing products available to consumers, however, dermatologists say that daily showering or bathing is still, for the majority of people, the most effective way to prevent unpleasant body odor. While antiperspirants and deodorants may work quite well, the products are still no substitute for simple soap and water. [Alison Mack]

Screening Tests for Genetic Disorders

If a fortune teller claimed to be able to predict your future, you'd probably be skeptical. But in the 1990's, scientists have made remarkable advances in helping people predict their medical futures with tests that reveal flawed genes that may cause disease.

Genetic fortunetelling is an inexact science, however, and gene tests often raise the fear of medical time bombs that may or may not explode. Nonetheless, discoveries of scores of disease-related genes and an ever-lengthening list of predictive tests for genetic disorders underscored the importance of learning about the potential impact of such tests on human lives.

During the 1990's scientists identified a number of disease-related genes. Along with these discoveries, researchers were able to devise tests, in many cases, to detect the presence of these genes. By 1998, at least 300 commercial and research laboratories in the United States were conducting genetic tests for more than 500 disorders. That number was expected to soar as work continues on the Human Genome Project, an international effort to map all human genes by the year 2005.

Most predictive tests were being offered only at university research laboratories or biotechnology companies, often as a part of scientific studies. However, tests for a number of genes responsible for inherited disorders were widely available. Those included tests for an inherited form of amyotrophic lateral sclerosis (Lou Gehrig's disease), cystic fibrosis, fragile X syndrome, Gaucher's disease, Huntington's disease, two types of muscular dystrophy, and polycystic kidney disease. Predictive gene tests were also available for many genes linked with various types of cancer that run in families, including breast, colorectal, ovarian, and thyroid cancers, melanoma, and retinoblastoma.

The first step in any sort of genetic testing is obtaining cells from a blood sample or other tissue, such as skin or hair cells. In *prenatal* (before birth) testing, a physician inserts a needle into the womb to extract a sample of the cell-rich fluid surrounding the fetus.

Housed within the nucleus of each cell is the material used for genetic tests: the person's genes. Genes are composed of a large, complex molecule called deoxyribonucleic acid (DNA), which contains encoded information. That information serves as an operating manual for the functioning of the body and for the determination of hereditary traits—such as eye and hair color—that are passed from generation to generation. A person's estimated 50,000 to 100,000 genes are housed in threadlike structures called chromosomes, which are located in the nucleus of nearly every cell in the body.

Each DNA molecule is made of two interconnected strands and has a structure resembling a twisted ladder. The "rungs" of the ladder are millions of pairs of chemical substances called bases. The four bases—adenine, cytosine, guanine, and thymine—in one strand pair up with the bases in the other strand. The sequence of bases in a gene tells a cell how to make a product, usually a protein. The genes encode tens of thousands of different proteins. Examples of these complex and versatile molecules are enzymes, disease-

Genetic counseling provides couples with advice on the possible risks of pregnancy—in particular, whether they are carrying certain defective genes that could cause health problems in their children. Such tests generally cost a few hundred to a few thousand dollars.

Selected conditions for which genetic tests are available*

Disease	What genetic testing may reveal	Possible action
Breast cancer	A woman's risk for this form of cancer, especially hereditary breast cancer.	Frequent mammograms and breast self-examinations to check for lumps.
Colorectal cancer	A person's risk for some forms of the cancer.	Colonoscopy once a year to check for growths, especially after age 30.
Cystic fibrosis	Whether a prospective parent is carrying a copy of the flawed gene.	Avoiding pregnancy if both members of a couple are found to be carriers of the defective gene and may thus produce a child with cystic fibrosis.
Fragile X syndrome	The presence of the defective gene that can result in this form of mental retardation.	Avoiding pregnancy if either member of a couple is found to be carrying the defect.
Gaucher's disease	Whether a prospective parent is carrying a copy of the flawed gene associated with this severe enzyme deficiency.	Enzyme replacement therapy for children who have inherited two copies of the gene and thus have developed the disease.
Huntington's disease	Whether a person has inherited the defective gene and will develop the disorder, which causes mental and physical deterioration.	No known cure, though drug therapy can lessen some symptoms.
Phenylketonuria (PKU)	Whether a child has inherited two defective copies of a gene, causing loss of a crucial enzyme, which—if left untreated—will lead to mental retardation.	A carefully restricted diet followed by an infant from birth to head off the disease.
Polycystic kidney disease	Whether a person has a copy of the flawed gene, which will eventually lead to the development of cysts in both kidneys.	No known cure, but condition can be treated with dialysis or a kidney transplant.
Sickle-cell anemia	Whether prospective parents are carriers of the gene causing this serious blood disorder, which afflicts primarily African Americans.	Avoiding pregnancy if both members of a couple carry the flawed gene.
Tay-Sachs disease	The presence of the gene responsible for the disease, a fatal hereditary brain disorder affecting the nervous system, which strikes chiefly Jews of eastern European descent.	Avoiding pregnancy if both members of a couple are found to be carriers of the gene.

* This is a partial list of conditions for which genetic testing is available. A physician referral was required for most genetic tests in 1998 because many of the tests were considered experimental and were not routinely available.

Critically reviewed by Professor David Haymer, Department of Genetics and Molecular Biology, University of Hawaii.

fighting antibodies, some hormones, and other substances that are essential to the body's activities.

Sometimes, however, a "genetic recipe" becomes garbled, usually due to an error during cell division or damage caused by an environmental factor such as radiation or chemicals. Some gene alterations, or mutations, involve a single "misspelling" that results in one base pair being altered. Other mutations involve losing or gaining a piece of DNA, ranging from a single base to an entire chromosome. Genetic instructions can also be disrupted when pieces of chromosomes are rearranged during cell division.

Some mutations have no discernible effects, while others result in altered or incomplete proteins that do not operate as well as their normal counterparts or do not function at all. For example, in sickle-cell anemia, a serious hereditary blood disease that afflicts primarily African Americans, a single base-pair change results in the production of abnormal molecules of hemoglobin, the oxygen-carrying protein in red blood cells. The altered hemoglobin causes red blood cells to become rigid, sickle shaped, and less capable of carrying oxygen. In contrast, a cancer of the eye called retinoblastoma is caused by a deletion of bases in a crucial gene, resulting in the absence of a protein that normally blocks tumor growth.

But inheriting a disease-related mutation doesn't always lead to illness. One reason is that genes and chromosomes come in pairs, with one copy inherited from each parent. For diseases such as cystic fibrosis, a disorder characterized by thick mucus that clogs the lungs, an individual must have two copies of a defective gene to develop the condition. Such genes are called *recessive.* People who have just one copy of a recessive disease-related gene are called carriers, because they are usually unaffected but can pass the flawed gene to their children. The children of carrier couples have a 1-in-4 risk of inheriting two copies of the defective gene—one from each parent—and thus developing the disease.

With some hereditary diseases, however, only one flawed copy of a gene is needed for the disorder to occur. In such cases, the mutated form of the gene is called *dominant,* because its effects override those of the other, normal copy of the gene. For example, if a person inherits just one flawed copy of the gene responsible for Huntington's disease, a disorder that causes progressive mental and physical deterioration, he or she will develop the illness.

Some genetic abnormalities, such as one form of Down syndrome and a type of leukemia, are caused by two chromosomes swapping pieces. This can be diagnosed by examining cells under a microscope. Others, like phenylketonuria (PKU), a disorder that causes mental retardation unless treated in early infancy, can be identified by the presence or absence of a key enzyme that points to a genetic flaw.

Diagnostic tests that require the direct detection of more subtle genetic abnormalities use *gene probes,* small pieces of DNA tagged with a radioactive or fluorescent compound. Each probe is designed to match the sequence of bases in a particular disease-related gene. If the patient's DNA contains the defective gene, the probe binds to the mutated portion of the DNA strand, and the radioactive tag signals that the mutation is present.

Some of the tests a physician may call for are designed to detect the presence of a genetic marker in order to determine a person's chances of developing a particular disease. A genetic marker is typically a telltale DNA sequence known to lie near a still-unidentified mutation. The marker, which does not itself cause disease, can be followed from generation to generation. Anyone found to have the marker in his or her genes is considered at high risk for developing the disease or passing on the gene.

Sometimes, a mutated gene is only one of several factors that may play a role in causing a disease. Scientists believe that in a number of disorders, such as heart disease, diabetes, and most cancers, illness results from interactions between one or more defective genes and various environmental factors. Such factors may include substances in food, exposure to intense sunlight, or cigarette smoke.

The most common use of genetic testing in 1998 was to screen newborns for certain disorders. The effects of several fairly common diseases caused by faulty genes can sometimes be prevented if the conditions are discovered at birth and treatment is begun immediately. For example, all 50 states require a test for PKU.

The screening test for PKU determines whether a child is lacking a critical enzyme because of a particular genetic defect. The absence of this enzyme ordinarily leads very quickly to mental impairment, but a carefully restricted diet, followed from birth, can head off the disease.

A number of prenatal tests were also available in 1998 to diagnose genetic disorders in an unborn baby. These included biochemical tests, which, like the test for PKU, detect the presence or absence of a substance that signals that a gene isn't working properly; DNA-based tests, which look for the genetic defect itself; and chromosome studies, which diagnose such conditions as Down syndrome, a disorder caused by an extra chromosome or a piece of it.

Another form of genetic testing is called carrier identification, which helps couples learn whether they carry certain disease-related genes they could pass on to their children. Although anyone can be a carrier, people who come from families with a history of recessive genetic disorders or from ethnic groups in which certain recessive disease-related genes are relatively common are prime candidates for carrier testing. Cystic fibrosis, the most common hereditary disease among people of northern European background, and Tay-Sachs disease, a fatal hereditary brain disorder affecting the nervous system that affects primarily Jewish children of eastern European ancestry, are two conditions for which carrier testing is used.

A procedure known as predictive gene testing seeks to determine whether a person may develop a particular disease in the future. For example, people who inherit the mutation that causes a condition called familial adenomatous polyposis develop hundreds of potentially cancerous growths called polyps in the colon and rectum. Individuals who learn from genetic testing that they have inherited the defective gene can undergo periodic examinations to detect and remove such polyps before they become cancerous.

The degree to which a genetic test can serve as a medical crystal ball depends on a number of factors, such as whether the disease in question is caused by a single gene or the interaction of several genes and whether environmental factors play a part. Furthermore, in some diseases, such as breast cancer and certain forms of Alzheimer's disease, mutations in several different genes can each produce a different form of the illness. As a result, the absence of a particular mutation does not eliminate the possibility of developing the disease.

On the other hand, not all disease-related mutations invariably cause disease. The presence of some mutations signals an increased risk for a disease, but does not guarantee that the disease will develop. Scientists estimate, for instance, that women who have inherited a defective gene known as BRCA1 have roughly an 80 percent risk of developing breast cancer in their lifetime—which means they also have a 20 percent chance of not developing the cancer.

Another complicating factor in genetic testing is that different mutations in the same gene can cause a wide range of effects in different individuals, ranging from no effects at all to severe disease. In cystic fibrosis, for example, some of the estimated 300 mutations in the gene associated with the disease can cause severe illness, while others produce only mild symptoms or none at all. In such cases, it's important to learn which mutation an individual carries.

Although most researchers agree that genetic screening offers patients many benefits, the information provided by such tests raises important ethical and social issues. One vital consideration is the right of individuals to keep genetic information private. Sensitive to this concern, many physicians and scientists have been seeking ways to protect the confidentiality of test results.

Many people's concerns about the confidentiality of genetic testing focus on the possibility of discrimination by insurance companies and employers. For example, insurers might deny coverage for a particular illness to people who have a gene linked to that disease. And employers might refuse to hire or promote healthy individuals with a known genetic defect that may or may not cause health problems in the future.

Such potential problems raise another important concern: the issue of informed consent and the need for high-quality genetic and psychological counseling before and after testing. Experts say that anyone considering genetic testing should first learn about both the potential benefits and possible risks of available tests, including their limitations, uncertainties, psychological impact, implications for childbearing decisions, and effect on other family members.

Genetic testing in 1998 was on its way to becoming a common practice and a valuable tool in health care. But such tests will not reach their full potential until medical science cannot only identify genetic defects but also prevent or cure the diseases they cause.

[Joan Stephenson]

WORLD BOOK Supplement

Seven new or revised articles reprinted from the 1998 edition of *The World Book Encyclopedia*

Climate

Climate is the weather of a place averaged over a length of time. The earth's climate varies from place to place, creating a variety of environments. Thus, in various parts of the earth, we find deserts; tropical rain forests; *tundras* (frozen, treeless plains); *conifer forests,* which consist of cone-bearing trees and bushes; prairies; and coverings of glacial ice.

Climate also changes with time. For example, a thousand years ago, northern latitudes were milder than they are today. The warmer climate enabled Vikings from Iceland to settle on the southern coast of Greenland. But the colder climate that developed over the following centuries wiped out the settlements.

One major environmental concern is that human activity may be changing the global climate. The burning of *fossil fuels*—coal, oil, and natural gas—to power motor vehicles, heat buildings, generate electric energy, and perform various industrial tasks is increasing the amount of carbon dioxide (CO_2) gas released into the atmosphere. Fossil fuels contain carbon, and burning them produces CO_2. This gas slows the escape of heat released by the earth into space. Thus, an increase in atmospheric CO_2 may cause *global warming*—a rise in the temperature of the air next to the earth's surface.

Global warming could change rainfall patterns, leading to shifts in plant and animal populations. It could also melt enough polar ice to raise the sea level, and it could increase the frequency and severity of tropical storms.

Why climates vary

Climates vary from place to place because of five main factors: (1) *latitude* (distance from the equator), (2) *altitude* (height above sea level), (3) *topography* (surface features), (4) distance from oceans and large lakes, and (5) the circulation of the atmosphere.

The role of latitude. The sun continually sends electromagnetic radiation into space. Most of the radiation is visible light, and it also includes *infrared* (heat) rays and ultraviolet rays. About 30 percent of the radiation that reaches the earth's atmosphere is reflected back into space, mostly by clouds. The remaining 70 percent is absorbed by the atmosphere and the earth's surface, heating them.

The intensity of the solar radiation reaching the atmosphere decreases with increasing latitude. The intensity depends on how high in the sky the sun climbs. The closer a place is to the equator, the higher the climb.

At latitudes between $23\frac{1}{2}°$ north and $23\frac{1}{2}°$ south, the sun is directly overhead at noon twice a year. In these cases, the sun's rays shine directly down toward the surface. The radiation that reaches the atmosphere is therefore at its most intense.

In all other cases, the rays arrive at an angle to the surface and are therefore less intense. The closer a place is to the poles, the smaller the angle and therefore the less intense the radiation. Due to decreases in the intensity of radiation, average temperatures decline from the equator to the poles. Seasonal changes in solar radiation and the number of hours of sunlight also vary with latitude.

Joseph M. Moran, the contributor of this article, is Professor of Earth Sciences at the University of Wisconsin at Green Bay.

© Martin Wendler, Okapia from Photo Researchers

A tropical wet climate supports a rain forest in South America. Rainfall is heavy, and the weather is always hot and muggy. As a result, trees and other plants grow throughout the year.

In tropical latitudes (those near the equator), there is little difference in the amount of solar heating between summer and winter. Average monthly temperatures therefore do not change much during the year.

In middle latitudes, from the Tropic of Cancer to the Arctic Circle and from the Tropic of Capricorn to the Antarctic Circle, solar heating is considerably greater in summer than in winter. In these latitudes, summers are therefore warmer than winters.

In high latitudes, north of the Arctic Circle and south of the Antarctic Circle, the sun never rises during large portions of the year. Therefore, the contrast in solar heating between summer and winter is extreme. Summers are cool to mild, and winters are bitterly cold.

The role of altitude. The higher a place is, the colder it is. Air temperature drops an average of about 3.5 Fahrenheit degrees per 1,000 feet of altitude (6.5 Celsius degrees per 1,000 meters). The temperature of the air determines how much precipitation falls as snow, rather than rain. Even in the tropics, it is not unusual for mountaintops to be snow-covered.

The role of topography. The surface features of the earth influence the development of clouds and precipitation. As humid air sweeps up the slopes of a mountain range, the air cools, and so clouds form. Eventually, rain or snow falls from the clouds. Some of the rainiest places on earth are on *windward* slopes, those facing the wind.

As winds blow down the opposite slopes, known as the *leeward* slopes, the air warms, and clouds thin out or vanish. Leeward slopes of mountain ranges are therefore dry. In addition, a *rain shadow* (dry area) may stretch hundreds of kilometers downwind of a mountain range.

Oceans and large lakes make the air temperature less extreme in places downwind of them. An ocean or lake surface warms up and cools down more slowly than a land surface. Thus, between summer and winter, the temperature of the water varies less than the temper-

© Uniphoto

A desert climate supports only sparse vegetation and gives rise to tremendous sand dunes in the southwest African nation of Namibia. The air is hot during the day but cools rapidly at night.

© Maxwell MacKenzie, Uniphoto

A humid continental climate supports a rich variety of plant life during a summer in Vermont. Precipitation falls fairly evenly all year. Snow remains on the ground much of the winter.

ature of the land. The temperature of the water strongly influences the temperature of the air above it. Therefore, air temperatures over the ocean or a large lake also vary less than air temperatures over land. As a result, places that are immediately downwind of the water have milder winters and cooler summers than places at the same latitude but well inland.

San Francisco and St. Louis, for example, are at about the same latitude and therefore receive about the same amount of solar radiation during the year. But San Francisco is immediately downwind of the Pacific Ocean, and St. Louis is well inland. Consequently, San Francisco has milder winters and cooler summers.

Atmospheric circulation influences climate by producing winds that distribute heat and moisture. Six belts of wind encircle the earth: (1) trade winds that blow between 30° north latitude and the equator, (2) trade winds that blow between the equator and 30° south latitude, (3) *westerlies* (winds from the west) that blow between 30° and 60° north of the equator, (4) westerlies blowing between 30° and 60° south of the equator, (5) polar winds north of 60° north latitude, and (6) polar winds south of 60° south latitude.

Trade winds north of the equator blow from the northeast. South of the equator, they blow from the southeast. The trade winds of the two hemispheres meet near the equator, causing air to rise. As the rising air cools, clouds and rain develop. The resulting band of cloudy and rainy weather near the equator is called the *doldrums*.

Westerlies blow from the southwest in the Northern Hemisphere and from the northwest in the Southern Hemisphere. Westerlies steer storms from west to east across middle latitudes.

Westerlies and trade winds blow away from the 30° latitude belt. Over broad regions centered at 30° latitude, surface winds are light or calm. Air slowly descends to replace the air that blows away. Descending air warms and is dry. The tropical deserts, such as the Sahara of Africa and the Sonoran of Mexico, occur under these regions of descending air.

Polar winds blow from the northeast in the Arctic and from the southeast in the Antarctic. In the Northern Hemisphere, the boundary between the cold polar easterly winds and the mild westerly winds is known as the *polar front*. A *front* is a narrow zone of transition, usually between a mass of cold air and a mass of warm air. Where the air masses overlap, storms can develop and move along the polar front, bringing cloudy weather, rain, or snow.

As the seasons change, the global wind belts shift north and south. In the spring, they move toward the poles. In the fall, they shift toward the equator. These shifts help explain why some areas have distinct rainy seasons and dry seasons. Parts of Central America, North Africa, India, and Southeast Asia have wet summers and dry winters. Southern California and the Mediterranean coast have dry summers and wet winters.

Kinds of climates

The earth's surface is a patchwork of climate zones. *Climatologists* (scientists who study the climate) have organized similar types of climates into groups. This article uses a modified version of a classification system introduced in 1918 by Wladimir Köppen, a German climatologist. Köppen based his system on a region's vegetation, average monthly and annual temperature, and average monthly and annual precipitation.

The modified version specifies 12 climate groups: (1) tropical wet, (2) tropical wet and dry, (3) semiarid, (4) desert, (5) subtropical dry summer, (6) humid subtropical, (7) humid oceanic, (8) humid continental, (9) subarctic, (10) tundra, (11) icecap, and (12) highland.

Tropical wet climates are hot and muggy the year around. They support dense tropical rain forests. Rainfall is heavy and occurs in frequent showers and thunderstorms throughout the year. Average annual rainfall varies from about 70 to 100 inches (175 to 250 centimeters).

Temperatures are high, and they change little during the year. The coolest month has an average temperature no lower than 64 °F (18 °C). The temperature difference

What the world's climate is like

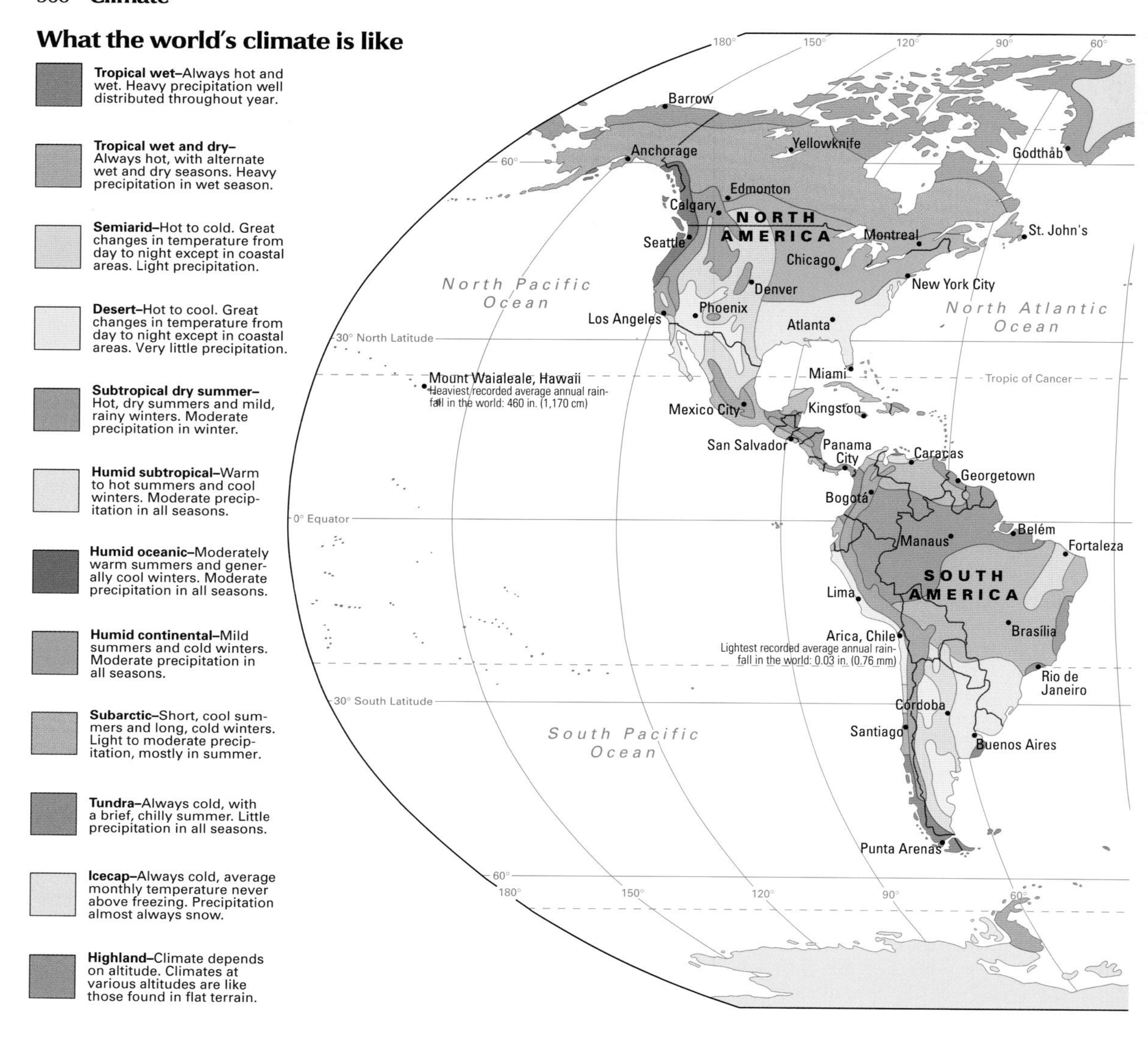

between day and night is greater than the temperature difference between summer and winter. Frost and freezing temperatures do not occur. Plants grow all year.

Tropical wet and dry climates occur in areas next to regions that have tropical wet climates. Temperatures in tropical wet and dry climates are similar to those in tropical wet climates, where they remain high throughout the year.

The main difference between the two climates lies in their rainfall. In tropical wet and dry climates, winters are dry, and summers are wet. Generally, the length of the rainy season and the average rainfall decrease with increasing latitude. Not enough rain falls in tropical wet and dry climates to support rain forests. Instead, they support *savannas*—grasslands with scattered trees.

Semiarid and desert climates occur in regions with little precipitation. Desert climates are drier than semiarid climates. Semiarid climates, also called *steppe climates,* usually border desert climates. In both climate groups, the temperature change between day and night is considerable. One reason for the wide swings in temperature is that the the skies are clear and the air is dry. Clouds would reflect much of the sun's intense radiation during the day, slowing the rate of heating of the air near the surface. At night, clouds and water vapor would absorb much of the earth's radiation—most of which consists of infrared rays—slowing the rate of cooling.

Semiarid and desert climates occur over a greater land area than any other climate grouping. They occur in both tropical and middle latitudes. They cover broad east-west bands near 30° north and south latitude.

Middle latitude semiarid and desert climates are in the rain shadows of mountain ranges. Winds that descend the leeward slopes of these ranges are warm and dry. Middle latitude semiarid areas and deserts differ from their tropical counterparts mainly in their seasonal

WORLD BOOK map

temperature changes. Winters are much colder in middle latitude semiarid areas and deserts.

Subtropical dry summer climates feature warm to hot, dry summers and mild, rainy winters. These climates, sometimes called *Mediterranean climates,* occur on the west side of continents roughly between 30° and 45° latitude. The closer to the coast the area is, the more moderate the temperatures and the less the contrast between summer and winter temperatures.

Humid subtropical climates are characterized by warm to hot summers and cool winters. Rainfall is distributed fairly evenly throughout the year. Winter rainfall—and sometimes snowfall—is associated with large storm systems that the westerlies steer from west to east. Most summer rainfall occurs during thunderstorms and an occasional tropical storm or hurricane. Humid subtropical climates lie on the southeast side of continents, roughly between 25° and 40° latitude.

Humid oceanic climates are found only on the western sides of continents where prevailing winds blow from sea to land. The moderating influence of the ocean reduces the seasonal temperature contrast so that winters are cool to mild and summers are warm. Moderate precipitation occurs throughout the year. Low clouds, fog, and drizzle are common. Thunderstorms, cold waves, heat waves, and droughts are rare.

Humid continental climates feature mild to warm summers and cold winters. The temperature difference between the warmest and coldest months of the year increases inland. The difference is as great as 45 to 63 Fahrenheit degrees (25 to 35 Celsius degrees). Precipitation is distributed fairly evenly throughout the year, though many locations well inland have more precipitation in the summer.

Snow is a major element in humid continental climates. Winter temperatures are so low that snowfall can be substantial and snow cover persistent. Snow cover has a chilling effect on climate. Snow strongly reflects solar radiation back into space, lowering daytime temperatures. Snow also efficiently sends out infrared radiation, lowering nighttime temperatures.

Subarctic climates have short, cool summers and long, bitterly cold winters. Freezes can occur even in midsummer. Most precipitation falls in the summer. Snow comes early in the fall and lasts on the ground into early summer.

Tundra climates are dry, with a brief, chilly summer and a bitterly cold winter. *Continuous permafrost* (permanently frozen ground) lies under much of the treeless tundra regions.

Icecap climates are the coldest on earth. Summer temperatures rarely rise above the freezing point. Temperatures are extremely low during the long, dark winter. Precipitation is meager and is almost always in the form of snow.

Highland climates occur in mountainous regions. A highland climate zone is composed of several areas whose climates are like those found in flat terrain. Because air temperature decreases with increasing elevation in the mountains, each climate area is restricted to a certain range of altitude.

A mountain climber may encounter the same sequence of climates in several thousand meters of elevation as he or she would encounter traveling northward several thousand kilometers. For example, the climate at the base of a mountain might be humid subtropical, and the climate at the summit might be tundra.

Causes of climate change

Climate changes over time. About 18,000 years ago, a sheet of glacial ice up to 10,000 feet (3,000 meters) thick covered much of what is now Greenland, Canada, and the northern United States. A warming trend gradually melted almost all the glaciers, except in Greenland. In Canada and the United States, the last large fields of ice had disappeared by about 11,500 years ago. The warming trend ended after a mild period from 7,000 to 5,000 years ago, when the global average temperature was higher than it is today.

Within the past 1,200 years, the period between about A.D. 950 and 1250 was mild. The years from about 1400 to 1850 were cool. Since then, global average temperatures generally have risen.

Many natural processes influence a region's climate. Some of these processes, such as volcanic eruptions, are short-lived and cause short-term changes. Other processes, such as mountain building, occur over long periods and cause long-term changes in climate. Human activity also may affect climate.

Volcanic eruptions can cause short-term cooling over large portions of the planet, especially if the eruptions throw large amounts of sulfur gases high into the atmosphere. The sulfur gases combine with moisture to produce droplets of sulfuric acid and tiny sulfate particles. The sulfur droplets and particles created by major volcanic eruptions absorb some solar radiation and reflect some back to space. As a result, less solar radiation reaches the earth's surface, and so air temperatures fall.

Because the droplets and particles are so small, they can remain suspended in the atmosphere for months or years. Meanwhile, winds carry them around the globe. Scientists believe that volcanic eruptions can cause a maximum global cooling of about 2 Fahrenheit degrees (1 Celsius degree).

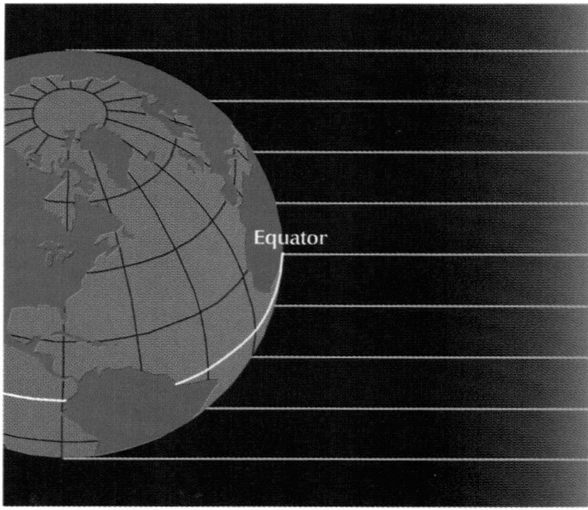

WORLD BOOK illustration

The sun heats the earth unevenly, producing cooler climates at increasing distances from the equator. The uneven heating occurs because the energy in the sun's rays spreads out over a greater surface area the farther the rays fall from the equator.

In 1991, the eruption of Mount Pinatubo in the Philippines threw large amounts of sulfur gases high into the atmosphere. This eruption likely caused a drop of 1.1 Fahrenheit degree (0.6 Celsius degree) in the global average temperature during the following few years.

Changes in ocean circulation can alter the climate. For example, changes in ocean currents that occur during El Niño can affect the climate for a year or two.

El Niño is a large-scale interaction between the tropical atmosphere and tropical oceans that happens about every two to seven years. Changes in the air pressure over the tropical Pacific Ocean cause the trade winds there to weaken or even reverse direction. This change enables the warm waters of the ocean surface to drift from the western tropical Pacific to the eastern tropical Pacific. This flow makes sea-surface temperatures lower than usual over the western tropical Pacific and higher than usual over the eastern tropical Pacific.

The sea-surface temperature changes, in turn, alter the circulation of the atmosphere in tropical and middle latitudes. These alterations cause weather extremes in various parts of the world. Heavy rains drench the normally arid coastal plain of western South America, drought is more likely in Hawaii and eastern Australia, and winters are wetter than normal along the Gulf of Mexico coast.

Activity on the sun's surface may affect the earth's climate for short periods. *Sunspots* are dark, relatively cool blotches that appear on the surface of the sun. *Faculae* are relatively bright, hot areas on the solar surface. The number of sunspots and faculae increases and decreases over a cycle of about 11 years. The average amount of energy given off by the sun is slightly higher during a sunspot maximum—when the number of sunspots and faculae is high. The average amount is slightly lower during a sunspot minimum—when the number is low.

Climatologists are not sure about the relationship between changes on the sun's surface and variations in the earth's climate. During the period from 1645 to 1715, the number of sunspots was unusually low. This episode corresponds to a portion of the Little Ice Age, a time of relatively cool conditions. Climatologists have not proved, however, that the reductions in the number of sunspots caused the cooling.

Changes in CO_2 concentration in the atmosphere may cause short-term and long-term variations in the climate. Atmospheric CO_2 slows the flow of heat from the earth to space. This gas absorbs heat that radiates from the earth's surface and radiates heat back to the surface.

Human activity is currently increasing the level of atmospheric CO_2, but this level has varied significantly throughout the history of the earth. With past variations in atmospheric CO_2, the global climate has warmed or cooled. For example, about 100 million years ago, volcanic activity on the floor of the Pacific Ocean released enough CO_2 to cause global warming of perhaps 18 Fahrenheit degrees (10 Celsius degrees).

Changes in the earth's orbit about the sun may cause climate changes over tens of thousands to hundreds of thousands of years. Milutin Milankovitch, a Serbian mathematician, proposed in the 1940's that three orbital variations change how sunlight is distributed seasonally and geographically over the planet: (1) a *precession* (wobble) of the earth's axis, (2) a variation in the tilt of the axis, and (3) a variation in the path of the orbit.

The precession of the axis varies over a period of 23,000 years. This cycle alters the times of the year when earth is closest to the sun and farthest from the sun.

The tilt of the axis changes from 22.1° to 24.5° over a period of 41,000 years. This cycle affects the contrast between winter and summer temperatures.

The path of the orbit varies over a period of 100,000 years. The orbit is always *elliptical*—that is, shaped like a flattened circle. But during the 100,000-year cycle, the

Orbital variations Three long-term variations in the earth's orbit may cause climate changes, according to a theory proposed by Serbian mathematician Milutin Milankovitch. These variations alter the amount of sunlight striking the earth at different latitudes at different times of year, perhaps triggering periods of warming and cooling.

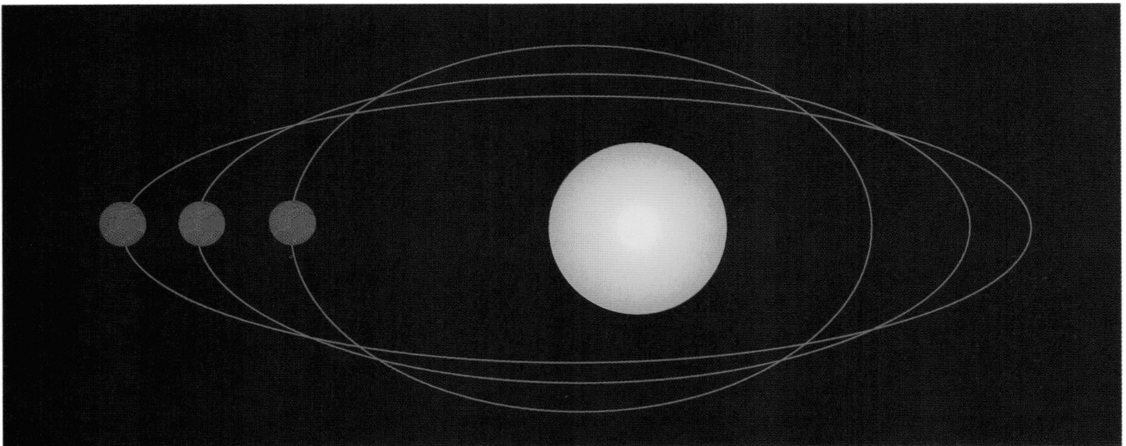

WORLD BOOK illustrations

The earth's orbit changes shape from oval to circular and back again in a cycle that lasts about 100,000 years. This change in orbit alters the distance from the earth to the sun at different times of year, varying the amount of sunlight hitting the earth. The diagram above is not drawn to scale. The orbit is much more circular than shown, and the sun is actually closer to the center of the orbit.

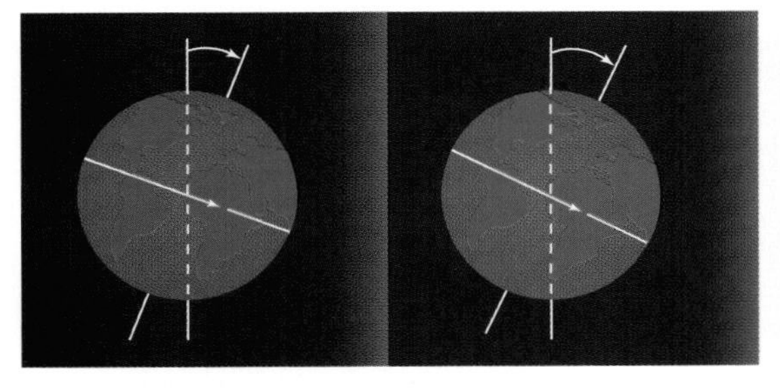

The tilt of the earth's axis changes from 21.1°, *above left,* to 24.5°, *above right,* and back every 41,000 years. When the tilt is 24.5°, the difference between the amount of sunlight received in summer and the amount received in winter is greater than at any other time. When the tilt is 21.1°, this difference is at a minimum.

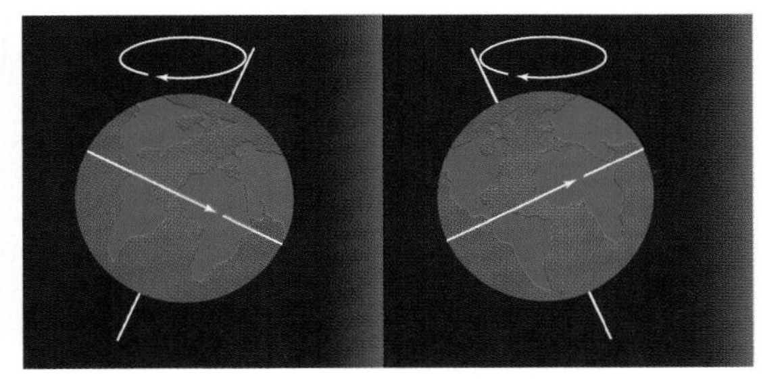

The earth's axis *precesses*—that is, wobbles—in a cycle that lasts about 23,000 years. As a result, the seasons occur at different times of year. For example, July is now summer in the Northern Hemisphere, *above left.* In 11,500 years, after a half-cycle of precession, that hemisphere will have winter in July, *above right.*

amount of flatness changes from a maximum to a minimum, then back to the maximum. This variation changes the distance between the earth and sun over the course of a year.

These three orbital cycles probably have altered global temperatures at regular intervals throughout the earth's history. For example, climatologists believe that the cycles may have governed the major fluctuations in the planet's glacial ice cover during the Pleistocene Epoch, the period from about 2 million years ago to 11,500 years ago. The three cycles probably produced temperature changes that caused the ice cover to expand and contract at regular intervals.

Continental drift is an extremely slow process that influences the climate over periods of tens of millions of years. The planet's solid outer shell, which is typically 60 miles (100 kilometers) thick, is divided into about 30 large plates that move slowly over the surface of the globe. The continents are embedded in these plates and slowly drift with them, a process known as *continental drift.* About 200 million years ago, there was just one huge continent, called Pangaea. Pangaea split into fragments that drifted apart and eventually reached their present locations as the continents we know today.

Continental drift helps explain the presence of coral reef fossils in Wisconsin and of tropical plant fossils in areas north of the Arctic Circle. When the coral and plants were alive, Wisconsin and the arctic regions were at much lower latitudes than they are today.

Mountain building, another extremely slow process, likely helped set the stage for the ice ages during the Pleistocene Epoch. About 30 million to 40 million years ago, the Himalaya and the adjacent Tibetan plateau began to rise in southern Asia. At about the same time, mountain building began in the western part of the United States. Half the total uplift of the Himalaya and the Colorado Plateau may have occurred within the past 10 million years. The rise of these massive landforms likely altered the wind belts that encircle the planet. As a result, the earth's climate became more diverse, wetter, and colder.

Human activity also affects the climate. The building of cities, the clearing of forests, and the burning of oil, coal, and natural gas can all cause climatic changes. Climatologists disagree, however, about the impact that human activity has had on climate, particularly on the recent global warming trend.

The construction of cities creates areas that are warmer and drier than the surrounding countryside. Cities are drier because they have storm sewer systems that quickly carry off rainwater and snowmelt.

Cities are warmer for several reasons. The use of storm drainage systems means that less solar radiation is used to evaporate water and more is used to heat the city surfaces and air. The brick, asphalt, and concrete surfaces readily radiate the heat they absorb and so raise urban air temperatures even more. In addition, cities themselves generate heat from a number of sources, including motor vehicles and heating and air conditioning systems.

Large urban areas also affect the climate in the areas downwind of them. Smokestacks and automobile tailpipes in cities emit water vapor and tiny particles that stimulate the formation of clouds. Heat from a city also spurs the growth of clouds. Thus, the climate downwind from many large urban areas is cloudier and wetter than the climate upwind from those same areas.

The burning of fossil fuels has contributed to recent increases in the amount of CO_2 in the atmosphere. Since the mid-1800's, the level of atmospheric CO_2 has risen about 25 percent, mainly because of an increased use of fossil fuels for transportation, space heating, and generation of electric energy. The clearing of forests also contributes to the buildup of atmospheric CO_2 by reducing the rate at which the gas is removed from the air. Trees and other green plants remove CO_2 from the air during *photosynthesis*—the process they use to produce food.

Climatologists disagree about the impact that the burning of fossil fuels and the clearing of forests have had on climate. Global temperature records indicate that a warming trend began in the late 1800's. It was interrupted by an episode of cooling from about 1940 to the late 1970's. The cooling effect of the Mount Pinatubo eruption also interrupted the trend in the early 1990's. Some scientists argue that the buildup of atmospheric CO_2 due to human activities has caused the warming trend. Others argue that recent warming is merely a natural fluctuation of the climate.

Determining past climates

Scientists study the climate record to learn about climate and its changes. The most reliable portion of the record is based on standardized measurements by weather instruments.

The reliable instrument-based record dates back only about 125 years and is too short to reveal all possible variations in climate. Climatologists have lengthened the record by studying historical documents, tree growth rings, fossil plants and animals, deposits of pollen, and cores drilled out of glacial ice and seafloor sediments.

Historical documents that contain information about climate include logs maintained by ships' captains and lighthouse keepers, diaries kept by farmers, and records of harvests. Another source is a record of the duration of ice cover in a harbor.

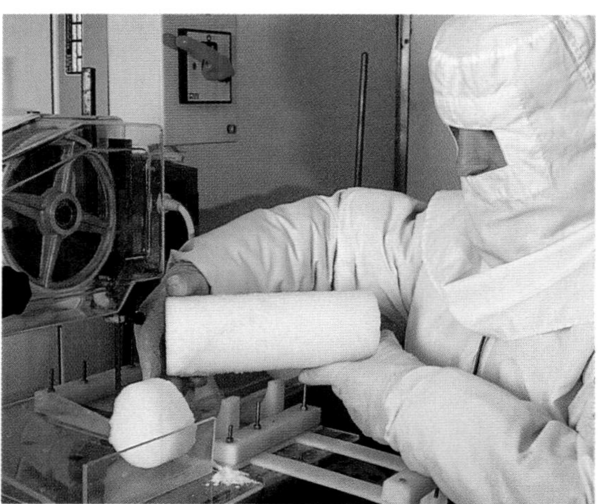

A core of ice drilled from a glacier in Greenland provides information about ancient climates. Scientists analyze the air and other substances that have been in the ice since the glacier formed.

Growth rings. A tree adds a growth ring each year. The thickness of those rings depends to some extent on seasonal weather conditions. By analyzing growth rings from living and dead trees, scientists can distinguish years of relatively favorable weather from years of stressful weather. In the southwestern part of the United States, tree growth rings provide information about the climate record dating back almost 8,000 years.

Fossils are the remains or imprints of plants and animals that lived in the past. Based on an understanding of the environmental conditions required by these organisms, scientists can reconstruct climatic conditions of the places where the fossils are found. Knowing the age of fossils, scientists can identify major shifts in climate.

Pollen consists of tiny grains that play an essential role in the reproduction of flowering and cone-bearing plants. Pollen is released by the plants and carried by the wind. Some pollen falls into lakes and settles to the bottom along with tiny bits of clay and other particles to form sediment.

Scientists can drill into a lake bottom and extract a *sediment core*—a sample of many layers of sediment. The core contains a record of changes in pollen, which reflect changes in a region's vegetation. Because climate largely determines the type of vegetation in a region, climatologists can use the pollen record to reconstruct an area's climate record.

For example, climatologists have used pollen records to determine that the average July temperature in western Europe was 3.5 Fahrenheit degrees (2 Celsius degrees) warmer 6,000 years ago. In parts of the midwestern United States, climatologists have reconstructed changes in climate from the pollen record as far back as 12,000 years ago.

Glaciers are composed of layers of ice created by the compression of winter snows. Each layer corresponds to one winter's snowfall. Through the chemical analysis of ice layers, scientists can determine winter temperatures. Tiny air bubbles trapped in the ice also provide clues about the chemical composition of the atmosphere when the snow fell. Scientists have obtained lengthy ice cores from ice sheets in Antarctica and Greenland. Cores extracted in Greenland in the early 1990's covered a period of almost 200,000 years.

Sea-floor sediments. The tiny shell and skeletal remains of organisms settle out of ocean water and accumulate with other sediment on the sea floor. A core extracted from sea-floor sediment provides a record of marine life through time. Scientists use chemical analysis of shells and skeletons removed from the sediment core to determine seawater temperatures. Analysis of deep-sea sediments revealed much of what is known about the climatic variations of the Pleistocene Epoch.

Climate models

Climatologists rely on computerized climate models to predict the earth's climate. A climatologist programs a computer with a numerical model of the climate. The model consists of a set of mathematical equations that describe how various factors influence climate. By altering one or more factors, the climatologist can use the model to predict changes in climate. One common application of climate models is to predict the impact of rising levels of atmospheric CO_2.

Joseph M. Moran

Related articles. See **Weather** and its list of *Related articles.* For information about the climate of individual states, provinces, and countries, see the *Climate* section in the state, province, and country articles, such as **India** (Climate). See also:

Acclimatization
Animal (Where animals live)
Arctic (Climate)
Biome
Desert
Food (Geographic reasons)
Global warming
Greenhouse effect
Gulf Stream
Ice age
Lake (Climate)
Latitude
Nuclear winter
Ocean (As an influence on climate)
Phenology
Plant (Where plants live)
Races, Human (Climatic adaptations)
Tropics

Outline

I. Why climates vary
A. The role of latitude
B. The role of altitude
C. The role of topography
D. Oceans and large lakes
E. Atmospheric circulation

II. Kinds of climates
A. Tropical wet climates
B. Tropical wet and dry climates
C. Semiarid and desert climates
D. Subtropical dry summer climates
E. Humid subtropical climates
F. Humid oceanic climates
G. Humid continental climates
H. Subarctic climates
I. Tundra climates
J. Icecap climates
K. Highland climates

III. Causes of climate change
A. Volcanic eruptions
B. Changes in ocean circulation
C. Activity on the sun's surface
D. Changes in CO_2 concentration
E. Changes in the earth's orbit
F. Continental drift
G. Mountain building
H. Human activity

IV. Determining past climates
A. Historical documents
B. Growth rings
C. Fossils
D. Pollen
E. Glaciers
F. Sea-floor sediments

V. Climate models

Additional resources

Bradley, Raymond S., and Jones, P. D., eds. *Climate Since A. D. 1500.* Routledge, 1992.
Linacre, Edward. *Climate Data and Resources.* Routledge, 1992.
Maunder, W. J., comp. *Dictionary of Global Climate Change.* Chapman & Hall, 1992.
Schotterer, Ulrich. *Climate: Our Future?* Rev. ed. Univ. of Minnesota Pr., 1992.

Questions

What is the difference between climate and weather?
How does latitude affect climate?
What can ice cores reveal about past climates?
What human activities contribute to a buildup of carbon dioxide in the atmosphere?
How do volcanic eruptions affect climate?
How does the building of cities affect climate?
What type of climate has the greatest temperature difference between summer and winter?
Where are highland climates found?
How does the level of carbon dioxide in the atmosphere affect climate?
What is the *El Niño* effect?

The Weather Channel® (WORLD BOOK photo by Steven Spicer)

© Gordon Garradd, SPL from Photo Researchers

Weather affects people's lives every day. Weather reports and forecasts help people decide what to wear, plan trips and other outdoor activities, and even avoid dangers, such as lightning storms, *above.* A television weather reporter explains current conditions in Texas and Oklahoma to a nationwide audience, *left.*

Weather

Weather is the state of the atmosphere at some place and time. We describe the weather in many ways. For example, we may refer to the temperature of the air, whether the sky is clear or cloudy, how hard the wind is blowing, or whether it is raining or snowing. At any given time, the weather is fair in some places, while it rains or snows in others. In some places it is warm, and in others it is cold.

Earth is not the only planet with a variety of weather conditions. Every planet in the solar system except Mercury and perhaps Pluto has enough of an atmosphere to support weather systems. In addition, Titan, a moon of the planet Saturn, has such an atmosphere. Pluto is so far away that little is known about its atmosphere. The remainder of this article discusses the weather on Earth.

Joseph M. Moran, the contributor of this article, is Professor of Earth Sciences at the University of Wisconsin at Green Bay.

The weather affects our lives every day. For example, it can have an impact on what type of clothing we wear and how we spend our free time. Weather also affects agriculture, transportation, and industry. Freezing temperatures can damage citrus crops in Florida or Spain, causing a rise in the price of oranges at the grocery store. Winter snows often create hazardous driving conditions. Thick fog may slow traffic on the roads and cause delays at airports. Our use of air conditioning during heat waves and heating during cold weather means that utility companies must supply more power at those times. Severe weather, such as tornadoes, hurricanes, and blizzards, can damage property and take lives.

Because of weather's importance, *meteorologists* (scientists who study the atmosphere and the weather) have developed ways to forecast weather conditions. Forecasts for the next 12 to 24 hours are correct more than 80 percent of the time. Long-range forecasts for the next week or month are less accurate. These forecasts indicate general trends, such as whether or not temperatures are expected to be warmer or colder than normal.

Closely related to weather is climate. Climate is the

weather of a place averaged over a length of time. Scientists determine a region's climate by examining its vegetation, average monthly and annual temperature, and average monthly and annual precipitation. The earth's surface is a patchwork of climate zones. For example, in various parts of the earth, we find deserts; tropical rain forests; prairies; forests of cone-bearing trees; frozen, treeless plains; and coverings of glacial ice. Unlike changes in the weather, which can occur in minutes, climate changes generally take many years. See **Climate**.

What causes weather

Weather takes place in the atmosphere, the layer of air that surrounds the earth. Air is a mixture of gases and tiny suspended particles.

The most plentiful gas is nitrogen, which accounts for about 78 percent of the air we breathe. The second most plentiful gas is oxygen, at 21 percent. The remaining 1 percent consists of a variety of gases. In spite of their low concentrations, some of these gases play vital roles. For example, the atmosphere contains little *water vapor*—an invisible gas produced when water evaporates. Yet, without water vapor there would be no clouds, no rain, no snow, and no plants or animals.

Most of the tiny particles floating in the atmosphere are too small to be visible. They are solid or liquid and come from a number of sources, such as the wind erosion of soil, volcanic eruptions, and the release of pollutants by smokestacks and automobile tailpipes.

Most weather occurs in the lowest portion of the atmosphere, called the *troposphere*. The troposphere extends from the earth's surface up to an altitude of about 6 to 10 miles (10 to 16 kilometers). Three key factors that determine the weather in the troposphere are air temperature, air pressure, and humidity.

Air temperature is a measure of the energy of motion of the air's gas molecules. The factors most responsible for the heating and cooling of the atmosphere are radiation arriving from the sun and radiation flowing from the earth.

The sun continually sends energy into space as electromagnetic radiation. One kind of solar radiation is visible light. The other forms of solar electromagnetic radiation are invisible to human beings. They include *infrared* (heat) rays and ultraviolet rays. About 30 percent of the solar electromagnetic radiation that reaches the atmosphere is reflected back into space, mostly by clouds. The atmosphere and the earth's surface absorb the remaining 70 percent, becoming warmer.

The warmed earth cools by radiating infrared rays. Some of this radiation travels directly into space. The atmosphere absorbs almost all the remainder as it streams off the surface of the planet. This absorption of radiation, known as the *greenhouse effect*, makes the air near the earth's surface about 59 Fahrenheit degrees (33 Celsius degrees) warmer than it would be otherwise.

The atmosphere also sheds heat energy by radiating infrared rays. Some of this infrared radiation flows down to the surface, while the remainder travels out into space.

Air temperature generally varies from day to night and from season to season because of changes in the amount of radiation heating the earth's atmosphere. For example, days usually are warmer than nights because the earth receives the heating rays of the sun only during the day. At night, infrared radiation from the planet streams off into space, and the air temperature drops.

Air temperature also changes with the seasons. Except near the equator, where temperatures remain fairly constant the year around, summers are warmer than winters. In the summer, the sun is higher in the sky, and days are longer. When the sun is higher above the horizon, the intensity of the sunlight striking the earth's surface increases. More hours of sunlight in summer also mean more solar heating.

Altitude also affects air temperature. Within the troposphere, the air temperature generally drops 3.5 Fahrenheit degrees per 1,000 feet of elevation (6.5 Celsius degrees per 1,000 meters of elevation). Thus, it is usually colder on top of a mountain than in the surrounding lowlands.

Air pressure is the weight per unit of area of a column of air that reaches to the top of the atmosphere. Air pressure always decreases with increasing altitude because as you move higher there is less and less air above you. Air pressure is, on average, highest at sea level and drops to about half its sea-level value at an average altitude of about 18,000 feet (5,500 meters).

Air pressure also changes from place to place across the earth's surface. Part of this change is due to differences in land elevation. Most of the remainder is caused by changes in air temperature. Cold air is relatively dense—that is, it has more air molecules per unit volume—and so it exerts relatively high pressure. Warm air is less dense and exerts relatively low pressure.

Regions where air pressure is relatively high usually experience fair weather, while regions where air pressure is relatively low experience cloudy, stormy weather. Generally, the weather stays fair or improves if air pressure rises. If the air pressure falls steadily, however, the weather may turn cloudy and rainy or snowy.

Air moves from areas where the air pressure is relatively high toward areas where the air pressure is relatively low. This movement of air is what we call wind.

Weather terms

Air mass is a large volume of air that is relatively uniform in temperature and humidity.

Front is a narrow zone of transition between air masses that differ in temperature or humidity. Most changes in the weather occur along fronts.

High-pressure area is an area in which the weight of the atmosphere on the earth is relatively high. High-pressure areas usually have clear skies.

Humidity is a measure of the amount of water vapor in the air.

Low-pressure area is an area in which the weight of the atmosphere on the earth is relatively low. Low-pressure areas usually have cloudy skies.

Precipitation is moisture that falls from clouds in the form of rain, snow, sleet, or hail.

Temperature is a measure of the heat energy of the oxygen and other gases in the air.

Wind is the movement of air. Air tends to move from a high-pressure area to a low-pressure area. Winds are named for the direction from which they blow. For example, a north wind blows from the north.

Weather extremes around the world

Highest temperature recorded was 136 °F (58 °C) at Al Aziziyah, Libya, on Sept. 13, 1922. The highest temperature recorded in North America was 134 °F (57 °C) in Death Valley, California, on July 10, 1913.
Lowest temperature observed on the earth's surface was −128.6 °F (−89.2 °C) at Vostok Station in Antarctica, on July 21, 1983. The record low in the United States was −80 °F (−62 °C) at Prospect Creek, Alaska, on Jan. 23, 1971.
Highest air pressure at sea level was recorded at Agata, in Siberia in the Soviet Union (now Russia), on Dec. 31, 1968. The barometric pressure reached 32.01 inches (81.31 centimeters or 108.4 kilopascals).
Lowest air pressure at sea level was estimated at 25.69 inches (65.25 centimeters or 87.00 kilopascals), during a typhoon in the Philippine Sea on Oct. 12, 1979.
Strongest winds measured on the earth's surface were recorded at Mount Washington, New Hampshire, on April 12, 1934. For five minutes, the wind blew at 188 miles (303 kilometers) per hour. One gust reached 231 miles (372 kilometers) per hour.
Driest place on earth is Arica, Chile. In one 59-year period, the average annual rainfall was $\frac{3}{100}$ inch (0.76 millimeter). No rain fell in Arica for a 14-year period.
Heaviest rainfall recorded in 24 hours was 73.62 inches (186.99 centimeters) on March 15-16, 1952, at Cilaos, on the island of Reunion, a part of France, in the Indian Ocean. The most rain in one year was at Cherrapunji, India. From August 1860 to July 1861, 1,401.78 inches (2,646.12 centimeters) fell. The wettest place is Mount Waialeale, on the island of Kauai in Hawaii, with an average annual rainfall of 460 inches (1,168 centimeters).
Heaviest snowfall recorded in North America in 24 hours—76 inches (193 centimeters)—fell at Silver Lake, Colorado, on April 14-15, 1921. The most snow recorded in North America in one winter—1,122 inches (2,850 centimeters)—fell at Rainier Paradise Ranger Station in Washington in 1971-1972.
Largest hailstone in the United States fell in Coffeyville, Kansas, on Sept. 3, 1970. The hailstone measured 17 $\frac{1}{2}$ inches (44.5 centimeters) in circumference, and it weighed 1$\frac{2}{3}$ pounds (0.76 kilogram).

Source: National Oceanic and Atmospheric Administration.

Humidity is a measure of the amount of water vapor in the air. There is an upper limit to this amount. Air that contains its maximum amount of water vapor is described as *saturated.* The amount of water vapor the air can hold increases as the air temperature rises and decreases as the temperature falls. Thus, saturated warm air has more water vapor than saturated cold air.

Weather reports commonly describe the amount of water vapor in the air in terms of the *relative humidity.* Relative humidity compares the actual amount of water vapor in the air with the amount of water vapor at saturation. Relative humidity is expressed as a percentage. If the relative humidity is 50 percent, the amount of water vapor in the air is half of what it would be if the air were saturated. Lowering the air temperature increases the relative humidity.

If the relative humidity is 100 percent, the air is saturated. When air becomes saturated, water vapor begins to condense into droplets of water. Condensation is the opposite of evaporation. It is a change from a gas to a liquid. If the air is cold enough, at saturation the water vapor develops into tiny ice crystals.

If condensation occurs on a cold surface such as the surface of an automobile window at night, dew or frost forms. Dew and frost do not fall from the sky like rain or snow. Rather, they form when air in contact with a relatively cold surface is chilled to saturation. The same process occurs when small drops of water appear on the outside of a cold soft-drink can on a hot summer day. The temperature to which air must be cooled to reach saturation and produce dew is known as the *dew point.*

When saturation occurs within the atmosphere, water vapor condenses into droplets (or develops into tiny ice crystals) that form clouds. A cloud that is in contact with the earth's surface is known as fog. Most clouds occur within the troposphere. Because air temperature drops with increasing altitude within the troposphere, high clouds are extremely cold and consist mostly of ice crystals. The crystals give these clouds a fuzzy appearance. Low clouds are warmer and generally are composed of droplets. These clouds appear to have sharper edges.

Clouds usually form where air moves upward. As air ascends, it encounters lower and lower pressure. Air responds to lower pressure by expanding. Whenever gases expand, they cool. As air cools, its relative humidity increases until it reaches saturation and clouds form.

Where air moves downward, clouds usually do not develop. Descending air is compressed, it warms up, and its relative humidity decreases. Saturation is not possible, and so clouds do not form.

Weather systems

Meteorologists classify weather systems according to their size and how long they last. The two largest and longest-lasting types of systems are *planetary-scale systems* and *synoptic-scale systems.* Planetary-scale systems are the belts of winds that circle the globe and may blow in the same direction for weeks at a time. Synoptic-scale systems cover a portion of a continent or ocean and last up to a week or so. The term *synoptic* comes from a Greek word meaning *a general view.*

Two briefer and smaller types of systems are *mesoscale systems* and *microscale systems.* Mesoscale systems may last an hour or less and are so small that they may affect the weather of only part of a city. Examples of this type include thunderstorms and sea breezes. Microscale systems, such as tornadoes, usually last only several minutes and affect an area not much larger than a few football fields.

Planetary-scale systems. Suppose that the earth did not rotate and that the noon sun was always directly above the equator. Air temperatures would be highest at the equator and decrease toward the poles. Cold air is denser than warm air. Thus, air pressure would be higher at the poles and lower at the equator. Because air moves from areas of high pressure to areas of low pressure, cold air would sweep toward the equator, where it would push the warm air upward. In the upper atmosphere, the warm air would move toward the poles, cool, and sink over the poles. Thus, the planetary-scale circulation of wind would consist of two huge cells, one in each hemisphere.

The real earth rotates. Rotation of the earth on its axis causes winds that blow long distances—thousands of miles or kilometers—to shift direction gradually. This shift is known as the *Coriolis effect,* which causes winds in the Northern Hemisphere to shift to the right and winds in the Southern Hemisphere to shift to the left. In

Cirrus clouds occur at heights that can exceed 35,000 feet (10,700 meters). The air is so cold at these altitudes that the clouds consist entirely of ice crystals.

© Joseph M. Moran

the Northern Hemisphere, for example, winds blowing southward shift to the west. Winds blowing northward shift to the east.

Rotation of the planet also causes winds near the earth's surface to split into three belts in each hemisphere. These three belts are (1) the trade winds, which blow near the equator, between 30° north latitude and 30° south latitude; (2) the *westerlies* (winds from the west), which blow in the middle latitudes between 30° and 60° north and south of the equator; and (3) the polar winds, which blow in the Arctic and Antarctic, from 60° latitude toward the poles.

Trade winds north of the equator blow from the northeast. South of the equator, they blow from the southeast. The trade winds of the two hemispheres meet near the equator, causing air to rise. Rising air cools, and its relative humidity therefore increases. Thus, a band of cloudy, rainy weather circles the globe near the equator.

Westerlies blow from the southwest in the Northern Hemisphere and from the northwest in the Southern Hemisphere. Westerlies and trade winds blow away from the 30° latitude belt. Over broad regions centered at 30° latitude, surface winds are light or calm, and air slowly descends. Air warms as it descends, and its relative humidity decreases, making clouds and precipitation unlikely. As a result, fair, dry weather characterizes much of the 30° latitude belt.

Polar winds are *easterlies* (winds from the east). They blow from the northeast in the Arctic and from the southeast in the Antarctic. In the Northern Hemisphere, the boundary between the cold polar easterlies and the mild westerlies is known as the *polar front.* A *front* is a narrow zone of transition, usually between a mass of cold air and a mass of warm air. Storms develop and move along the polar front, bringing cloudiness, rain, or snow.

Important seasonal changes take place in the earth's wind patterns. Wind belts shift toward the poles in spring and toward the equator in fall. For example, during the fall, the polar front in the Northern Hemisphere often moves from Canada down to the continental United States.

Planetary-scale winds control the direction of movement of smaller-scale weather systems. For example, in the tropics, trade winds generally steer hurricanes and other weather systems from east to west. In middle latitudes, westerlies move weather systems from west to east. The westerlies are particularly vigorous near the top of the troposphere and just over the polar front. This corridor of exceptionally strong winds is known as a *jet stream.* The jet stream of the polar front supplies energy to developing storms and then moves them rapidly along the front.

Synoptic-scale systems include air masses, fronts, lows, and highs. The movement of these systems causes the day-to-day changes in the weather of Europe, the continental United States, and other regions in the middle latitudes.

Air masses. An *air mass* is a huge volume of air covering thousands of square miles or kilometers that is relatively uniform in temperature and humidity. The properties of an air mass depend on where it forms. Air masses that develop at high latitudes are colder than air masses that form over low latitudes. Air masses that form over the ocean are relatively humid, and those that form over land are relatively dry. The four basic types of air masses are (1) cold and dry, (2) cold and humid, (3) warm and dry, and (4) warm and humid.

Across North America, warm air masses move north and northeastward, and cold air masses move south and southeastward. Maritime polar air, which is cool and humid, forms over the North Pacific and North Atlantic. This air mass brings low clouds and precipitation to the Pacific Northwest, New England, and eastern Canada. Continental polar air, which is dry and cold in winter and dry and mild in summer, forms in north-central Canada. Arctic air, which is dry and much colder than continental polar air, forms over the snow-covered regions north of about 60° latitude in the Northern Hemisphere. The movement of Arctic air to the south causes the bone-numbing cold waves that sweep across the Great Plains and Northeast in winter.

Most of the maritime tropical air that invades North America originates over the Gulf of Mexico and the tropical Atlantic. This warm, humid air mass brings oppressive summer heat waves to areas east of the Rocky Mountains. Continental tropical air forms over the deserts of Mexico and the southwest United States.

How air masses affect the weather

The weather in the Northern Hemisphere is greatly influenced by the movements of *air masses.* These enormous bodies of air form over areas in which the temperature is fairly constant. The air masses take on the temperature of these areas. As the masses move across great distances, they influence the weather below. Arctic air can cause dangerously cold weather as it sweeps south. Continental polar air is dry and cold in the winter and dry and mild in the summer. Maritime polar air masses, which are cool and humid, bring low clouds and precipitation to coastal areas. Warm and humid maritime tropical air can cause oppressive heat waves in the summer.

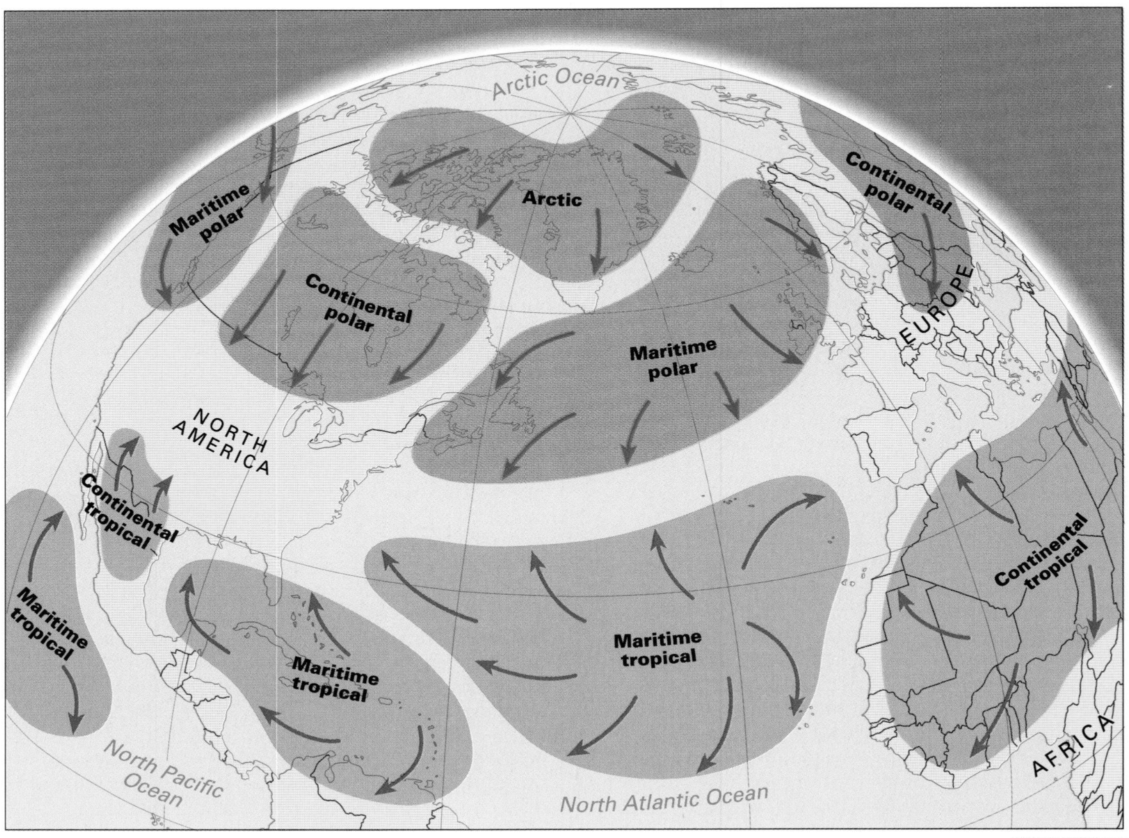

WORLD BOOK map

In summer, this hot, dry air mass surges over Texas and other parts of the American Southwest.

As an air mass travels from place to place, its temperature and humidity can change. For example, air over the Pacific Ocean west of North America is mild and humid. If that air mass moves eastward, it is forced up the slopes of the coastal mountain ranges. Air temperature drops, the relative humidity increases to saturation, clouds form, and rain or snow develops. As the air travels down the opposite slopes of the mountain ranges, the air temperature rises, the relative humidity decreases, and clouds thin out or vanish.

This process repeats with each mountain range the air mass encounters as it moves eastward. By the time it reaches the Western Plains, the air mass has become considerably drier and milder. This modified air mass, known as Pacific air, brings mild, dry weather to much of the central and eastern regions of the United States and Canada.

Fronts form where air masses meet. A front is a narrow zone of transition between air masses that differ in temperature or humidity. In most cases, the air masses differ in temperature, so that the fronts are either warm or cold.

A *warm front* is the leading edge of an advancing warm air mass. Warm air is less dense than cold air, so warm air advances by riding up and over the retreating cold air. As the warm air ascends, its temperature drops, relative humidity increases, and clouds and perhaps precipitation form. In North America, clouds can extend hundreds of miles or kilometers to the north and northwest of a warm front. Rain or snow is usually light to moderate and may last 12 to 24 hours or longer.

A *cold front* is the leading edge of an advancing cold air mass. Cold air is denser than warm air, so that cold air advances by moving under and pushing up the retreating warm air. As warm air ascends, its temperature falls, relative humidity rises, and clouds and often precipitation develop. Clouds associated with a cold front typically form a narrow band along the front. Rain or snow falls in brief showers. If the cold front is fast-moving and well-defined by considerable temperature contrast across the front, thunderstorms are likely. Some of these thunderstorms could become severe and produce hail, torrential rains, or strong winds. Tornadoes also may develop from severe thunderstorms.

A front that stalls is known as a *stationary front*. The weather along a stationary front often consists of considerable cloudiness and light rain or snow.

Cold fronts move faster than warm fronts, so a cold front may catch up to and merge with a warm front. The warm air is lifted off the earth's surface, and the merged front is known as an *occluded front*. An occluded front sometimes moves very slowly and causes several days of considerable cloudiness and light precipitation.

Lows are areas of relatively low air pressure. The winds in a low pressure system bring contrasting air masses together to form fronts. For this reason, lows are sometimes described as the chief weather-makers of regions in the middle latitudes. Scientists use the term *cyclone* to refer to a synoptic-scale low-pressure area. They also use the term to mean a hurricane in some parts of the world.

Viewed from above in the Northern Hemisphere, surface winds in a low-pressure area blow in a counterclockwise and inward direction. Surface winds converging in the low cause air to rise, cool, and reach saturation. Clouds and precipitation usually develop. Air ascends mostly along fronts that develop as winds in the low bring cold and warm air masses together.

Lows generally travel from southwest to northeast across North America and may complete a journey from Colorado to New England in three or four days. As a rule, temperatures are lower to the left (north) of the path followed by the low-pressure area and higher to the right (south). In winter, the heaviest snows usually fall about 90 to 150 miles (150 to 250 kilometers) to the north and west of the moving low-pressure area.

Highs, also known as *anticyclones*, are areas of relatively high air pressure. A high, which brings fair weather, often follows in the wake of a low. Viewed from above in the Northern Hemisphere, surface winds in a high blow in a clockwise and outward direction. As winds blow out and away from a high, air descends near

The formation of fronts

A *front* forms where air masses meet. An air mass is a huge volume of air that is relatively uniform in temperature and humidity. A front is a zone between masses that differ in temperature or humidity. In most cases, the temperatures differ, so the zones are either *cold fronts* or *warm fronts*.

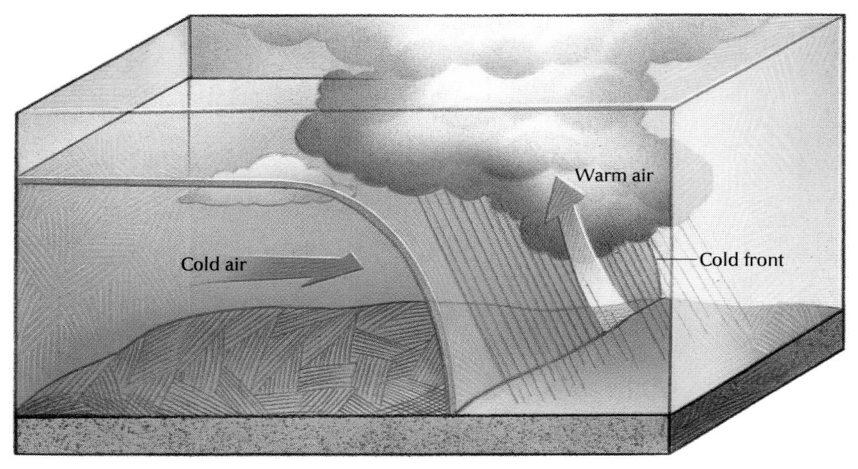

A cold front is the leading edge of a mass of cold air that advances into a region occupied by warm air. Cold air weighs more than warm air, so the cold air moves under the retreating warm air and pushes it upward. The rising warm air cools, so its relative humidity increases, often bringing precipitation. The diagram at the left represents a region about 300 miles (500 kilometers) long.

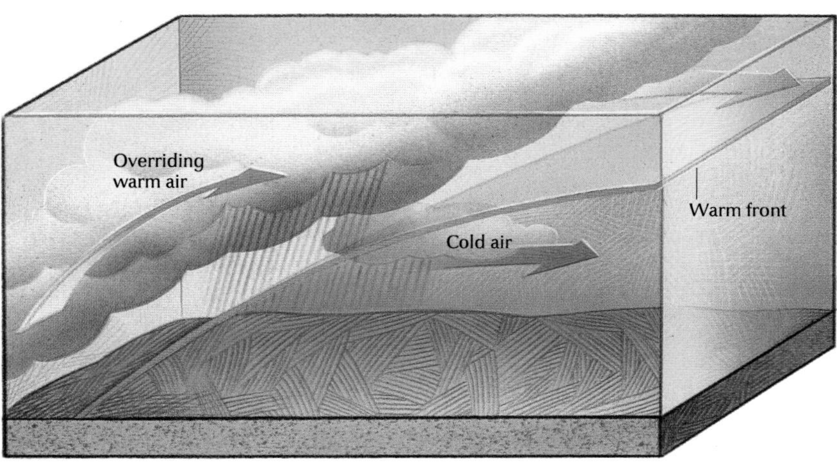

A warm front is the leading edge of a mass of warm air that advances into a region occupied by cold air. Warm air weighs less than cold air, so the warm air rides up and over the cold air. As the warm air rises, it cools, so its relative humidity increases, bringing clouds and precipitation. The diagram at the left represents a region about 900 miles (1,500 kilometers) long.

WORLD BOOK illustrations by Paul Turnbaugh

National Center for Atmospheric Research

A supercell thunderstorm is a violent storm dominated by a single gigantic *cell*—a weather system made up of storm clouds and the winds associated with them. Rain and hail may fall for hours.

the center of the system. Descending air warms, and the relative humidity decreases.

Highs are either warm or cold. Warm highs form south of the polar front and are characterized by high temperatures and low relative humidities. Such highs are massive weather systems that extend from the earth's surface to the top of the troposphere. In the summer, a warm high sometimes stalls over North America. If the high remains stationary for several weeks, it creates a drought.

Cold highs form north of the polar front. They are shallow masses of cold, dry air that develop mostly in the winter over the snow-covered regions of northern Canada, Alaska, and Siberia. As cold highs move southeastward over Canada and into the continental United States, they bring fair but cold weather.

Mesoscale and microscale systems result from the development and movement of synoptic and planetary-scale systems. Mesoscale systems, which may last an hour or less and affect only part of a city, include thunderstorms and sea breezes. A tornado is an example of a microscale system, the smallest and briefest of significant weather systems.

Measuring the weather

Because no single country can constantly measure and report on conditions in every part of the atmosphere, the world's nations must cooperate to monitor the weather effectively. To this end, they founded the International Meteorological Organization in 1873, renaming it the World Meteorological Organization (WMO) in 1950. Members of the WMO participate in the worldwide observation of the atmosphere and in the exchange of weather data and forecasts. Weather observations come from many different sources, including land-based observation stations, radar systems, weather balloons, airplanes, ships, and satellites.

Land-based observation stations. About 10,000 land-based weather stations—also known as *surface stations*—monitor the weather worldwide. In the United States, the National Weather Service coordinates weather observations at about 1,000 land-based stations with information obtained from about 10,000 volunteer weather observers. The Atmospheric Environment Service of Canada operates a similar observation network.

Observation stations use a variety of instruments to monitor the state of the atmosphere. An electronic thermometer checks air temperature and registers the highest and lowest temperature of the day. A *hygrometer* measures the amount of water vapor in the air. A *barometer* shows the air pressure. A weather vane indicates the direction of the wind, and an anemometer monitors wind speed. Rain gauges measure rainfall or snowfall. For more information, see the separate articles on weather instruments in *World Book*.

Weather radar. Some observation stations use radar to track storms. Weather radar can operate in either a *reflectivity mode* or a *velocity mode*.

In the reflectivity mode, weather radar locates areas of rain or snow. The system sends out a radar signal, which consists of microwave energy pulses. If the radar signal encounters rain, snow, or hail, the falling precipitation reflects some of that signal back to the radar. The reflected radar signal, called a *radar echo,* appears as a blotch on a television-type screen. As the radar rotates, it generates a map of radar echoes that represents the pattern of precipitation surrounding the radar. In the reflectivity mode, weather radar can detect precipitation more than 250 miles (400 kilometers) away.

In the velocity mode, weather radar determines the circulation of air from the motion of raindrops, snowflakes, or dust particles. This radar is also known as Doppler radar because it uses the Doppler effect to calculate how the air in a weather system is moving.

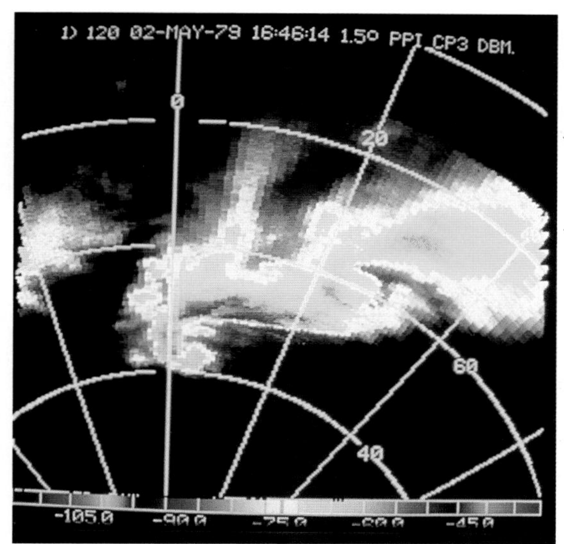

National Center for Atmospheric Research

A Doppler radar image is color-coded to indicate the speed and direction at which rain clouds and other masses of air are moving. Doppler radar can provide warning of severe weather.

The Doppler effect is the change in frequency of sound or radiation waves caused by the motion of the source of the waves relative to their observer. For example, the *pitch* (frequency) of a train whistle seems higher as a train approaches and lower as the train moves away. Similarly, as raindrops, snowflakes, or dust particles move through the atmosphere, the radar signals they reflect change in frequency. The radar monitors these frequency changes and then uses them to calculate the speed at which the drops, flakes, or particles are advancing or receding.

By enabling meteorologists to monitor the motion of air in a weather system, rather than merely track areas of precipitation, Doppler radar has improved their ability to provide advance warning for severe weather. For example, meteorologists can use Doppler radar to detect the development of a tornado before it descends from its parent thunderstorm and strikes the earth's surface. In the velocity mode, radar can detect the speed of precipitation or dust particles more than 120 miles (190 kilometers) away. A network of more than 150 Doppler radar stations across the United States called NEXRAD began operation in 1997 to improve the forecasting of severe weather.

Weather balloons, airplanes, and ships. To obtain information on the state of the atmosphere above the earth's surface, meteorologists routinely use an instrument package called a *radiosonde.* The radiosonde, which is carried aloft by a weather balloon, measures changes in temperature, pressure, and humidity. A small radio transmitter beams these data back to a weather station, where they are recorded by computer. At an altitude of about 19 miles (30 kilometers), the balloon bursts, and a parachute carries the instrument package back to the earth's surface.

Meteorologists use a special antenna to track a radiosonde and thereby measure wind speed and direction at different altitudes within the atmosphere. Such an observation is known as a *rawinsonde.* Radiosonde and rawinsonde observations are made every 12 hours.

Meteorologists sometimes use a *dropwindsonde* to obtain atmospheric measurements over the oceans. A dropwindsonde is a radiosonde attached to a parachute and dropped from an aircraft. As the instrument package falls toward the sea, it radios back to the aircraft measurements of temperature, pressure, and humidity.

Ships also report on weather conditions at sea. Some launch weather balloons, and others release special ocean buoys that record and transmit information about weather at sea level.

Weather satellites play a major role in worldwide weather observation. They monitor clouds associated with weather systems, track hurricanes and other severe weather systems, measure winds in the upper atmosphere, and obtain temperature measurements.

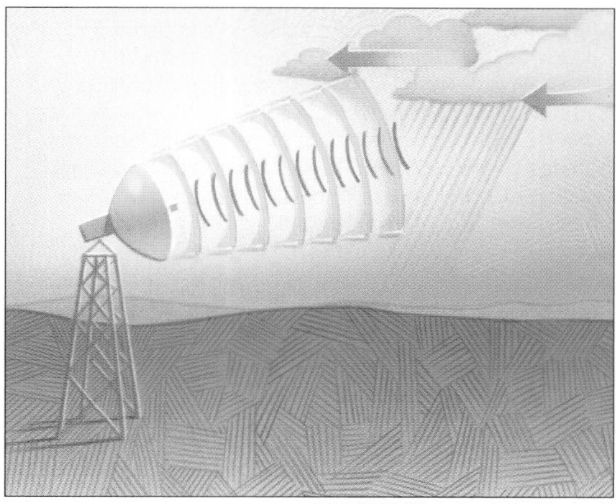

WORLD BOOK illustrations by Paul Turnbaugh

Doppler radar enables meteorologists to see how fast a storm is moving and in what direction. The radar sends out microwave pulses that bounce off raindrops or ice particles. The frequency of the pulses as they return to the antenna shows whether the storm is retreating, *left,* or advancing, *right.*

National Oceanic and Atmospheric Administration

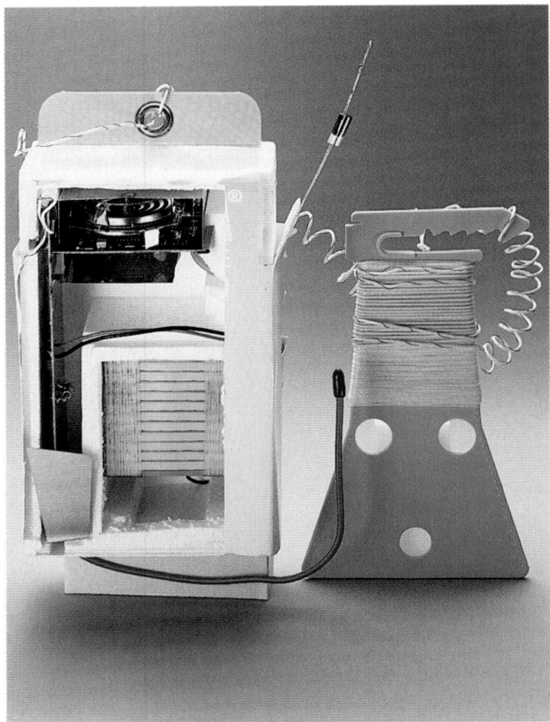
Vaisala Inc. (WORLD BOOK photo by Steven Spicer)

Weather instruments measure characteristics of the atmosphere that determine the weather. A weather balloon, *left*, carries aloft a package of instruments. One such package, a *radiosonde, above,* measures temperature, pressure, and humidity. A built-in radio transmitter sends the data to a weather station on the ground, where they are recorded by a computer.

Weather satellites offer significant advantages over the network of surface weather stations. They can observe weather over a broad and continuous field of view, whereas surface stations are widely spaced and may not observe some weather systems directly. Furthermore, satellites provide valuable data from the oceans, which cover about 70 percent of the globe. Land-based weather stations provide little information about these vast regions.

Sensors aboard weather satellites detect two types of radiation signals coming from the planet. One signal consists of reflected sunlight. These satellite images, which resemble black-and-white photographs of the planet, reveal cloud patterns.

The second signal recorded by weather satellites is infrared radiation (IR). The intensity of the infrared radiation emitted by an object depends on the object's temperature. For example, low clouds and fog, which are relatively warm objects, give off more intense infrared radiation than do high clouds, which are relatively cool. Thus, an IR image reveals not only cloud patterns but also cloud temperatures. Weather satellites can record IR images at any time—day or night—because objects continually emit infrared radiation.

There are two main types of weather satellites—*geostationary* and *polar-orbiting.* A geostationary satellite orbits about 22,300 miles (35,900 kilometers) above the equator and travels eastward at the same rate as the earth rotates. Thus, a geostationary satellite remains above the same point on the equator. Because geostationary satellites orbit at such a high altitude, they can record images that cover a wide area. For example, two of them cover most of the United States and Canada. Most satellite images shown on televised weather reports come from this type of satellite.

A polar-orbiting satellite follows a north-south path that takes it over the polar regions. Because the satellite does not rotate east as the earth does, the rotation of the earth causes the satellite to pass over different areas of the earth each orbit. Polar-orbiting satellites travel at a much lower orbit than geostationary satellites, and so they record more detailed images.

Weather forecasting

After meteorologists have gathered weather data from around the world, they can try to predict the development and movement of future weather systems. To forecast the weather, meteorologists use such tools as weather maps and mathematical models.

Weather maps. Meteorologists use weather observation data to create weather maps that represent the state of the atmosphere at a particular time. To capture

How to read a weather map

Maps like the one below appear in many daily newspapers and are based on government weather service reports. This map shows the weather conditions expected across the United States at a certain time of the day. It also predicts the high and low temperature for the day in various cities.

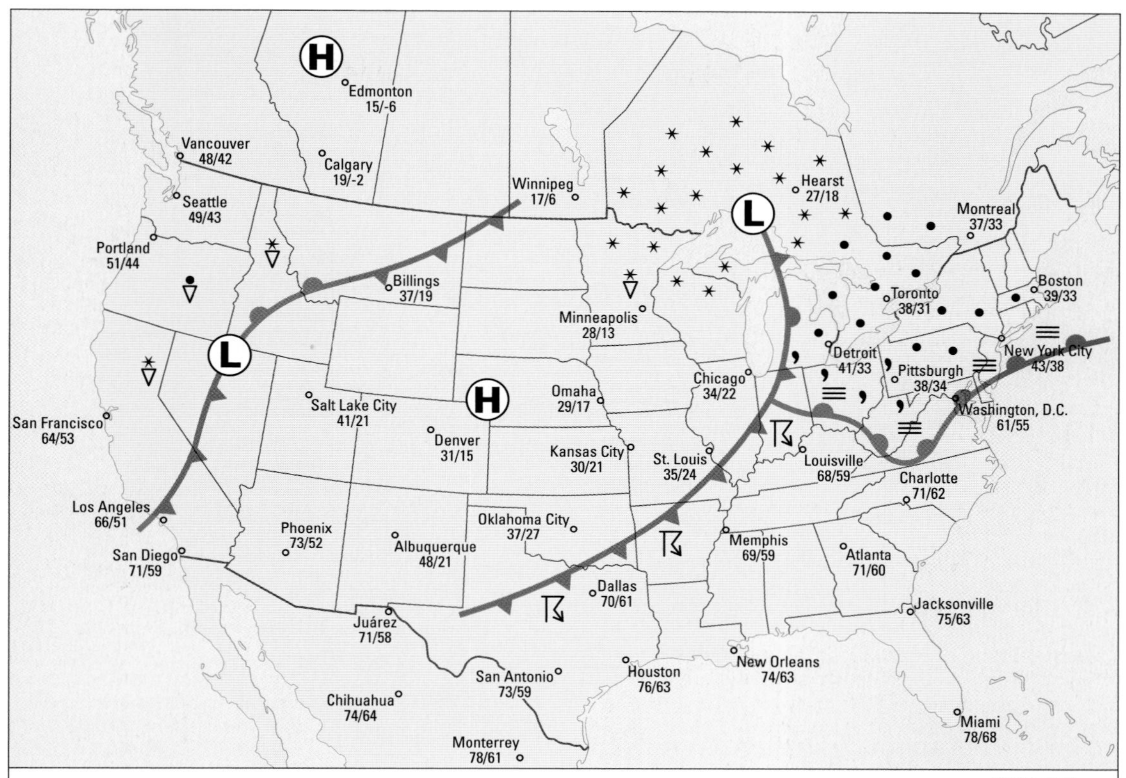

WORLD BOOK map

Weather map symbols

- 71/59 **High/Low daily temperature (°F)**
- **Rain**
- **Drizzle**
- **Rain shower**
- **Thunderstorm**
- **Snow**
- **Snow shower**
- **Fog**

High pressure system: A region where air pressure is relatively high. This system usually brings fair weather. Viewed from above in the Northern Hemisphere, surface winds blow clockwise and outward around the center of a high.

Low pressure system: A region where air pressure is relatively low. This system usually brings cloudy skies and a chance of precipitation. Viewed from above in the Northern Hemisphere, surface winds blow counterclockwise and inward around the center of a low.

Cold front: A narrow zone of transition between a mass of relatively cold air and a mass of relatively warm air. The cold air is advancing as the warm air is retreating. A narrow area of cloudiness and showers often accompanies the passage of a cold front.

Warm front: A narrow zone of transition between a mass of relatively warm air and a mass of relatively cold air. The warm air is advancing as the cold air is retreating. A broad area of cloudiness and light to moderate precipitation often occurs ahead of a warm front.

Stationary front: A narrow zone of transition between warm and cold air masses that have stalled. The weather along a stationary front often consists of considerable cloudiness and light precipitation.

Occluded front: A narrow zone of transition between contrasting air masses formed when a fast-moving cold front catches up to and merges with a slower-moving warm front. An occluded front sometimes moves very slowly, bringing a lengthy period of clouds and light precipitation.

National Oceanic and Atmospheric Administration

A hurricane approaches the Bahamas and Florida in this image taken from a weather satellite in orbit about 22,300 miles (35,900 kilometers) above the earth. Only satellites can provide images of the weather over vast expanses of the earth's surface, making satellites an essential part of modern storm detection.

the three-dimensional nature of weather, they draw maps for conditions at the earth's surface and at various levels within the atmosphere. By examining a sequence of weather maps, forecasters can determine how the weather changes through time and then make predictions about the future state of the atmosphere.

Mathematical models. Since the mid-1950's, weather forecasters have used mathematical models of the atmosphere, processed by computers, to improve the accuracy of their predictions. A mathematical model of the atmosphere consists of a set of equations intended to approximate the atmospheric processes that drive the development and movement of weather systems. Mathematical models are based on scientific laws, and through the years scientists have developed increasingly sophisticated models. Improvements in computer technology have greatly enhanced meteorologists' ability to use mathematical models effectively because computers can process enormous amounts of observation data and perform a multitude of calculations extremely rapidly.

A mathematical model begins with the current state of the atmosphere, as determined by the most recent weather observations. The model uses these data to predict the state of the atmosphere for a specific time interval—for example, the next 10 minutes. Using this predicted state as a new starting point, the model then forecasts the state of the atmosphere for another 10-minute period. This process repeats over and over again until the model produces short-range weather forecasts for the next 12, 24, 36, and 48 hours.

The accuracy of weather forecasts generated by mathematical models declines steadily over time for two main reasons. First, the weather observation data initially fed into the model can never provide a complete picture of the present state of the atmosphere. Not all the data are reliable, due to both technical and human error, and data are missing from vast stretches of the atmosphere over the oceans. Second, mathematical models of the atmosphere are only approximations of the way the atmosphere actually works, and errors in the models tend to grow with each repetition.

Meteorologists understand the limitations of mathematical models. They base their forecasts on observations of how the weather has changed over the past several days and on their understanding of atmospheric processes. Experience and even intuition play important roles, along with the cautious interpretation of the output of mathematical models.

Meteorologists also use mathematical models to produce long-range weather forecasts, such as 6- to 10-day forecasts and monthly (30-day) and seasonal (90-day) outlooks. Long-range forecasts and outlooks typically are less accurate than short-range forecasts, but they can provide an indication of general trends, such as whether conditions will be wetter or drier than normal.

Reporting the weather

The National Weather Service of the United States and the Atmospheric Environment Service of Canada issue weather forecasts, watches, warnings, and advisories to the public through regional forecast offices. The public accesses this information through radio and television broadcasts, newspapers, and the Internet.

When hazardous weather threatens, forecasters issue outlooks, watches, warnings, and advisories. An outlook provides advance notice of a general weather trend. For example, the outlook for spring flooding due to expected snowmelt is usually available many weeks in advance. A weather watch is issued when hazardous weather is possible based on current or predicted atmospheric conditions. A weather warning applies when hazardous weather is taking place nearby. Watches and warnings are issued for severe thunderstorms, tornadoes, floods, hurricanes, and winter storms, such as blizzards and ice storms.

Weather advisories refer to expected weather hazards that are less serious than those covered by a warning. An example is a winter weather advisory.

Weather advisories are also issued for low wind chill temperatures and for a high heat index. *Wind chill* is a measure of the cooling power of a combination of low air temperature and strong winds. Even if the air temper-

(Text continued on page 328)

National Oceanic and Atmospheric Administration

A National Weather Service meteorologist works with computer images of weather patterns at the Weather Service control center near Washington, D.C.

Some kinds of storms

Storms are periods of strong wind accompanied by rain, snow, hail, or thunder and lightning. Violent storms can kill people and destroy property. The most common types of storms include (1) thunderstorms, (2) winter storms, (3) tornadoes, and (4) hurricanes.

Robert H. Glaze, Artstreet

A thunderstorm brings lightning, thunder, and rain.

Robert H. Glaze, Artstreet

A winter storm can be a blinding blizzard.

E. R. Degginger, Earth Scenes

A tornado is a deadly, spinning funnel cloud.

E. R. Degginger, Earth Scenes

A hurricane's swirling winds form over tropical seas.

A *World Book* science activity

A cloud in a bottle

This activity has two purposes: (1) to demonstrate that air warms when it is compressed and cools when it expands; and (2) to demonstrate that, if air cools sufficiently, water vapor in the air will condense to form a cloud.

What you need:

To carry out this activity, you will need a two-liter plastic soft-drink bottle, masking tape, matches, and a strip thermometer designed to mount on the outside of a home-aquarium tank. The thermometer is made of film and contains special materials called *liquid crystals* that indicate the temperature. It is available at aquarium-supply stores.

Caution:

This activity involves lighting a match, which can cause burns or fire. Do not use matches unless you have a responsible adult help you.

What to do:

1. Bend the thermometer into a half-circle, with the numbers facing outward. Use a short piece of the tape to join the ends of the thermometer, forming a "D." Preparing the thermometer in this way will enable you to read it easily when it is inside the bottle.

2. Put the thermometer inside the bottle and replace the cap. Let the bottle stand for two minutes. Record the temperature of the air inside it.

3. Squeeze the bottle—either between your hands or around the edge of a desk. After two minutes, again record the temperature. You will see that squeezing the bottle—thereby compressing the air in the bottle—increases the temperature of the air. If you squeeze hard enough, the temperature will rise about 2 Celsius degrees.

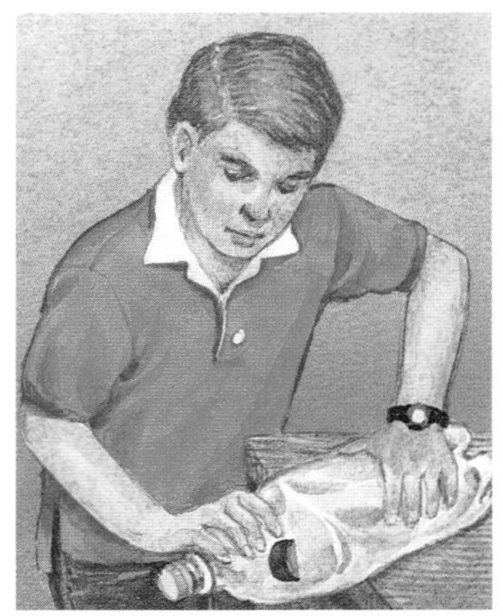

4. Release the bottle. It will pop back to almost its original shape. You may have to "help" it by pushing against large dents. Again, wait two minutes and record the temperature. The expansion of the air causes the temperature to fall back to its original value.

5. Pour about $\frac{3}{4}$ inch (2 centimeters) of water into the bottle and replace the cap. Swirl the water to wet all the inside of the bottle. Let the bottle stand for two minutes, then pour out the excess water. The bottle will now contain a large amount of *water vapor*—a clear, colorless gas resulting from evaporation.

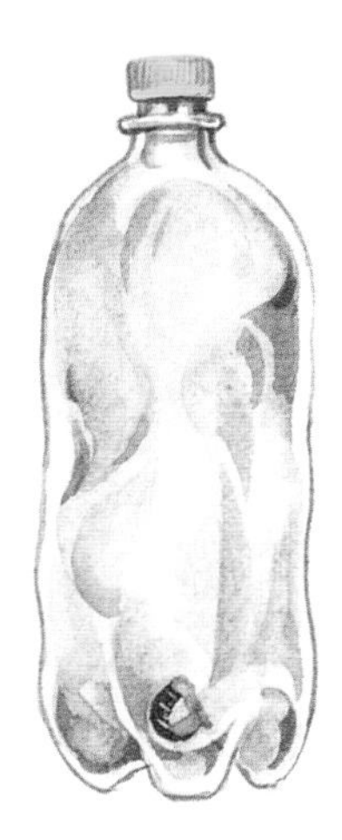

6. Quickly turn the bottle upside down and squeeze it just a little. Light a match and quickly blow it out. While easing your grip on the bottle—enabling it to expand—allow smoke from the match to enter the bottle. Quickly put the cap back on the bottle. The bottle will now contain the ingredients needed to form a cloud: water vapor and tiny particles from the smoke on which the vapor can *condense* (turn into liquid droplets).

7. Water vapor will condense rapidly on the smoke particles, forming a cloud. The cloud will be a thin, foglike mist filling the bottle—rather than a separate object floating in the air inside the bottle.

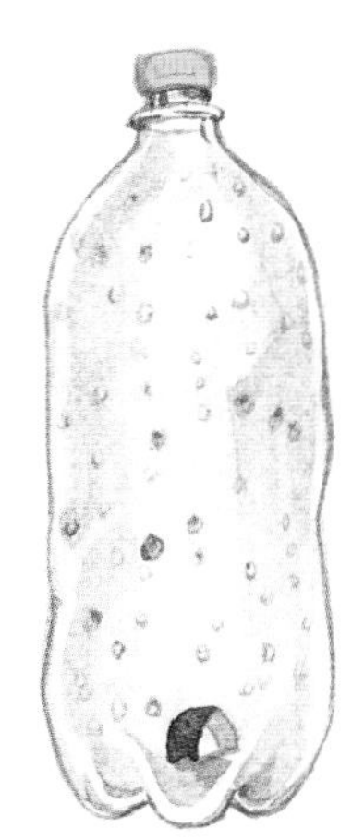

8. Squeeze the bottle as you did at the beginning of the experiment. You will see the cloud vanish. Compressing the air raises the temperature of the air, increasing its ability to hold water vapor. As a result, water droplets in the cloud evaporate.

9. Release the bottle, enabling the bottle and the air inside it to expand. You will see the cloud reappear. The expansion of the air causes the temperature of the air to fall, decreasing its ability to hold water vapor. As a result, water vapor in the bottle condenses.

10. Quickly squeeze and release the bottle several times. The cloud will vanish and reappear each time.

11. You can "set the cloud free" by taking the cap off the bottle, then squeezing the bottle. The cloud will puff out of the top of the container.

WORLD BOOK illustrations by Yoshi Miyake

ature remains constant, the human body loses increasing amounts of heat to the environment as wind speed increases. At low wind chill temperatures, people need to take special precautions to prevent *frostbite* (the freezing of skin tissue) and *hypothermia* (a dangerous drop in body temperature).

Heat index is a measure of the stress produced by a combination of high air temperature and high relative humidity. During excessively hot and muggy weather, the human body may not be able to release sufficient heat to prevent *hyperthermia* (a dangerous rise in body temperature). High humidity reduces the rate at which perspiration evaporates from the skin's surface. The cooling that accompanies this evaporation represents one of the body's main ways to release heat.

Private weather forecasters provide weather information tailored to a special need. For example, prior to pouring concrete, a building contractor may consult a private forecaster to find out the probability of rainfall during specific hours of the day. Department stores may hire a private forecaster to advise them of prospects for hot summer weather to ensure the stores have enough air conditioners and fans in stock.

How people affect the weather

Human activities affect the weather both intentionally and unintentionally. For example, the construction of cities creates areas that are drier and warmer than the surrounding countryside. Cities are drier because they have storm sewer systems that quickly carry off rainwater and snowmelt.

Cities are warmer for several reasons. The use of storm drainage systems means that less solar radiation is used to evaporate water and more is used to heat the city surfaces and air. The brick, asphalt, and concrete surfaces of city buildings, sidewalks, and streets readily transmit the heat they absorb and so raise urban air temperatures even more. In addition, cities themselves generate heat from a number of sources, including motor vehicles and heating and air conditioning systems.

Large urban areas also affect the weather in the areas downwind of them. Smokestacks and automobile tailpipes in cities emit water vapor and tiny particles that stimulate the formation of clouds. Heat energy rising from a city also spurs the growth of clouds. Thus, the weather downwind from many large urban areas is cloudier and rainier than the weather upwind from those same areas.

Urban and industrial areas also produce air pollutants, such as carbon monoxide, nitrogen oxides, and hydrocarbons. Although improved controls on factories and motor vehicles have reduced considerably the amount of these gases released into the atmosphere, air quality problems persist. For example, many large cities still have problems with smog—a mixture of gases and tiny particles that reduces visibility and poses serious health hazards.

Smog and other air pollution problems are particularly serious in areas where winds are light and a *temperature inversion* occurs in the lower atmosphere. In a temperature inversion, warm air overlies cold air, so that the air temperature rises with increasing altitude, which is the opposite of the usual situation in the troposphere. Under such circumstances, smokestack and tailpipe emissions do not rise and disperse, and so emissions may build up to unhealthy concentrations.

From time to time, scientists have tried to alter the weather. Most of these efforts—including attempts to modify hurricanes, suppress hailstorms, and clear fog—have not worked. Today, scientists focus their weather modification efforts primarily on *cloud seeding,* an attempt to stimulate the natural precipitation-forming process.

Most clouds do not produce rain or snow. This is because few clouds have just the right combination of tiny ice crystals and *supercooled water droplets* (droplets that remain liquid even at subfreezing temperatures). In such clouds, ice crystals grow at the expense of water droplets and eventually become snow crystals. If the temperature is below freezing all the way to the earth's surface, precipitation falls as snow. If the air temperature is above freezing, the snow crystals melt into raindrops.

Cloud seeding is intended to stimulate precipitation in clouds that do not have enough ice crystals. Scientists use a small aircraft to introduce into the clouds either dry ice or silver iodide crystals. Dry ice, which is frozen carbon dioxide, lowers the temperature so that cloud droplets freeze into ice crystals. Silver iodide crystals resemble ice crystals and cause supercooled water to form crystals of ice around them. When cloud seeding succeeds, precipitation is increased by perhaps as much as 20 percent.

Careers in weather

High school students interested in meteorology as a career should take college-preparatory classes, including courses in mathematics, computer science, physics, and chemistry. Most entry-level jobs in meteorology require at least a bachelor's degree, and many require a master's degree. A limited number of U.S. and Canadian colleges and universities offer degrees in meteorology, and the U.S. armed forces offer meteorological training as well.

Meteorologists specialize in a number of different areas. For example, research meteorologists study some subfield of meteorology, such as tropical weather systems. Regional forecasters prepare weather forecasts for portions of one or more states or provinces. Consulting meteorologists provide weather information tailored for specific industrial, business, or government needs. Broadcast meteorologists have skills in both meteorology and television or radio broadcasting. This type of meteorologist informs the public of current and expected weather conditions. Specialists called *synoptic meteorologists* analyze weather observations, interpret the output of computer models, and monitor weather systems. *Physical meteorologists* study the physical and chemical properties of the atmosphere. *Dynamic meteorologists* focus on creating models of atmospheric circulation.

Joseph M. Moran

Related articles in ***World Book*** include:

Storms

Blizzard
Cloudburst
Cyclone
Dust devil
Dust storm
Hail
Hurricane
Lightning
Rain
Sandstorm
Sleet
Snow
Squall
Storm
Thunder
Tornado
Typhoon
Waterspout

A tornado roars across the Oklahoma plain. A deadly funnel has descended from a dark, heavy *wall cloud* and touched the ground, its violent, rotating winds hurling debris in all directions.

Tornado is the most violent of all storms. A tornado, sometimes called a *twister*, consists of a rapidly rotating column of air that forms under a thundercloud or a developing thundercloud. Tornado winds swirl at speeds that may exceed 300 miles (480 kilometers) per hour. A powerful tornado can lift cattle, automobiles, and even mobile homes into the air and destroy almost everything in its path. Fortunately, most tornadoes are relatively weak, and only a few are devastating.

Scientists use the word *cyclone* to refer to all spiral-shaped windstorms that circulate in a counterclockwise direction in the Northern Hemisphere or in a clockwise direction in the Southern Hemisphere. The term *cyclone* comes from the Greek word for *circle*. Cyclones come in many sizes. Among the largest such storms are hurricanes and typhoons, which may reach 250 miles (400 kilometers) across.

Most tornadoes are small, intense cyclones. On rare occasions, the winds whirl in the direction opposite that of a cyclone—for example, clockwise in the Northern Hemisphere.

Most tornadoes have damage paths less than 1,600 feet (500 meters) wide, move at less than 35 miles (55 kilometers) per hour, and last only a few minutes. Extremely destructive twisters may reach 1 mile (1.6 kilometers) in diameter, travel at 60 miles (100 kilometers) per hour, and blow for more than an hour.

The United States has the highest incidence of tornadoes in the world. Most of these storms occur in a belt known as Tornado Alley that stretches across the Midwestern and Southern states, especially Texas, Oklahoma, Kansas, Nebraska, and Iowa. However, tornadoes also strike many other parts of the world. Australia ranks second to the United States in number of twisters, and many damaging tornadoes strike Bangladesh. Tornadoes occur most often during the spring and early summer in the late afternoon and early evening.

A tornado over a body of water is called a *waterspout*. Waterspouts occur frequently in summer over the Florida Keys. Waterspouts also form elsewhere in the Gulf, along the Atlantic and Pacific coasts, over the Great Lakes, and even over the Great Salt Lake in Utah.

The story of a tornado. The majority of tornadoes develop from severe thunderstorms. A hurricane, when it makes landfall, can also generate tornadoes.

The most damaging tornadoes form in large, powerful thunderstorms called *supercells.* For a supercell to form, and perhaps spawn a tornado, several conditions must exist. There must be an adequate supply of moisture to feed the storm. In Tornado Alley, air from the Gulf of Mexico provides the moisture. There must be a layer of warm, moist air near the ground and a layer of much cooler air above. Often, a *front* (the boundary between two air masses at different temperatures) powers an upward flow of warm air. As the warm air rises, it begins to cool, and the moisture it holds condenses into raindrops. The air stops rising at high levels and spreads sideways to form a characteristic anvil-shaped storm cloud.

For a supercell to develop, the winds at higher elevations must differ markedly from those at lower levels in speed, direction, or both. Such a large difference in wind speed or direction is called *wind shear.* Wind shear makes the column of rising air begin to rotate, forming a broad, horizontal tube of swirling air. As the storm continues, this tube turns on its end, producing a rotating column of air called a *mesocyclone.* Studies show that most supercells containing mesocyclones eventually produce tornadoes.

A low, dark, heavy cloud called a *wall cloud* forms underneath the mesocyclone. Tornado funnels develop out of the wall cloud.

The first sign of an approaching tornado may be light rain, followed by heavier rain, then rain mixed with hail. The hailstones may grow to the size of golf balls or even baseballs. After the hail ends, a tornado may strike. In most tornadoes, a funnel-shaped cloud forms and descends from the wall cloud until it touches the ground. However, there might be a tornado even if the funnel does not touch the ground or if the air is too dry for a funnel cloud to form. Sometimes, the first sign of a tornado is dust swirling just above the ground.

A few small tornadoes begin near the ground and build upward with no apparent connection to the storm aloft. Many of these storms occur without mesocyclones and lack a funnel cloud.

Damage by tornadoes. Most tornado damage results from the force of the wind. Each time the wind speed doubles, the force of the wind increases four times. For example, the force of the wind at 220 miles per hour is four times as great as the force at 110 miles per hour. This tremendous strength may knock over buildings and trees. Other damage occurs when the wind picks up objects and hurls them through the air.

Scientists estimate the wind speed of a tornado by the damage it inflicts, using a gauge called the Fujita scale. The scale was developed by the Japanese-born weather scientist T. Theodore Fujita. On the Fujita scale, F0 is the weakest rating and F5 is the strongest. An F5 tornado can remove a house from its foundation.

A tornado sucks up air when passing over a building. For this reason, some people think they should open windows to help equalize the pressure if a tornado threatens. They fear that the air pressure outside the building might drop so suddenly that the structure would explode outward. Safety experts know, however, that air moves in and out of most buildings so quickly that air pressure remains nearly equal inside and out, even during a tornado. Open windows do not reduce damage and may even increase it because wind blowing in may hurl loose objects through the air.

Tornado damage is often localized. A tornado may demolish one house and leave an adjacent house untouched.

Some tornadoes consist of smaller rotating columns of air called *suction spots* or *suction vortices.* The suction spots revolve around the central axis of the tornado and can inflict tremendous damage to small areas in the tornado's path.

Forecasting tornadoes. *Meteorologists* (scientists who study weather) can predict possible severe weather 12 to 48 hours in advance. They make such forecasts using data from weather balloons, satellites, and conventional weather radar. Computers help meteorologists an-

WORLD BOOK illustration by Bruce Kerr

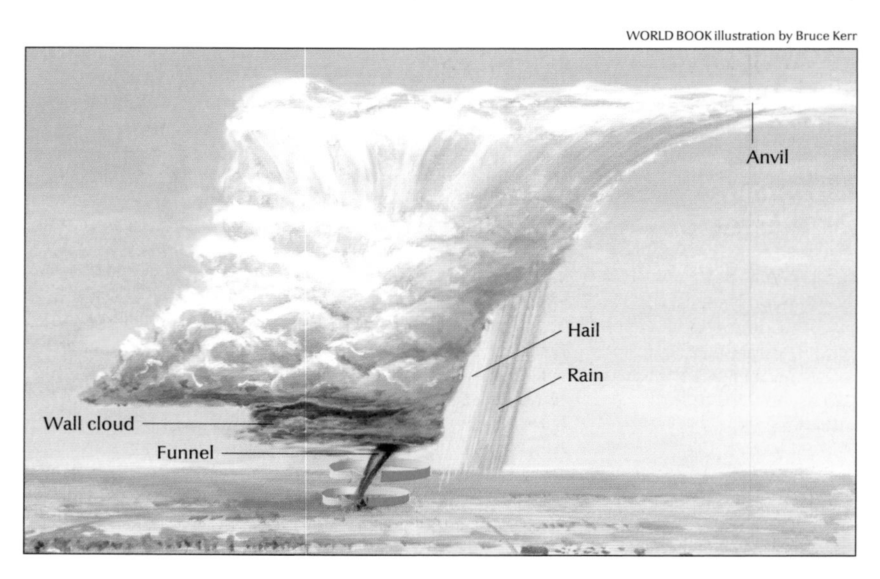

A tornado is a rapidly rotating column of air that can develop under a large, anvil-shaped thundercloud. First, a dark *wall cloud* forms underneath the thundercloud. In most cases, a twisting, funnel-shaped cloud then descends from the wall cloud and touches the ground. Almost all tornadoes in the Northern Hemisphere rotate as shown in the diagram at the left—counterclockwise when viewed from above.

Where tornadoes occur in the United States

Tornadoes most often hit the midwestern and southern United States. This map shows the number of tornadoes that occur yearly in each 10,000 square miles (25,900 square kilometers) of area.

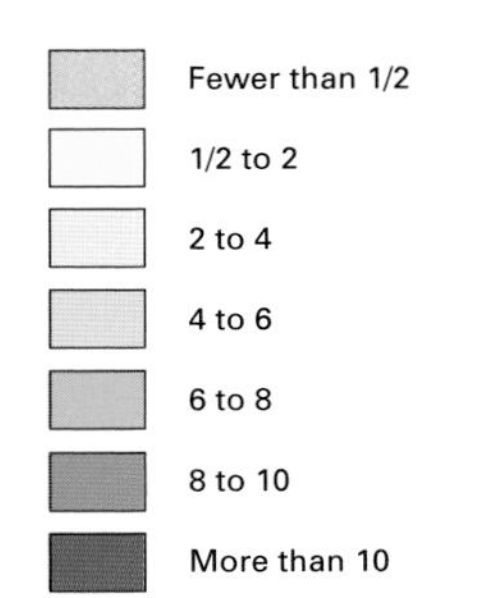

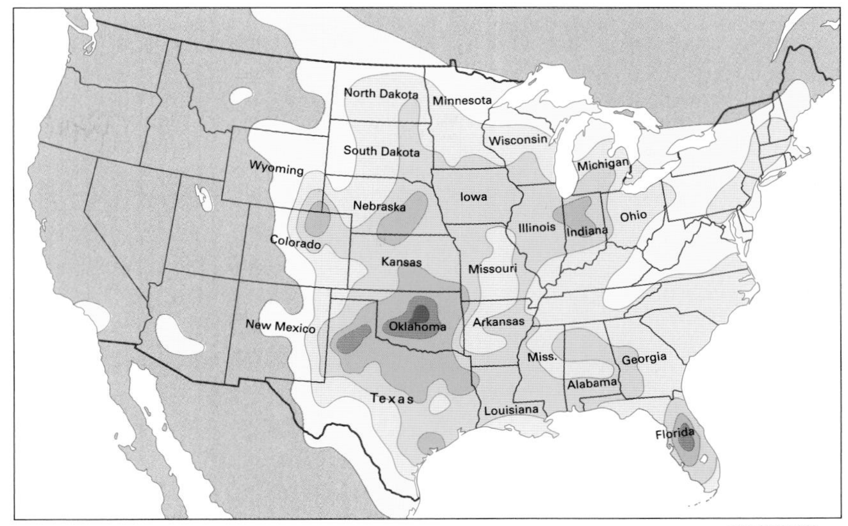

WORLD BOOK map

The Fujita scale

The chart below shows the Fujita scale for estimating the speed, in miles per hour (mph) and kilometers per hour (kph), of tornadoes and other violent winds based on the damage caused.

Scale	Damage	Wind speed: Mph	Kph
F0	**Light** Tree branches broken, damage to chimneys and large signs.	40-72	64-116
F1	**Moderate** Trees snapped, surface of roofs peeled off, windows broken.	73-112	117-180
F2	**Considerable** Large trees uprooted, roofs torn off frame houses, mobile homes demolished.	113-157	181-253
F3	**Severe** Roof and some walls torn off well-constructed houses, cars overturned.	158-206	254-332
F4	**Devastating** Well-constructed houses leveled, cars and large objects thrown.	207-260	333-419
F5	**Incredible** Strong frame houses lifted off foundation and destroyed, car-sized objects thrown more than 300 feet (90 meters).	261-318	420-512

alyze the data and recognize when conditions favor the formation of thunderstorms.

Meteorologists use a special type of radar called Doppler radar to look for mesocyclones. This type of radar works because radar waves change frequency depending on whether the objects they bounce off, such as raindrops or dust particles, are advancing or receding. This change in frequency, called the Doppler effect, can reveal the rotating pattern of a mesocyclone.

In the United States, the National Weather Service issues a *tornado watch* when conditions in the atmosphere promote the formation of tornadoes. If a tornado watch is issued for your area, you should watch for threatening weather and listen to weather bulletins on the radio or television for additional information.

If Doppler radar detects a mesocyclone in a thunderstorm, the National Weather Service issues a *tornado warning.* The Weather Service also issues a tornado warning if someone actually sees a funnel cloud. If a tornado warning is issued for your location, take cover immediately. The safest place is a basement or other underground shelter. If no underground shelter is available, an interior bathroom or closet is best.

Studying tornadoes. Meteorologists study tornadoes both outdoors and in the laboratory. Tornadoes are difficult to study outdoors because they form and vanish quickly and occupy a small area. Another problem is that scientists do not know exactly what causes tornadoes, so they find it difficult to get to the right place at the right time to gather data.

Many meteorologists form mobile teams of "storm chasers" to study tornadoes outdoors. The storm chasers travel in specially equipped automobiles, vans, trucks, and aircraft, trying to get near tornadoes without coming too close. The researchers drop instruments in or near the paths of tornadoes to measure wind, temperature, and air pressure. They also record flying debris on film or videotape so they can analyze wind patterns. Doppler radar enables storm chasers to map wind speed and direction and to study the changes that take place in a thunderstorm before a tornado forms.

Meteorologists make computer *models* (mathematical representations) of thunderstorms and tornadoes. They also simulate tornadoes using rotating air currents in chambers. Scientists hope to learn why tornadoes form, to know what happens inside a tornado, and to better forecast these destructive storms. Howard B. Bluestein

See also **Cloud; Cyclone; Waterspout.**

Hurricane is a powerful, swirling storm that begins over a warm sea. When a hurricane hits land, it can cause great damage through fierce winds, torrential rain, flooding, and huge waves crashing ashore. A powerful hurricane can kill more people and destroy more property than any other natural disaster.

The winds of a hurricane swirl around a calm central

NASA

Hurricane winds swirl about the *eye,* a calm area in the center of the storm. The main mass of clouds shown in this photograph measures almost 250 miles (400 kilometers) across. The hurricane, named Andrew, struck the Bahamas, Florida, and Louisiana in 1992, killing 54 people and causing billions of dollars in damage.

zone called the *eye* surrounded by a band of tall, dark clouds called the *eyewall.* The eye is usually about 10 to 20 miles (16 to 32 kilometers) in diameter and is free of rain and large clouds. In the eyewall, large changes in pressure create the hurricane's strongest winds. These winds can reach nearly 200 miles (320 kilometers) per hour. Damaging winds may extend 250 miles (400 kilometers) from the eye.

Hurricanes are referred to by different labels, depending on where they occur. They are called *hurricanes* when they happen over the North Atlantic Ocean, the Caribbean Sea, the Gulf of Mexico, or the Northeast Pacific Ocean. Such storms are known as *typhoons* if they occur in the Northwest Pacific Ocean, west of an imaginary line called the International Date Line. Near Australia and in the Indian Ocean, they are referred to as *tropical cyclones.*

Hurricanes are most common during the summer and early fall. In the Atlantic and the Northeast Pacific, for example, August and September are the peak hurricane months. Typhoons occur throughout the year in the Northwest Pacific but are most frequent in summer. In the North Indian Ocean, tropical cyclones strike in May and November. In the South Indian Ocean, the South Pacific Ocean, and off the coast of Australia, the hurricane season runs from December to March. Approximately 85 hurricanes, typhoons, and tropical cyclones occur in a year throughout the world. In the rest of this article, the term *hurricane* refers to all such storms.

Hurricane conditions

Hurricanes require a special set of conditions, including ample heat and moisture, that exist primarily over warm tropical oceans. For a hurricane to form, there must be a warm layer of water at the top of the sea with a surface temperature greater than 80 °F (26.5 °C).

Warm seawater evaporates and is absorbed by the surrounding air. The warmer the ocean, the more water evaporates. The warm, moist air rises, lowering the *atmospheric pressure* of the air beneath. In any area of low atmospheric pressure, the column of air that extends from the surface of the water—or land—to the top of the atmosphere is relatively less dense and therefore weighs relatively less.

Air tends to move from areas of high pressure to areas of low pressure, creating wind. In the Northern Hemisphere, the earth's rotation causes the wind to swirl into a low-pressure area in a counterclockwise direction. In the Southern Hemisphere, the winds rotate clockwise around a low. This effect of the rotating earth on wind flow is called the *Coriolis effect.* The Coriolis effect increases in intensity farther from the equator. To produce a hurricane, a low-pressure area must be more than 5 degrees of latitude north or south of the equator. Hurricanes seldom occur closer to the equator.

For a hurricane to develop, there must be little *wind shear*—that is, little difference in speed and direction between winds at upper and lower elevations. Uniform winds enable the warm inner core of the storm to stay intact. The storm would break up if the winds at higher elevations increased markedly in speed, changed direction, or both. The wind shear would disrupt the budding hurricane by tipping it over or by blowing the top of the storm in one direction while the bottom moved in another direction.

The life of a hurricane

Meteorologists (scientists who study weather) divide the life of a hurricane into four stages: (1) tropical disturbance, (2) tropical depression, (3) tropical storm, and (4) hurricane.

Tropical disturbance is an area where rainclouds

are building. The clouds form when moist air rises and becomes cooler. Cool air cannot hold as much water vapor as warm air can, and the excess water changes into tiny droplets of water that form clouds. The clouds in a tropical disturbance may rise to great heights, forming the towering thunderclouds that meteorologists call *cumulonimbus clouds*.

Cumulonimbus clouds usually produce heavy rains that end after an hour or two, and the weather clears rapidly. If conditions are right for a hurricane, however, there is so much heat energy and moisture in the atmosphere that new cumulonimbus clouds continually form from rising moist air.

Tropical depression is a low-pressure area surrounded by winds that have begun to blow in a circular pattern. A meteorologist considers a depression to exist when there is low pressure over a large enough area to be plotted on a weather map. On a map of surface pressure, such a depression appears as one or two circular *isobars* (lines of equal pressure) over a tropical ocean. The low pressure near the ocean surface draws in warm, moist air, which feeds more thunderstorms.

The winds swirl slowly around the low-pressure area at first. As the pressure becomes even lower, more warm, moist air is drawn into the system, and the winds blow faster.

Tropical storm. When the winds exceed 38 miles (61 kilometers) per hour, a tropical storm has developed. Viewed from above, the storm clouds now have a well-defined circular shape. The seas have become so rough that ships must steer clear of the area. The strong winds near the surface of the ocean draw more and more heat and water vapor from the sea. The increased warmth and moisture in the air feed the storm.

A tropical storm has a column of warm air near its center. The warmer this column becomes, the more the pressure at the surface falls. The falling pressure, in turn, draws more air into the storm. As more air is pulled into the storm, the winds blow harder.

Each tropical storm receives a name. The names help meteorologists and disaster planners avoid confusion and quickly convey information about the behavior of a storm. The World Meteorological Organization (WMO), an agency of the United Nations, issues four alphabetical lists of names, one for the North Atlantic Ocean and the Caribbean Sea, and one each for the Eastern, Central, and Northwestern Pacific. The lists include both men's and women's names that are popular in countries affected by the storms.

Except in the Northwestern and Central Pacific, the first storm of the year gets a name beginning with *A*—such as Tropical Storm Alberto. If the storm intensifies into a hurricane, it becomes Hurricane Alberto. The second storm gets a name beginning with *B*, and so on through the alphabet. The lists do not use all the letters of the alphabet, however, since there are few names beginning with such letters as Q or U. For example, no At-

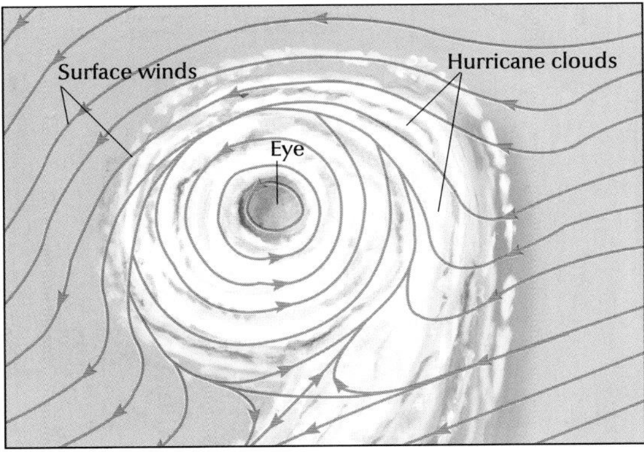

WORLD BOOK illustrations by Bruce Kerr

Hurricane winds on the ocean surface swirl counterclockwise around a calm eye, *left,* in the Northern Hemisphere. Surface winds feed heat energy and water vapor into the storm, *below.* A core of warm air maintains a zone of low pressure near the surface, drawing more air into the hurricane. Much rain falls from tall bands of clouds called the *eyewall* and *rainbands.* The illustration below is not to scale. The cloud structure is about 150 miles (240 kilometers) across and 8 miles (13 kilometers) high.

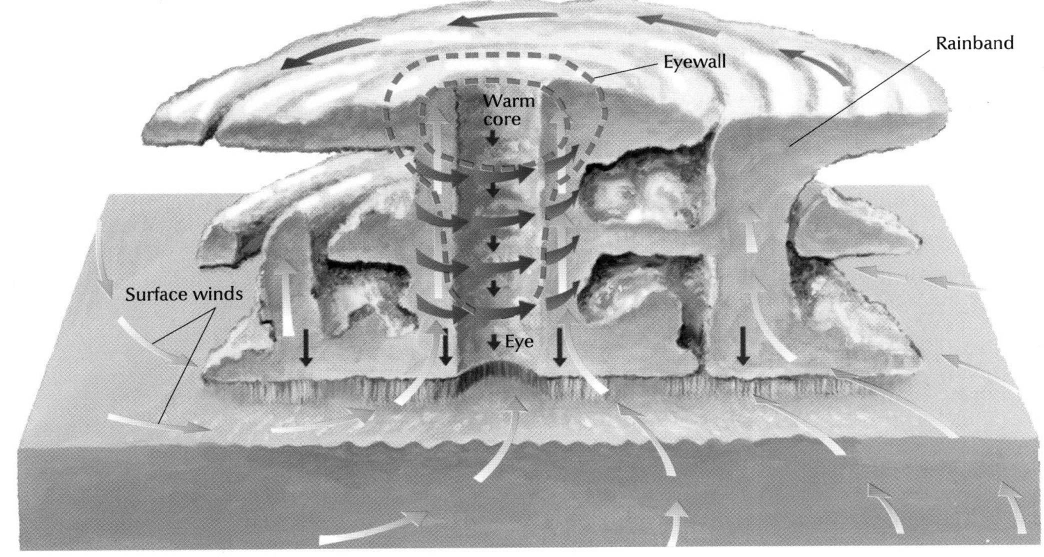

Where hurricanes strike

Hurricanes, typhoons, and tropical cyclones form in warm tropical seas, then move to higher latitudes. The storms can die out quickly if they move over land because the tropical water can no longer provide them with energy and moisture.

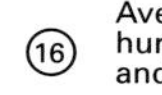

Average number of hurricanes, typoons, and tropical cyclones per year

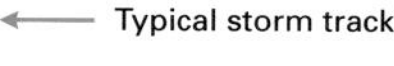
Typical storm track

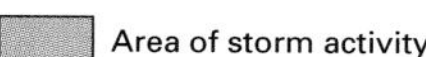
Area of storm activity

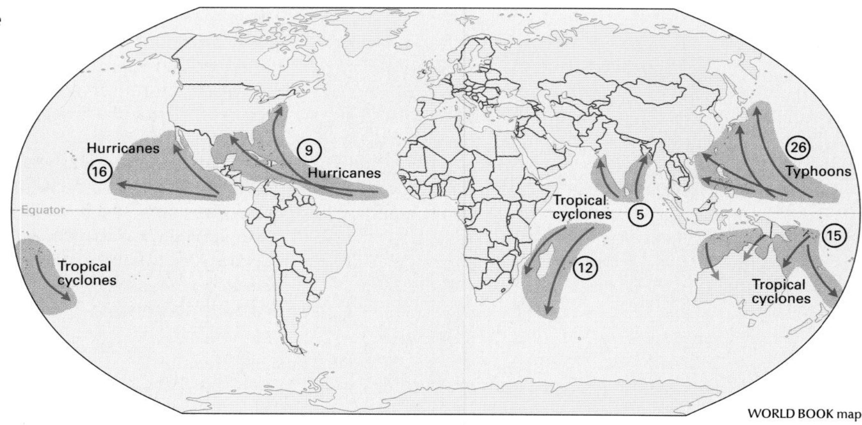

WORLD BOOK map

lantic or Caribbean storms receive names beginning with Q, U, X, Y, or Z.

Because storms in the Northwestern Pacific occur throughout the year, the names run through the entire alphabet instead of starting over each year. The first typhoon of the year might be Typhoon Nona, for example. The Central Pacific usually has fewer than five named storms each year.

The system of naming storms has changed since 1950. Before that year, there was no formal system. Storms commonly received women's names and names of saints of both genders. From 1950 to 1952, storms were given names from the United States military alphabet—Able, Baker, Charlie, and so on. The WMO began to use only the names of women in 1953. In 1979, the WMO began to use men's names as well.

Hurricane. A storm achieves hurricane status when its winds exceed 74 miles (119 kilometers) per hour. By the time a storm reaches hurricane intensity, it usually has a well-developed eye at its center. Surface pressure drops to its lowest in the eye.

In the eyewall, warm air spirals upward, creating the hurricane's strongest winds. The speed of the winds in the eyewall is related to the diameter of the eye. Just as ice skaters spin faster when they pull their arms in, a hurricane's winds blow faster if its eye is small. If the eye widens, the winds decrease.

Heavy rains fall from the eyewall and bands of dense clouds that swirl around the eyewall. These bands, called *rainbands,* can produce more than 2 inches (5 centimeters) of rain per hour. The hurricane draws large amounts of heat and moisture from the sea.

The path of a hurricane

Hurricanes last an average of 3 to 14 days. A long-lived storm may wander 3,000 to 4,000 miles (4,800 to 6,400 kilometers), typically moving over the sea at speeds of 5 to 20 miles (8 to 32 kilometers) per hour.

Hurricanes in the Northern Hemisphere usually begin by traveling from east to west. As the storms approach the coast of North America or Asia, however, they shift to a more northerly direction. Most hurricanes turn gradually northwest, north, and finally northeast. In the Southern Hemisphere, the storms may travel westward at first and then turn southwest, south, and finally southeast. The path of an individual hurricane is irregular and often difficult for weather forecasters to predict.

All hurricanes eventually move toward higher latitudes where there is colder air, less moisture, and greater wind shears. These conditions cause the storm to weaken and die out. The end comes quickly if a hurricane moves over land, because it no longer receives heat energy and moisture from warm tropical water. Heavy rains may continue, however, even after the winds have diminished.

Famous recent hurricanes, typhoons, and cyclones

1900 A hurricane and storm surge killed about 6,000 people in the Galveston, Texas, area.

1928 About 1,800 people died in a hurricane and floods in the Lake Okeechobee area of Florida. The storm also killed 300 people in Puerto Rico.

1938 A hurricane that became known as the New England Hurricane tore through the northeastern United States, killing about 600 people.

1944 A typhoon in the Philippine Sea sank three destroyers and wrecked over 100 aircraft of the U.S. Pacific fleet; 778 lives were lost.

1963 Hurricane Flora killed about 5,000 people in Haiti, over 1,700 in Cuba, and over 400 in the Dominican Republic.

1969 Hurricane Camille killed more than 250 people in seven states from Louisiana to Virginia.

1970 The storm surge from a tropical cyclone drowned about 266,000 people in East Pakistan (now Bangladesh).

1972 Floods well inland caused by Hurricane Agnes killed 117 people from Florida to New York.

1974 Hurricane Fifi struck Honduras, killing about 8,000 people.

1975 Floods caused by Typhoon Nina killed over 100,000 people in China.

1979 Sea-level pressure dropped to a record 870 millibars as Typhoon Tip passed through the Philippine Sea. (Normal sea-level pressure is about 1,013 millibars.)

1980 Tropical Cyclone Hyacinthe looped around Reunion Island in the Indian Ocean, delivering 252 inches of rain in just 15 days.

1988 Hurricane Gilbert, the most intense hurricane ever recorded in the Western Hemisphere, struck the West Indies and Mexico, causing about 300 deaths.

1989 Hurricane Hugo struck the West Indies and the southeastern United States, causing more than 60 deaths.

1992 Hurricane Andrew raked the Bahamas, Florida, and Louisiana, killing 54 and causing about $22 billion in damage.

1995 Hurricane Ismael struck northwestern Mexico, causing 107 deaths.

© Gary Williams, Gamma/Liaison

A storm surge, a rapid rise in sea level, is the most dangerous part of a hurricane. A storm surge occurs when fierce hurricane winds drive a tremendous volume of ocean water ashore.

The Saffir-Simpson hurricane scale

The chart below shows the Saffir-Simpson scale of hurricane intensity. The scale is based on wind speed in miles per hour (mph) and kilometers per hour (kph) and the height of the resulting *storm surge*—that is, how much the sea level rises above normal high tide.

Hurricane level		Wind speed		Storm surge	
		Mph	Kph	Feet	Meters
Level 1	(weak)	74-95	119-153	4-5	1.2-1.5
Level 2	(moderate)	96-110	154-177	6-8	1.8-2.4
Level 3	(strong)	111-130	178-209	9-12	2.7-3.7
Level 4	(very strong)	131-155	210-250	13-18	4.0-5.5
Level 5	(devastating)	156-	251-	19-	5.8-

Effects

Level 1	Minimal damage to trees, shrubbery, and mobile homes.
Level 2	Considerable damage to trees, mobile homes, and piers; some damage to roofs.
Level 3	Trees blown down or stripped of leaves; mobile homes destroyed; some damage to other buildings.
Level 4	Extensive damage to windows, doors, and roofs, especially near shore; possible flooding.
Level 5	Small buildings overturned or blown away; severe structural damage to other buildings.

Hurricane damage

Hurricane damage results from wind and water. Hurricane winds can uproot trees and tear the roofs off houses. The fierce winds also create danger from flying debris. Heavy rains may cause flooding and mudslides.

The most dangerous effect of a hurricane, however, is a rapid rise in sea level called a *storm surge.* A storm surge is produced when winds drive ocean waters ashore. Storm surges are dangerous because many coastal areas are densely populated and lie only a few feet or meters above sea level. A 1970 cyclone in East Pakistan (now Bangladesh) produced a surge that killed about 266,000 people. A hurricane in Galveston, Texas, in 1900 produced a surge that killed about 6,000 people, the worst natural disaster in United States history.

Hurricane watchers rate the intensity of storms on a scale called the Saffir-Simpson scale, developed by American engineer Herbert S. Saffir and meteorologist Robert H. Simpson. The scale designates five levels of hurricanes, ranging from Level 1, described as weak, to Level 5, which can be devastating. Level 5 hurricanes have included Hurricane Camille, which hit the United States in 1969, and Hurricane Gilbert, which raked the West Indies and Mexico in 1988.

Forecasting hurricanes

Meteorologists use weather balloons, satellites, and radar to watch for areas of rapidly falling pressure that may become hurricanes. Specially equipped airplanes called *hurricane hunters* investigate budding storms.

If conditions are right for a hurricane, the National Weather Service issues a *hurricane watch.* A hurricane watch advises an area that there is a good possibility of a hurricane within 36 hours. If a hurricane watch is issued for your location, check the radio or television often for official bulletins. A *hurricane warning* means that an area is in danger of being struck by a hurricane in 24 hours or less. Keep your radio tuned to a news station after a hurricane warning. If local authorities recommend evacuation, move quickly to a safe area or a designated hurricane shelter. Gary Barnes

See also **Cyclone; Safety** (During a hurricane); **Tornado; Typhoon.**

Paleobotany, *PAY lee oh BAHT uh nee,* is the study of ancient plants. Paleobotany is a branch of *paleontology,* the study of ancient plants, animals, and other organisms. Specialists called *paleobotanists* investigate the evolution of plant life and the origins and relationships of plant groups. They also examine the link between vegetation and the earth's changing climate. Paleobotany includes the study of such simple organisms as ancient algae, fungi, and bacteria. In addition, it involves searching for the earliest evidence of life in rocks more than 3 billion years old.

Paleobotanists interpret the earth's history by examining plant fossils. These fossils have been preserved in *sedimentary rocks* (rocks formed from deposits laid down by ancient rivers, lakes, and seas). Paleobotanists have found the earliest land plants in sedimentary rocks that are more than 430 million years old. Remains of early forests are abundant in rocks 350 million years old. The ancestors of all major groups of land plants lived in these forests. Today, the most numerous plants on earth are *angiosperms* (flowering plants). Paleobotanists have discovered that the first angiosperms appeared about 140 million years ago, during the age of the dinosaurs.

Paleobotanists study the features that plants have developed to survive in their environment. These scientists are thus able to describe the type of climate that existed millions of years ago. Paleobotany contributes to an understanding of how and why the earth's climate changes. This understanding is important in predicting the changes in climate that humans may cause through the *greenhouse effect* (a gradual warming of the earth's surface). James F. Basinger

Global warming is an increase in the average temperature of the earth's surface. Since the late 1800's, the average temperature has increased about 0.5 to 1.5 Fahrenheit degrees (0.3 to 0.8 Celsius degree). Scientists disagree about the causes and potential impact of this warming trend.

Many scientists believe that human activities have caused the trend by enhancing the earth's natural *greenhouse effect.* This effect warms the lower atmosphere and surface of the earth through a complex process involving sunlight, gases, and particles in the atmosphere.

Since the mid-1800's, human activities—chiefly the burning of *fossil fuels* (coal, oil, and natural gas) and the clearing of land—have increased the amounts of heat-trapping atmospheric gases, called *greenhouse gases.* The burning of fossil fuels produces the greenhouse gas carbon dioxide. Most of this burning takes place in automobiles, electric power plants, and industrial facilities. The clearing of land reduces the amount of carbon dioxide that trees and other plants remove from the atmosphere in a process called *photosynthesis.*

Although researchers have not yet fully proved that the increase in greenhouse gases has raised the surface temperature, many scientists consider such a relationship likely. A much smaller number of scientists argue that the increase in greenhouse gases has not made a measurable difference in the climate. These scientists say that the warming trend is a normal change in the climate system. They argue that natural processes, such as increases in the energy given off by the sun, could have caused global warming. But the greater weight of evidence suggests that an unusual climate change has occurred and that human activities are at least partly responsible for it.

The impact of continued global warming. Global warming affects many aspects of the environment, including sea levels, coastlines, agriculture, forestry, and wildlife. Continued global warming could have a beneficial impact in some areas and a harmful impact in others. For example, people could begin to farm in regions where it is currently too cold. At the same time, global warming could cause sea levels to rise and thereby increase the threat of flooding in low-lying coastal areas, many of which are densely populated.

To determine the impact of global warming, scientists must consider the rate of climate change. Many experts estimate that, due to human activities, the average surface temperature will rise between 3 and 8 Fahrenheit degrees (1.5 and 4.5 Celsius degrees) by 2100. But even a warming of 2 Fahrenheit degrees in a century would be several times faster than the typical natural rate.

A rapid and large-scale climate change could severely harm the earth's *ecosystems* (the living organisms and physical environment in particular areas). For example, such a change could make it difficult for many species to survive in the regions they now inhabit. Some species could be forced to migrate, while others could become extinct.

Efforts to prevent global warming. Because global warming might do much harm, many scientists recommend a reduction in the emissions of greenhouse gases. The least controversial way to reduce such emissions would be to use gains in efficiency to decrease the use of energy. For example, manufacturers would adopt more efficient processes to produce goods. Makers of electrical equipment would introduce more efficient motors, light bulbs, and other devices. New manufacturing equipment and the new devices might be more costly than those they replaced. However, the savings resulting from decreased energy consumption might help make up for the extra costs. As a result, the new technologies could reduce emissions and save money.

Experts have also offered a number of more controversial strategies. For example, governments could create regulations that specify the type of technologies used or the amount of fossil fuels burned. Many policy analysts recommend that governments consider a tax on emissions of greenhouse gases.

Another option would limit the amount of greenhouse gases each nation emits. Under this plan, each country would receive emissions "permits" that could be bought and sold. Richer nations with higher emissions per person could purchase permits from poorer nations with lower emissions per person. Thus, the more developed countries would have incentives to create and use more efficient technologies, and less developed countries would receive money that could be used to aid their development. Stephen H. Schneider

See also **Greenhouse effect.**

Human Genome Project, *JEE nohm,* is an international scientific program to analyze the complete chemical instructions that control heredity in human beings and certain other organisms. All living things contain such instructions, which are carried on long coils of a chemical called DNA (deoxyribonucleic acid). DNA contains four types of simpler chemicals called *bases.* The four types of bases are cytosine, adenine, guanine, and thymine. Scientists estimate that the human *genome*—that is, one complete set of instructions—contains about 3 billion bases. Scientists worldwide are sharing information to determine the exact order in which these bases occur.

Although the bases can combine chemically in any order, they occur in almost the same order in all people. In any two human genomes, only one base in every few thousand differs. These rare differences in the order of bases control such inheritable characteristics as eye color, skin color, and blood type.

Instructions for a particular characteristic are encoded in one or more sections of bases called *genes.* Genes are arranged on 23 pairs of rod-shaped structures called *chromosomes.* Scientists hope that identifying the order of bases in the human genome will lead to development of highly detailed *gene maps.* These maps will show where each gene lies on a chromosome and reveal each gene's role in normal body processes or disease.

The Human Genome Project began formally in 1990. Knowledge gained in the project may someday enable scientists to analyze an individual's genome. Because genetic variations cause or contribute to certain diseases, analysis of a person's genome could reveal health information that should remain private. Scientists in the project are studying how to protect people from misuse of such information.

Maynard V. Olson

See also **Cell; Gene; Gene mapping; Heredity; Nucleic acid.**

How to use the index

This index covers the contents of the 1997, 1998, and 1999 editions.

Each entry gives the last two digits of the edition year, followed by a colon and the page number or numbers. For example, this entry indicates that information on gene probes may be found on page 303 of the 1999 edition.

The indication (il.) after a page number means that the reference is to an illustration only. For example, this entry refers to the picture of gene mapping on page 244 of the 1999 edition.

When there are many references to a topic, they are grouped alphabetically by clue words under the main topic. For example, the clue words under **Genetic engineering** group the references to that topic under the main heading and several subtopics.

The "see" and "see also" cross-references indicate that references to a topic are listed under another entry in the index.

An entry in all capital letters indicates that there is a Science News Update article with that name in at least one of the three volumes covered by this index. References to the topic in other articles may also be listed in the entry.

An entry that only begins with a capital letter indicates that there are no Science News Update articles with that title but that information on this topic may be found in the editions and on the pages listed.

An entry followed by *WBE* refers to a new or revised *World Book Encyclopedia* article in the supplement section. This entry means that there is a *World Book Encyclopedia* article on global warming on page 336 of the 1999 edition.

Index

A

B

C

Index

G

H

I

N

O

P

Q

R

Index

S

Index

ACKNOWLEDGMENTS

The publishers gratefully acknowledge the courtesy of the following artists, photographers, publishers, institutions, agencies, and corporations for the illustrations in this volume. Credits are listed from top to bottom, and left to right, on their respective pages. All entries marked with an asterisk (*) denote illustrations created exclusively for this yearbook. All maps, charts, and diagrams were staff-prepared unless otherwise noted.

2 NASA; © John Reader/SPL from Photo Researchers
3 Reuters/Archive Photos; Frederick R. Blattner, *Science*, vol. 277, pp.1453-62; NASA/JPL/CALTECH
4 © M.P.L. Fogden, Bruce Coleman, Ltd.; © Alain Nogues, Sygma
5 NASA/JPL
10 © Institute for Molecular Manufacturing; Applied Precision, Inc.
11 *Tess* from California Science Center; © *New Zealand Herald*
12 Giacomo Giacobini, University of Turin, Italy
19 © Javier Trueba, Madrid Scientific Films
20 © John Reader/SPL from Photo Researchers
22 © Javier Trueba, Madrid Scientific Films
26-27 NASA/JPL/CALTECH
29 Schiaparelli Observatory, Varese, Italy; © Lowell Observatory; Lowell Observatory; Stock Montage; NASA/JPL
30 NASA/JPL
32-33 NASA /JPL/CALTECH
36 NASA/JPL/MSSS; NASA/JPL/MSSS; NASA/JPL/MSSS; NASA/JPL/CALTECH
38 ISAS
39 NASA/JPL
40 © Michael Carroll
42 © Stephen Dalton, NHPA; © Joseph M. Kiesecker, Oregon State University; © M.P.L. Fogden, Bruce Coleman, Ltd.; © John D. Cunningham, Visuals Unlimited
47 © Daniel Heuclin, NHPA; © Deb Rose, Minnesota Pollution Control Agency
48 U.S. Fish and Wildlife Service; U.S. Fish and Wildlife Service; © G. Prance, Visuals Unlimited
49 © Andrew Blaustein, Oregon State University; © Andrew Blaustein, Oregon State University; © E.A. Janes, NHPA; © Cabisco, Visuals Unlimited; © Daniel Heuclin, NHPA
51 © A.N.T. from NHPA; Corel Corporation
54 © Michael J. Lannoo, Indiana University
56 Eric Lessing from Art Resource
59 E.M. Winkler, *Stone Properties, Durability in Man's Environment; Self Portrait* (1554) oil on panel by Sofonisba Anguissola; Kunsthistorisches Museum, Vienna, Austria (Eric Lessing from Art Resource)
62 *Young Lady in a Tricorn Hat* (1755) oil on canvas by Giovanni Battista Tiepolo; National Gallery of Art, Washington, D.C., Samuel H. Kress Collection © Board of Trustees, National Gallery of Art; *Detail of the Abduction of the Sabine Women* (1675) oil on canvas by Luca Giordano; Major Acquisitions Endowment; Charles H. and Mary F.S. Worcester Collection © The Art Institute of Chicago. All rights reserved.; © Board of Trustees, National Gallery of Art, Washington, D.C.; © The Art Institute of Chicago. All rights reserved.
63 Eric Lessing from Art Resource; Leonardo Secca, Anna Maria Ianucci, Nicola Santopuli and Barbara Vernia, CIRAM (L.S.,B.V.) Research Centre of Applied Mathematics at the University of Bologna directed by Prof. Tommaso Ruggeri, and at the Superintendence of Ravenna (A.M.I.,N.S.); Leonardo Secca, Anna Maria Ianucci, Nicola Santopuli and Barbara Vernia, CIRAM (L.S., B.V.) Research Centre of Applied Mathematics at the Univerity of Bologna directed by Prof. Tommaso Ruggeri, and at the Superintendence of Ravenna (A.M.I.,N.S.)
65 © The Art Institute of Chicago. All rights reserved; Getty Conservation Institute © The J. Paul Getty Trust; © The Art Institute of Chicago. All rights reserved; © The Art Institute of Chicago. All rights reserved.
68 *Lady Reading the Letters of Heloise and Abelard* (1758) oil on canvas by Jean-Baptiste Greuze; Mrs. Harold T. Martin Fund; Lacy Armour Endowment; Charles H. and Mary F.S. Worcester Collection © The Art Institute of Chicago. All rights reserved.; The Minneapolis Institute of Arts; © The Art Institute of Chicago. All rights reserved; The Minneapolis Institute of Arts.
69 Folger Shakespeare Library, Washington, D.C.; © John Asmus; Folger Shakespeare Library, Washington, D.C.
72 Cari Biamonte*
75 Roberta Polfus*.
77 Roberta Polfus*; Image copyright © 1997 PhotoDisc, Inc..
78 Peg Gerrity and Roberta Polfus*
79 Roslin Institute; ABS Global
82 Wyeth-Ayerst Pharmaceuticals; © John Chadwick; AP/Wideworld
84 University of Bath, England
86 Granger Collection
90 National Museum of Natural History © Smithsonian Institution
93 Granger Collection
94 *New Zealand Herald*
95 © Glen Loates
97 © John M. Francis/National Geographic Television
98 © Ingrid Roper
99 © John Anderson/National Geographic Television
102 © Peter Menzel
103 © Institute for Molecular Manufacturing and Xerox
105 © Institute for Molecular Manufacturing ; Tom Herzberg*
106-107 Joyce Benna*
108 Space Studies Institute
109 Foresight Institute
110 Joyce Benna*
111 Joyce Benna*; © Institute for Molecular Manufacturing
112-114 Joyce Benna*
117 COSI Toledo
119 Deutsches Museum von Meisterwerken der Naturwissenschaft und Technik, Munich, Germany
120 © The Exploratorium
121 © S. Schwartzenberg, The Exploratorium
122 New York Hall of Science; © A. Douyon, New York Hall of Science
124 Andrew Brilliant, Museum of Science, Boston
125 Museum of Science, Boston
126 © Teri Bloom, New York Hall of Science
127 *Tess* from California Science Center; Franklin Institute
128 New York Hall of Science
130 NASA
132 Deutsches Museum von Meisterwerken der Naturwissenschaft und Technik, Munich. Gemany
133 NASA; Energia Ltd.
134 The Boeing Company
135 NASA; Spar Aerospace
138 The Boeing Company; Jean-Pierre Wery and Eli Lilly Co.; Jean-Pierre Wery and Eli Lilly Co.
141-142 NASA
144 © David Scharf. All rights reserved.
146 Elaine Fitzsimmons & B.R. Smith, Center for InVivo Microsopy, Duke University Medical Center
148 © David Scharf. All rights reserved.
150 Biophysical Imaging & Opto-Electronics, Cornell University
151 Charles Lieber, et al., Harvard University, *Nature* Vol. 391, pp.62-64; Z.L. Wang. Georgia Institute of Technology
152 Applied Precision, Inc.
153 Richard Tuft, University of Massachusetts Medical School
154 D. Mueller & A. Engel, Universitat Basel, Switzerland
155 PHI Evans, Redwood City, CA from Digital Instruments
157 © Jurgen Scriba/SPL from Photo Researchers; © Demange Francis, Gamma/Liaison; © Demange Francis, Gamma/Liaison; © Alain Nogues, Sygma
158 Archive Photos
159 Corbis-Bettmann
160 Archive Photos
162 Forensic Analytical
164 © Demange Francis, Gamma/Liaison
165-166 Forensic Analytical
167 © Demange Francis, Gamma/Liaison
168-169 Forensic Analytical

170	© Mark C. Burnett, Photo Researchers
171	© Jurgen Scriba/SPL from Photo Researchers; © Alain Nogues
172	© A. Jason, Gamma/Liaison
173	© A.H. Bingen, Sygma
174	ImageWare Software, Inc.
176	NASA; NASA/Goddard Space Flight Center
177	Duke University Primate Center; NASA
179	Agricultural Research Center, USDA
181	© Kenneth Garrett, National Geographic Society
182	© Gen Suwa, National Museum of Ethiopia from Brill Atlanta
183	NASA
184	Agence France-Presse; © Thomas J. Abercrombie, National Geographic Society
187	© Guiseppe Avallone, NYT Pictures; © Guiseppe Avallone, NYT Pictures; NYT Graphics
189-195	NASA/JPL.
197	NASA/Goddard Space Flight Center
201	© Marc Van Roosmalen
202	© Paul D. Heideman
204	Jeffrey Plautz & Steve Kay, Scripps Research Institute
209	W.E. Moerner, University of California at San Diego
210	Keiki-Pua Dancil, University of California at San Diego
213	Ames Research Center
215	Chequemate Technologies
217	© Gary Braasch; © Paul Gier, Visuals Unlimited
219	Duke University Primate Center
220	AP/Wide World; Carl A. Kroch Library, Cornell University; Woods Hole Oceanographic Institution.
221	AP/Wide World; AP/Wide World; Argonne National Laboratory
223	Merck & Company
224	David Read, University of Sheffield, England
225	© Dan L. Perlman
227	*Daily Southtown*
228	Peter Duine, Philips Research Laboratories
231	Bechtel Corporation
232	C.C. Lin & M.A. Schmidt, MIT
233	© Xinhua/C. Nouvelle from Gamma/Liaison
234	Sky City
235	AP/Wide World
236	Reuters/Archive Photos
239	AP/Wide World
241	M.A. Fedonkin & Benjamin M. Waggoner, *Nature*, vol. 388, pp.868-71
242	Stefan Bengtson, *Science*, Vol. 277, pp.1645-48
244	Frederick R. Blattner, *Science*, Vol. 277, pp.1453-62; © David M. Phillips, Visuals Unlimited
247	Sara Figlio*
249	AP/Wide World
250	NOAA
251	Ken Jezek, Byrd Polar Research Center, Ohio State University
252	Yang Shen & C.J. Wolfe, Woods Hole Oceanogaphic Institution
255	The Cleveland Clinic Foundation
257	N.L'Heureux, et al., *The FASEB Journal* (1988), Vol. 12, pp.47-56
259	AP/Wide World; © Remsberg, Gamma/Liaison; AP/Wide World
262	SeaWiFS, Goddard Space Flight Center
263	Charles Fisher, Penn State University
265	Carol Brozman*; NASA/JPL
267	Dustin Carr & Harold Craighead, Cornell University
269	MacroSonix Corporation
271	© Joe Bergeron
273	Kim, Relkin, Lee & Hirsch, *Nature*, Vol. 388, pp.171-74
275	Reuters/Archive Photos; © Thomas S. England, NYT Pictures
278	AP/Wide World
281	NASA
282	AP/Wide World
283-284	NASA
285	© Sargent, Gamma/Liaison; Corel Corporation
286	© Sargent, Gamma/Liaison
290	© John Barr, Gamma/Liaison
294	Corel Corporation
295	Top-Flight Golf
296	Paul Perrault*
297	© Joe Polillio, Gamma/Liaison
298	Paul Perrault*
301	© Will & Deni McIntyre, Photo Researchers
305	NASA; © Howard P. Bluestein